Contamination of Groundwater: Analysis and Remediation

Contamination of Groundwater: Analysis and Remediation

Editor: Jonah Murray

Callisto Reference
www.callistoreference.com

Callisto Reference,
118-35 Queens Blvd., Suite 400,
Forest Hills, NY 11375, USA

Visit us on the World Wide Web at:
www.callistoreference.com

ISBN: 978-1-64116-833-5 (Hardback)

Cataloging-in-Publication Data

Contamination of groundwater : analysis and remediation / edited by Jonah Murray.
p. cm.
Includes bibliographical references and index.
ISBN 978-1-64116-833-5
1. Groundwater--Pollution. 2. Groundwater--Quality. 3. Groundwater--Purification I. Murray, Jonah.
TD426 .C66 2023
628.168--dc23

Table of Contents

Preface

Groundwater refers to the fresh water that seeps into the soil. It is stored in the minuscule spaces between soil particles and rocks known as aquifers. Entry of substances such as road salts, gasoline, chemicals and oil in the groundwater leads to groundwater contamination. Contaminated groundwater is hazardous for human consumption. Other possible sources of groundwater contamination include uncontrolled hazardous waste, atmospheric contaminants, landfills and septic systems. Groundwater remediation is the process of treating polluted groundwater by either eliminating the contaminants or transforming them into harmless products. Abiotic redaction, chemical precipitation, biotic reduction, and adsorption and ion exchange are the four biological, chemical, and physical uptake mechanisms used in the remediation of contaminated groundwater. This book is a compilation of chapters that discuss the most vital concepts and emerging trends in the analysis and remediation of groundwater contamination. It will serve as a valuable source of reference for graduate and postgraduate students.

The information shared in this book is based on empirical researches made by veterans in this field of study. The elaborative information provided in this book will help the readers further their scope of knowledge leading to advancements in this field.

Finally, I would like to thank my fellow researchers who gave constructive feedback and my family members who supported me at every step of my research.

Editor

1

Remediating Contaminated Groundwater with an Aerated, Direct-Push, Oxidant Delivery System

James Reece [1], Mark Christenson [1], Ann Kambhu [2,3], Yusong Li [2], Clifford E. Harris [4] and Steve Comfort [1,5,*]

[1] AirLift Environmental, LLC, 5900 N. 58th, Suite 5, Lincoln, NE 68507, USA; james@airliftenvironmental.com (J.R.); mark@airliftenvironmental.com (M.C.)
[2] Department of Civil Engineering, University of Nebraska, Lincoln, NE 68588-6105, USA; ann.k@ku.ac.th (A.K.); yli7@unl.edu (Y.L.)
[3] Department of Environmental Technology and Management, Kasetsart University, Bangkok 10900, Thailand
[4] Department of Chemistry, Albion College, Albion, MI 49224, USA; charris@albion.edu
[5] School of Natural Resources, University of Nebraska, Lincoln, NE 68583-0915, USA
* Correspondence: scomfort@unl.edu

Abstract: One of the biggest challenges to treating contaminated aquifers with chemical oxidants is achieving uniform coverage of the target zone. In an effort to maximize coverage, we report the design and installation of a novel aerated, slow-release oxidant delivery system that can be installed by direct-push equipment. By continuously bubbling air beneath a slow-release oxidant in situ, an airlift pump is created that causes water and oxidant to be dispersed from the top of the outer screen and drawn in at the bottom. This continuous circulation pattern around each drive point greatly facilitates the spreading of the oxidant as it slowly dissolves from the wax matrix (i.e., oxidant candle). Given that the aeration rate controls the outward flow of oxidant from the outer screen in all directions, the radius of influence around each drive point is largely a function of the outward velocity of the oxidant exiting the screen and the advection rate opposing the upgradient and lateral spreading. Temporal sampling from three field sites treated with the aerated oxidant system are presented and results show that contaminant concentrations typically decreased 50–99% within 6–9 months after installation. Supporting flow tank experiments that demonstrate oxidant flow patterns and treatment efficacy are also presented.

Keywords: airlift pump; groundwater remediation; modular oxidant delivery system; oxidant cylinders; oxidant delivery device

---

1. Introduction

A critical review of two decades of contaminated groundwater sites treated by in situ chemical oxidation (ISCO) has uncovered many examples where liquid oxidant injections have failed to achieve desired cleanup goals [1,2]. Unlike the positive results obtained when chemical oxidants are used in well-mixed reactors, such as during the treatment of municipal wastewaters, direct-push delivery techniques of liquid oxidants into groundwater can fail to deliver uniform coverage. The reason for this is the uncontrolled, unmixed and stratified nature of subsurface environments. Moreover, the volume of oxidant delivered is often too small to allow for complete coverage of the target zone.

Huling et al. [1] systematically reviewed a series of ISCO performance surveys and concluded that poor ISCO performance often corresponds with sites treated with low oxidant volumes (i.e., percentage of contaminant plume). A review of 40 ISCO sites showed that nearly 50% of the ISCO treatments injected liquid oxidant volumes equivalent to 10% of the contaminant plume [2–4]. While diffusion and mass transport will ultimately increase the percentage of the plume impacted following ISCO injections, the heterogeneous and anisotropic nature of aquifers prevents uniform mixing and can leave significant portions of the contaminated aquifer untouched [2].

Advancements in high-resolution site characterizations (e.g., coupled direct-push membrane interface with hydraulic profiling) combined with prescription-type injections has advanced injection technologies by providing for more targeted treatments. While improved profile characterization can direct better oxidant placement, the injected liquid oxidant is still subject to advection and density-driven flow. These two processes, coupled with the presence of residual non-aqueous phase liquids (NAPLs) within low permeable zones often necessitates additional injections as contaminants desorb back (i.e., rebound) into the transmissive regions following declines in contaminant concentrations from ISCO injections [5,6]. McGuire et al. [7] reported that in a review of 23 NAPL sites treated with liquid oxidants, 81% of the wells exhibited rebound with 30% bouncing back to concentrations greater than those observed before treatments began.

While the chemistry of treating groundwater contaminants with chemical oxidants such as permanganate or persulfate is proven, crucial challenges remain. The biggest challenges to contaminant plume treatment appear to be how to more uniformly deliver the chemical oxidants to targeted zones and determining how to make the oxidant remain in place long enough to mitigate rebound. To overcome the limitations associated with liquid injections, researchers began encapsulating oxidants as slow-release treatments for contaminated aquifers several years ago [8–10]. Since these initial reports, increased interest has resulted in a number of publications that have documented the efficacy of slow-release oxidants to remove groundwater contaminants at the laboratory-scale and in larger flow-tank systems [11–23]. In their analysis of peer-reviewed research involving controlled-release materials for groundwater remediation, O'Connor et al. [24] indicated that of the 30 publications identified, most have been published in the last five years.

Christenson et al. [25] was the first to report the use of slow-release oxidants in the field. They treated a former unregulated landfill with known trichloroethene (TCE) contamination by manufacturing and deploying slow-release oxidant candles (oxidant-wax cylinders). Installation involved inserting half of the oxidant candle mass into the aquifer with direct-push technology and the other half by placing the candles in screens and then dropping them down designated wells. For the candles placed in wells, it was observed that the permanganate dissolving from the oxidant-wax matrix was subject to downward density flow and thus producing an uneven oxidant distribution across the well screen. To counteract the downward flow, Christenson et al. [25] began bubbling air at the base of the oxidant. This bubbling action creates what is analogous to an airlift pump where the injection of air into a submerged pipe (i.e., well screen) causes water to rise up and discharge near the top of the screen and enter at the bottom. Kambhu et al. [26] recently simulated these flow patterns by coupling a two-phase bubbly flow model with the Darcy flow model in a wide flow tank (Figure S1). By combining aeration with the slow-release oxidants, Christenson et al. [25,27] created a continuous (i.e., 24 h per day, 7 days per week; 24/7) circulation pattern around each drive point, which over time, greatly facilitated the distribution of the oxidant to the target zone. Results from the initial field study showed that after more than seven years of deployment, the aerated oxidant candles were still functioning and had reduced downgradient TCE concentrations and associated contaminants by 90%. The initial success of combining aeration with oxidant candles to decrease contaminant concentrations subsequently led to a reloadable design that could be installed by direct-push equipment [28]. This modular oxidant delivery system is currently treating a variety of aquifers contaminated with petroleum and chlorinated solvents.

In this paper, we report the groundwater monitoring results from three field sites treated with the direct-push, aerated, oxidant delivery system. Supporting flow tank experiments that demonstrate oxidant flow patterns and treatment efficacy are also presented. This new method of treating contaminated groundwater is a novel combination of an airlift pump with slow-release oxidants. Merging aeration with slow-release oxidants creates a technology that differs from typical ISCO treatments and avoids some of the problems associated with injecting liquid oxidants. By creating a circulation pattern around each drive point, density-driven flow of the oxidant is minimized. Given that the oxidant is added as a solid, no surfacing (i.e., daylighting) of the oxidant will ever occur. Moreover, the stationary and sustained release of an oxidant to a targeted aquifer zone greatly diminishes the chances of rebound. Finally, the reloadable design allows more oxidant to be added as needed with minimal effort or safety concerns; this allows the technology to provide an effective groundwater treatment for months to years.

2. Materials and Methods

2.1. *Treatability Studies, Oxidant Manufacturing and Field-Scale Installation and Sampling Procedures*

Prior to installing the aerated oxidant delivery system in the field, treatability studies using miniature candles were performed with groundwater from the field sites. Details of the procedures used in the treatability studies are provided in the supplemental materials (SM). The identification of acetone and 2-butonone in one of the treatability studies and at one of the field sites prompted additional investigations to determine the source of the acetone. Methodologies used to manufacture the field-scale wax-based oxidant candles have previously been reported [17,25,27]. Supplementary Materials also contains detailed schematics of the oxidant delivery devices (Figure S2), direct-push installation procedures of the oxidant drive points, groundwater sampling and analysis methodologies, and the methods used to measure the longevity of the persulfate candles under field conditions.

2.2. *Generalized Plot Plan and Technology Components*

The placement and spacing of drive points containing the aerated oxidant delivery devices are site specific and tailored to the contaminant plume. Regardless of the plume geometry, the treatment technology consists of three components. These include: (i) two or more rows of individual drive points that are offset and spaced 0.9–2.4 m apart (Figure 1A); (ii) a dual manifold system that supplies an aeration tube to the drive points and a ventilation line away from the drive points to a central outlet and (iii) a rocking piston compressor (EasyPro Pond Products, Grant, MI, USA). Rocking piston compressors have routinely been used to aerate lakes and are compact, quiet and able to run 24/7 with minimal maintenance. This three-part modular design is scalable to accommodate most contaminant plumes. Width can be accommodated by adding more drive points and depth can be adjusted by direct-push placement of the oxidant screens.

2.3. *Oxidant Delivery System*

The drive points used to deliver oxidant to the aquifer consist of an outer screen that is emplaced with direct-push equipment and an inner screen that holds the slow-release oxidant [28]. To cover the treatment zone, the outer screen extends from the targeted depth to slightly above the water table; this allows potential smear zones to be treated as the water table fluctuates. An aeration tube is placed in the gap between the outer and inner screens. For the reloadable design, the outer screen consists of multiple 6.0325 cm OD by 1.52 m SCH 40 PVC pipes with 0.0254 cm screens. The inner screen, which holds the 2.54-cm diameter oxidant candles, consists of 4.2164 cm OD SCH 80 PVC with 0.0254 cm screens (Figure 1B). Compressed air supply and vent line headers were constructed of propylene pipe installed with a hydraulic vibratory plow to a depth of 30–40 cm. Self-tapping branch connections tapped into the headers to connect to each insertion point with 6.35 mm polyethylene tubing (Figure 1C).

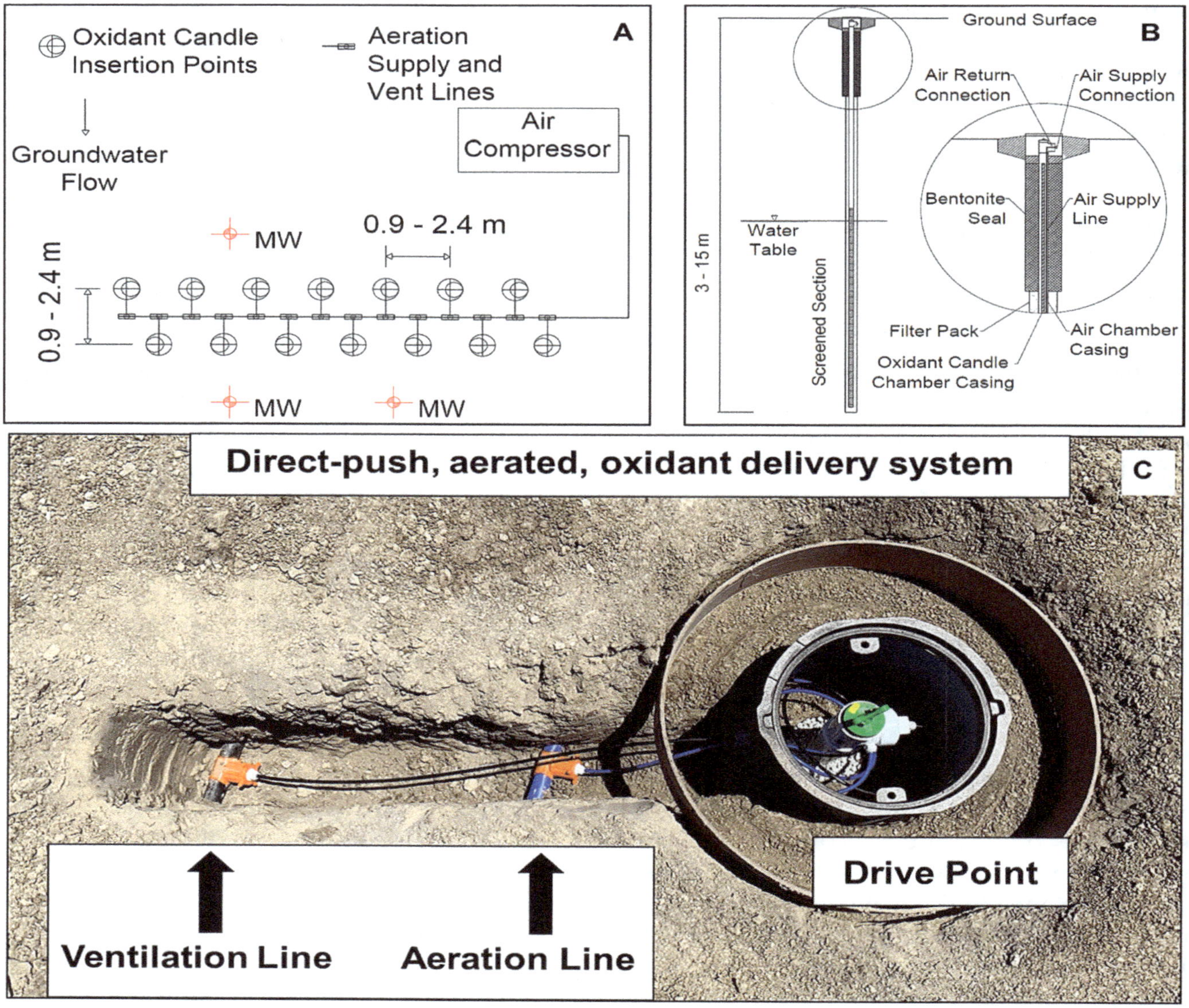

Figure 1. (**A**) Generic plot plan (plan view); (**B**) vertical view of drive points used in deploying the direct-push, aerated, oxidant delivery system and (**C**) photograph of direct-push, aerated, oxidant delivery system.

2.4. Visualizing Oxidant Dispersal from Aerated Oxidant Candles

To understand the flow patterns created by aerating the oxidant candles in the double screened assembly (i.e., airlift pump), a large acrylic tank was filled with water and field-scale 1.52-m segments of the outer and inner screen containing permanganate oxidant candles were lowered into the tank and aerated. Temporal photographs were taken as the permanganate dissolved and circulated around the outer screen. To aid in understanding the flow dynamics of the system, we used similarly dimensioned clear outer and inner screens and added dye to specific locations between the outer and inner screen (i.e., screen gap) and recorded water flow directions.

2.5. Field Testing of Aerated, Direct-Push, Oxidant Candles

Results from three field sites are presented. The site names and locations include: Freeman (Sargent, NE, USA), Textron (Lincoln, NE, USA) and Hoover (Beatrice, NE, USA). All three sites were treated with aerated oxidant candles that were installed with direct-push equipment. The Freeman site used a non-reloadable design whereas the Textron and Hoover site used a reloadable design that is currently being used at other locations. The number of drive points installed at each site is shown in the individual plot plans provided below; additional plot details and brief site histories are provided in SM.

The aerated oxidant candle drive points were placed in the contaminant source areas for the Freeman and Textron sites while the drive points at the Hoover site were positioned to test a pilot-scale barrier design (Table 1). Each field site was located in a different geographical location within the state of Nebraska (USA) and thus site and groundwater characteristics differed. Individual field site characteristics are presented and include: depth to groundwater; screen depths of oxidant candles and monitoring wells; soil textures; seepage velocity; groundwater quality characteristics and major contaminants of concern (Table 1).

Table 1. General site characteristics of three field sites treated with aerated oxidant candle drive points.

Site Characteristics	Freeman	Textron	Hoover
Treatment type	Source Zone	Source Zone	Permeable Reactive Barrier
Depth to Groundwater (m)	2.0	4.1	4.4
Screened Interval of Oxidant Drive Points (m)	1.5–6.1	1.5–6.1	1.5–6.4
Screened Interval of Monitoring wells (m)	1.2–4.3	1.5–6.1	3.4–6.4
Texture	Layers of Clay and Fine to Coarse Sand	Layers of Clay, Sandy Clay, and Medium to Fine Sand	Lean Clay with various amounts of Fine Sands
Seepage Velocity (cm/day)	16.5	26.5	0.21
Contaminant(s) of Concern	BTEX	BTEX	1,4-dioxane, TCE, BTEX
Dissolved Oxygen (mg/L) Initial Range [1]	0.76–1.64	0.34–3.30	0.32–1.80
ORP (mV) Initial Range	−248.6 to −207.1	−118.9 to 48.0	−167.4 to −48.5
pH Initial Range	6.45–6.78	6.15–6.98	6.47–7.06

[1] Initial range of water quality parameters measured in site monitoring wells (Freeman: SG-10, SG-19 and SG-20; Textron: TX-1, TX-2, TX-3 and TX-4 and Hoover: HV-1, HV-2 and HV-3) prior to initiating the aerated oxidant candle treatment.

2.6. *ISCO with Liquid Oxidant versus Aerated Oxidant Candle*

Given that the Freeman site had been treated by ISCO using a liquid oxidant in a previous investigation and aerated oxidant candles from this research, we compared the efficacy of using oxidant candles versus liquid oxidant in a controlled wide flow box experiment. Tank experiments used Ottawa graded sand (grain size 0.150–0.425 mm) that was packed into the tank by uniformly adding the sand in layers while raising the water level during packing. Additional details of the flow box construction (inlet, soil and outlet chambers) are presented in Supplementary Materials (Figure S4). To compare the two forms of oxidant delivery, we used phenol as the surrogate contaminant and flooded the tank with phenol (50 mg L^{-1}) under constant advection. Once inlet and outlet phenol concentrations stabilized, the tank inlet chamber was switched to tritiated water (^3H_2O) and pumped into the inlet chamber to produce the same advection rate. Effluent samples were collected from the outlet chamber to produce a tritiated water breakthrough curve. For the control treatment, the rise in ^3H_2O corresponded with a decrease in phenol as it was physically flushed out of the tank.

To determine phenol destruction by either the addition of liquid oxidant (injected at 8% of plume volume, with a permanganate concentration of 20,000 or 25,000 mg/L) or the aerated oxidant candles, we repeated this experiment and compared the mass of phenol recovered from the outlet chamber. In brief, the control treatment provided the baseline removal of phenol by advection or simple displacement (by ^3H_2O) whereas the oxidant treatments represented phenol removal by advection and chemical oxidation. We also compared the mass of permanganate used by both the liquid and oxidant candle treatments.

Samples for phenol and ^3H_2O analysis were taken from the outlet reservoir every 6 h while the total mass of effluent eluted from the flow tank was determined by collecting all the effluent in a single reservoir and periodically weighing. One milliliter effluent samples were taken for phenol analysis by high performance liquid chromatography (HPLC; Shimadzu Scientific Instruments Columbia, MD, USA) using a photodiode array detector and AQUASIL C18 column (Thermo Fisher Scientific, Waltham, MA, USA, dia. 250 mm $\times$ 4.6 mm, particle size 5 µm) that was coupled to a guard column. An isocratic mobile phase of 50% acetonitrile was used at a 1 mL min^{-1} flow rate. The samples were analyzed at 256 nm by an external calibration curve. A second 2 mL sample was taken and thoroughly mixed with 12 mL of Ultima Gold™ liquid scintillation cocktail (PerkinElmer, Inc., Waltham, MA, USA) to measure tritium (^3H) activity via liquid scintillation counting (Packard 1900TR scintillation counter, Downers Grove, IL, USA).

3. Results and Discussion

3.1. Visualizing Oxidant Dispersal from Aerated Oxidant Candles

Airlift pumps are historical devices (circa 1800) that have proven to be a reliable and low maintenance way of moving liquids in the petrochemical and oil industries [29,30]. In its simplest form, a compressor was used to inject air into the bottom of a submerged pipe and because air has a lower density than liquid water, it moves upward. In the context of our modular oxidant delivery system, the air bubbles placed in the screen gap created a buoyancy force that lifted and caused water and oxidant to dispel from the upper portions of the screen (Figure 2A). With water exiting the upper half of the screen, equilibrium is maintained by water being drawn in at the bottom of the screen (Figure 2A). Thus, the airlift pump, which has no mechanical or moving parts, provided a simple way of circulating and treating contaminated groundwater in the field.

Using clear screens and colored dye, we observed four distinct flow patterns created in the aerated oxidant delivery system. The flow patterns detected included: (1) water moved upward in the screen gap on the side where the aeration tube was located but moved (2) partially downward on the opposite side, causing an internal circulation within the screen gap; (3) water and oxidant were expelled from the top half of the outer screen and (4) water was drawn into the screen gap in the lower sections of the outside screen (Figure 2A).

Aerating the oxidant candles provides two important functions to the treatment. First, the induced flow patterns around the candle broke the intermolecular forces holding the oxidant together and allowed the oxidant to solvate with water, thus reducing downward density driven flow. Secondly, by aerating the candles inside the reloadable screen, a continuous, 24/7 circulation pattern around the drive point was created. Kambhu [31] simulated this action around miniature aerated double-screened oxidant candles by coupling a two-phase bubbly flow model with the Darcy flow model in a wide flow tank (Figure 2B; Figure S1). Kambhu et al. [26] showed that the aeration rate controls the outward flow of the oxidant from the outer screen in all directions and that the radius of influence around each drive point is largely a function of the outward velocity of the oxidant exiting the screen and the advection rate opposing the upgradient and lateral spreading of the oxidant (Figure 2B). For this reason, site-specific seepage velocities will be an important consideration in determining the spacing between oxidant drive points. Modeling by Kambhu et al. [26] also predicted the oxidant would move upgradient or against groundwater flow near the upper half of the screen. This upgradient flow pattern has been detected in our flow box experiments (Figure 2B,C) and as presented below, in some of our upgradient monitoring wells at the field sites. In the absence of aeration, our narrow flow box shows that much less coverage of the target zone occurred during advection due to density-driven flow of the permanganate (Figure 2C).

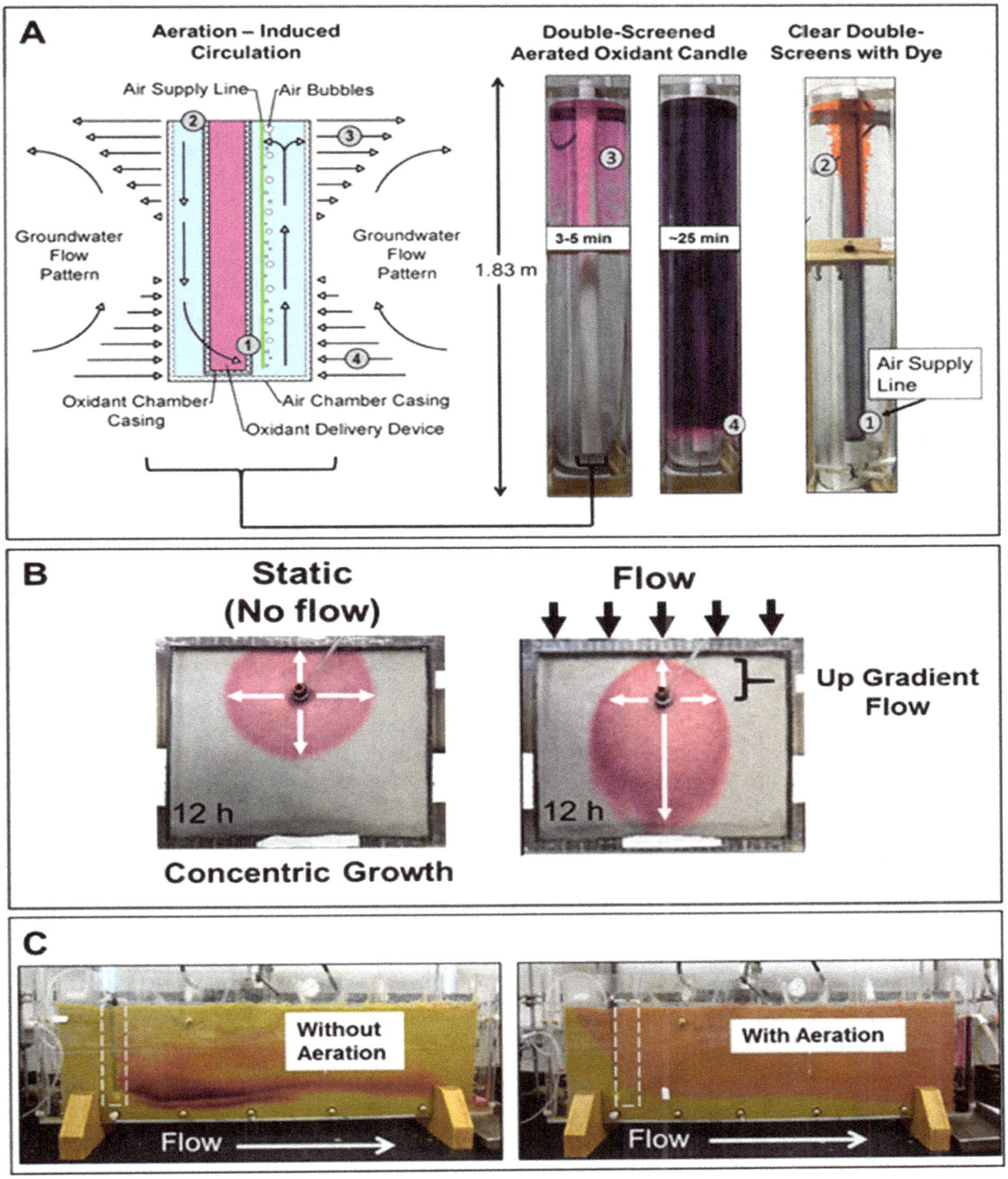

Figure 2. (**A**) Schematic illustrating observed flow patterns and temporal side-view photographs of oxidant dispersal patterns from field-scale, aerated oxidant candles; (**B**) top view of wide flow box (35.6 cm W × 30.5 cm L, see Figure S3) showing permanganate spreading from miniature double-screened candles after receiving similar aeration rates for 12 h, one with advection and one without (static) and (**C**) photographs of narrow flow box (122 cm W × 45 cm H) containing permanganate candle under advection, with and without aeration.

3.2. *Field Testing of Aerated, Direct-Push, Oxidant Candles*

Each field site was treated with multiple drive points containing permanganate or persulfate candles. Research that went into developing these field-scale oxidant candles has previously been reported [17,18,21,25,27]. These publications provide details on oxidant-wax ratios, oxidant candle compositions; oxidant release rates, pairing of persulfate candles with zerovalent iron candles and contaminant degradation rates via oxidant candle exposure. The chemistries of the two oxidants used in the slow-release candles (i.e., permanganate and persulfate) have also been extensively studied from both a kinetic and mechanistic perspective (e.g., [32–36]). In brief, permanganate is widely accepted as an efficient oxidant for ISCO applications and is extremely efficient in oxidizing chlorinated ethenes to CO_2 [37–40]. Likewise, several researchers have reported positive results in treating BTEX compounds with persulfate in contaminated soil and water [35,41–45].

When treating a contaminated plume, traditional ISCO injections using liquid oxidants must typically account for soil oxidant demand (SOD) of the aquifer solids. While soil oxidant demand is still a consideration when using aerated oxidant candles, the direct-push, aerated oxidant candles are stationary and continuously release oxidant to the same zone around the drive points. This means that the SOD within the radius of influence will eventually decrease with time. Thus, as oxidant candles are reloaded into the drive points, less oxidant would be consumed by the SOD and more would be available to treat the contaminants. For sites with high soil oxidant demands, using the aerated oxidant candles in a barrier design may be preferred because it would provide a stationary source of oxidant to the same treatment zone for months to years and avoid having to account for the soil oxidant demand of the entire plume.

3.3. *Freeman Site*

The Freeman site was a typical abandoned gas station that had experienced leaking petroleum from its underground storage tanks. This site was responsible for vapor intrusion in a resident's basement approximately 21 m downgradient (Figure 3). Following several years of investigation and remedial treatments, which included two ISCO injections, the Nebraska Department of Environment and Energy (NDEE) approved the use of aerated, direct-push, oxidant candles. Before the oxidant candles were installed at the site, a treatability study was first performed where miniature oxidant candles were allowed to react with the groundwater for 14 days prior to quenching and sending off for analysis. Based on this treatability test, the persulfate + zerovalent iron (ZVI) candles worked best followed by the permanganate candles (Table S1). For comparison, we placed the field-scale persulfate + ZVI candles upgradient of two monitoring wells (SG-10, SG-19, Figure 3) and permanganate candles upgradient of monitoring well SG-20 (Figure 3).

While the chemistries of permanganate and persulfate are effective in degrading various contaminants, the key to successful treatment has always been achieving adequate coverage of the target zone. By including aeration in the technology to induce a circulation pattern around each oxidant drive point (Figure 2), oxidant coverage was greatly enhanced.

Temporal samplings of Freeman monitoring wells showed significant reductions (>99%) in groundwater contaminants following oxidant candle installation. By plotting the historical xylene concentrations (which were in most abundance), results showed contaminant concentrations decreased following the two ISCO injections, but concentrations eventually rebounded (Figure 4). By contrast, all BTEX concentrations decreased after installation of the oxidant candles with no sustained rebound (Figure 4). The continued reduction in BTEX concentrations combined with the lack of vapor intrusion in the nearby resident's basement prompted NDEE regulators to close the site to further treatment. These results provided support for combining aeration with slow-release oxidants.

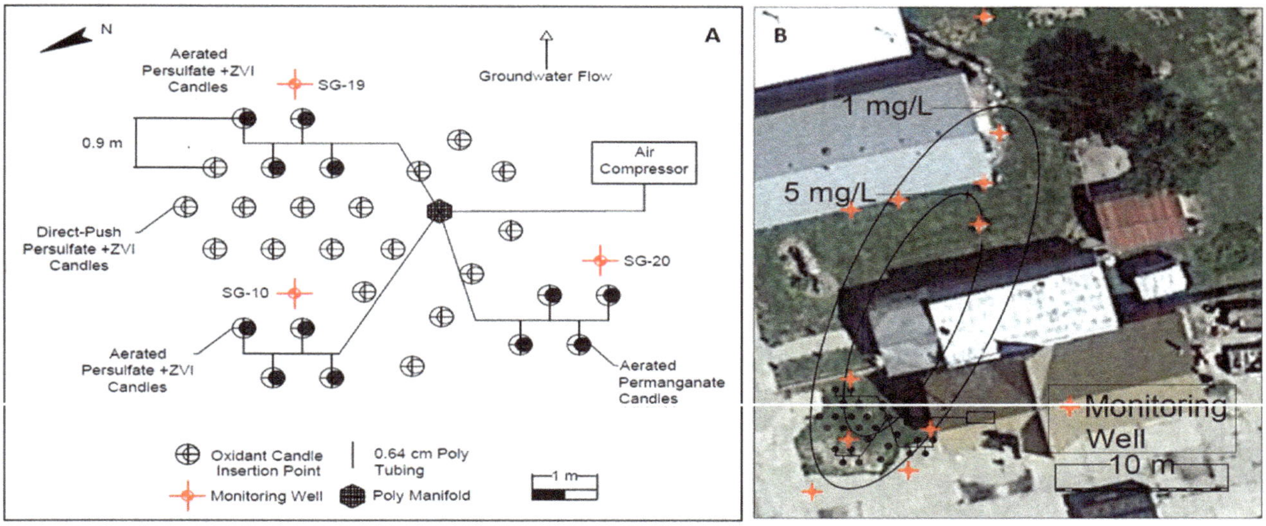

Figure 3. (**A**) Plot plan of Freeman site showing location of monitoring wells and direct-push oxidant candles, with and without aeration and (**B**) plan view of Freeman site showing locations of xylene plume and oxidant drive points (circled).

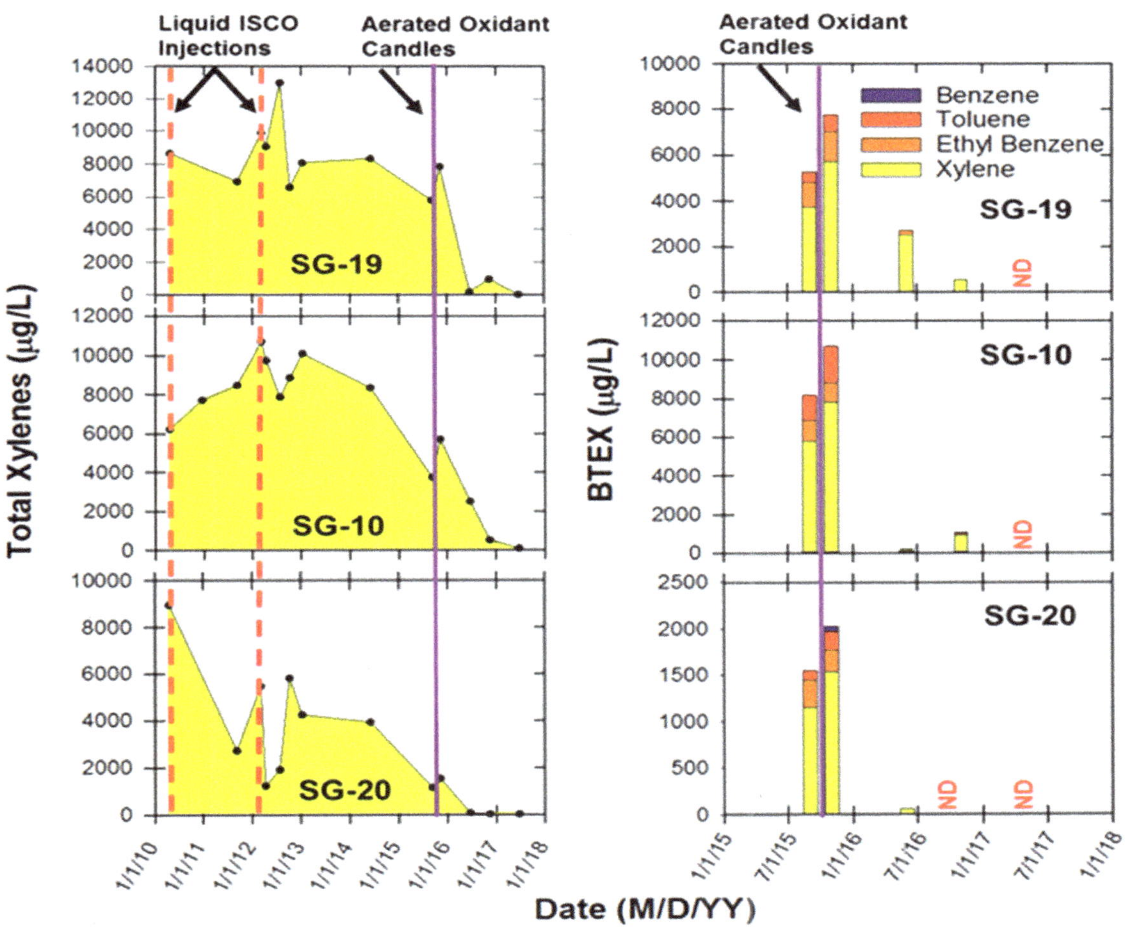

Figure 4. (**Left**) Historical changes in total xylene concentrations at the Freeman site following treatments with two liquid oxidant injections and aerated oxidant candles and (**Right**) temporal changes in BTEX concentrations following installation of aerated oxidant candles.

3.4. *ISCO with the Liquid Oxidant versus Aerated Oxidant Candle*

While traditional ISCO (i.e., injected liquid oxidants) and oxidant candles were both used on the Freeman site (Figure 4), these two field technologies were not set up to be directly compared due to differences in time (years) and space (injection points) and the mass of oxidant deployed. To provide a more objective comparison, a larger flow tank experiment was conducted where similar masses of oxidant were added (via liquid vs. oxidant candles) under the same advection rate. This series of transport experiments demonstrated the removal of the surrogate contaminant (i.e., phenol) from the flow tank by three means, namely: (i) physical displacement (control); (ii) physical displacement plus oxidation by liquid oxidant injection and (iii) physical displacement plus oxidation by aerated oxidant candles.

Differences in the mass of phenol recovered between the control and liquid injections were between 34 and 39% (for the two concentrations) whereas the difference between the control and aerated oxidant candles was 66% (Figure 5). Therefore, the aerated oxidant candles removed significantly more phenol than the liquid oxidant injections. Given that the mass of the oxidant added was similar between the liquid injections (12 and 15 g for the two concentrations) and oxidant candles (12.5 g via mass released), differences in the percent of phenol removed were due to differences in the spatial distribution of the permanganate throughout the flow box, which ultimately impacted contaminant contact. Although the liquid oxidant was added as 8% of the flow tank pore volume and injected at three locations and multiple depths, the density of the permanganate eventually caused it to sink to the bottom of the flow tank (Figure 5). By contrast the aerated oxidant candles produced a more even vertical distribution of the permanganate in the flow tank, with permanganate visibly present at the surface (Figure 5). The greater oxidant spreading produced by the aerated system resulted in a more efficient removal of the phenol during transport.

3.5. *Textron Site*

The Textron site is a targeted brownfield and previous environmental assessments have confirmed the presence of petroleum products in the groundwater, namely benzene, ethylbenzene, toluene and xylene (i.e., BTEX). Using the reloadable drive points at the Textron site (Figure 6), we observed that the aerated oxidant candles dramatically decreased BTEX concentrations (Figure 7). The highest initial BTEX concentrations were located in monitoring wells TX-1 (6909 μg BTEX/L) and TX-2 (7687 μg BTEX/L). Monitoring well TX-1 was located just upgradient from the first line of drive points and TX-2 was just downgradient of the drive points (Figure 6). Within 3 months, we observed significant BTEX decreases in all four monitoring wells (81–99%) and by 12 months, decreases ranged between 96% and below detection limits (Figure 7).

The decrease in BTEX concentrations in the upgradient well (TX-1) is perhaps curious but as mentioned above (Figure 2), the circulation pattern created around the drive point causes water and oxidant to be expelled and move out concentrically away from the drive point (Figure 2B). This can cause upgradient concentrations to decrease, as observed in well TX-1, where traces of persulfate were observed (<0.7 mg/L) and BTEX concentrations declined but at a slower rate than the downgradient wells where higher persulfate concentrations were recorded (TX-2, 2.1 mg/L; TX-3, >70 mg/L; TX-4, 21 mg/L). Supporting flow tank experiments verified that the oxidant can move upgradient (Figure S1) with the radius of influence around each drive point largely a function of the outward velocity of the oxidant exiting the screen. This outward velocity would be a function of the aeration rate and the depth of water above the aeration tube, with greater heads producing more velocity [21,26].

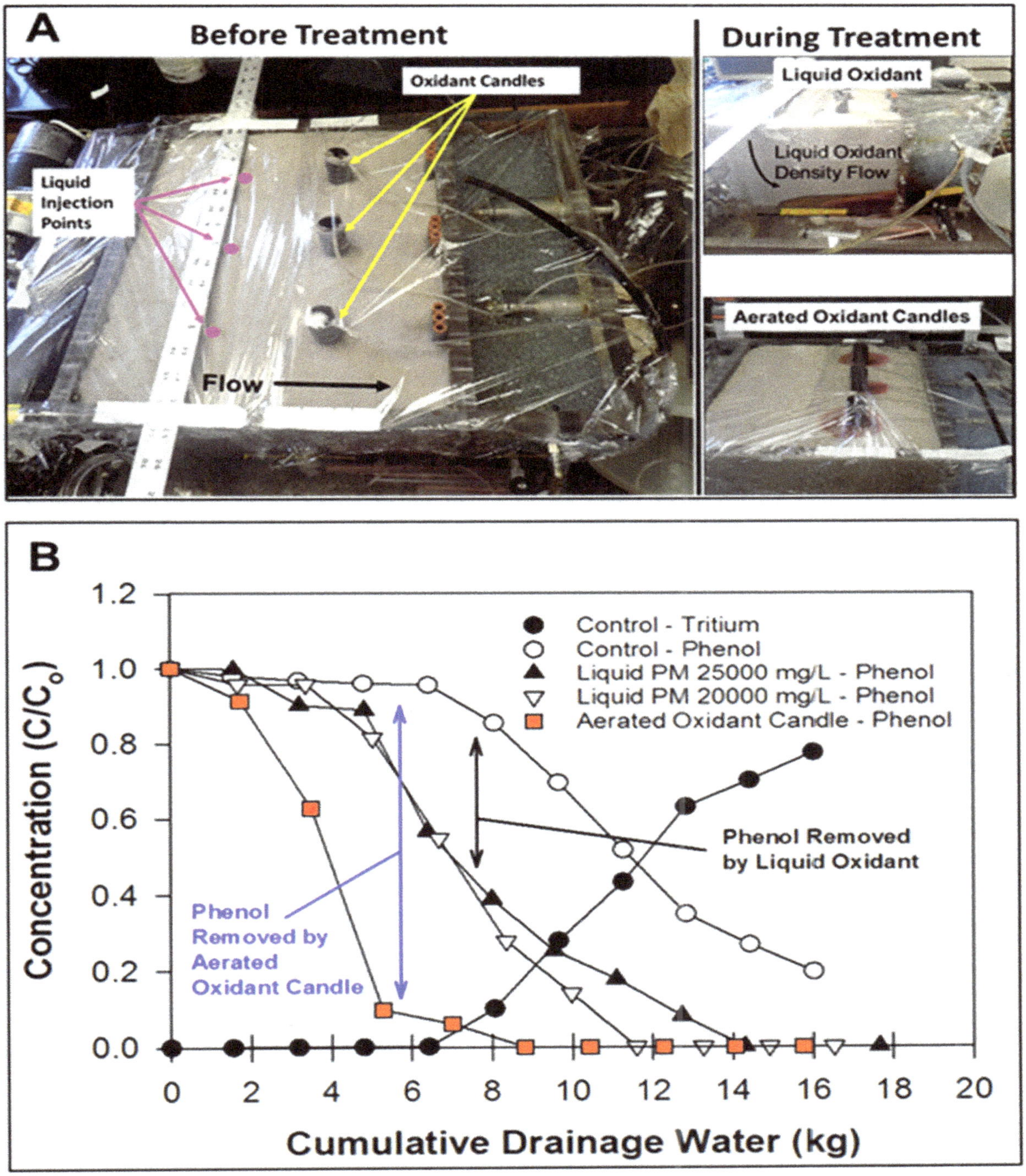

Figure 5. (**A**) Photographs of the flow box used to compare liquid oxidant versus aerated oxidant candles during transport and (**B**) breakthrough curves of ^{3}H2O and phenol removal during transport experiments. Transport experiments were repeated four times and included the following treatments: control (^{3}H20 only); liquid permanganate (PM) injection (20,000 mg/L); PM injection (25,000 mg/L) and aerated oxidant candles.

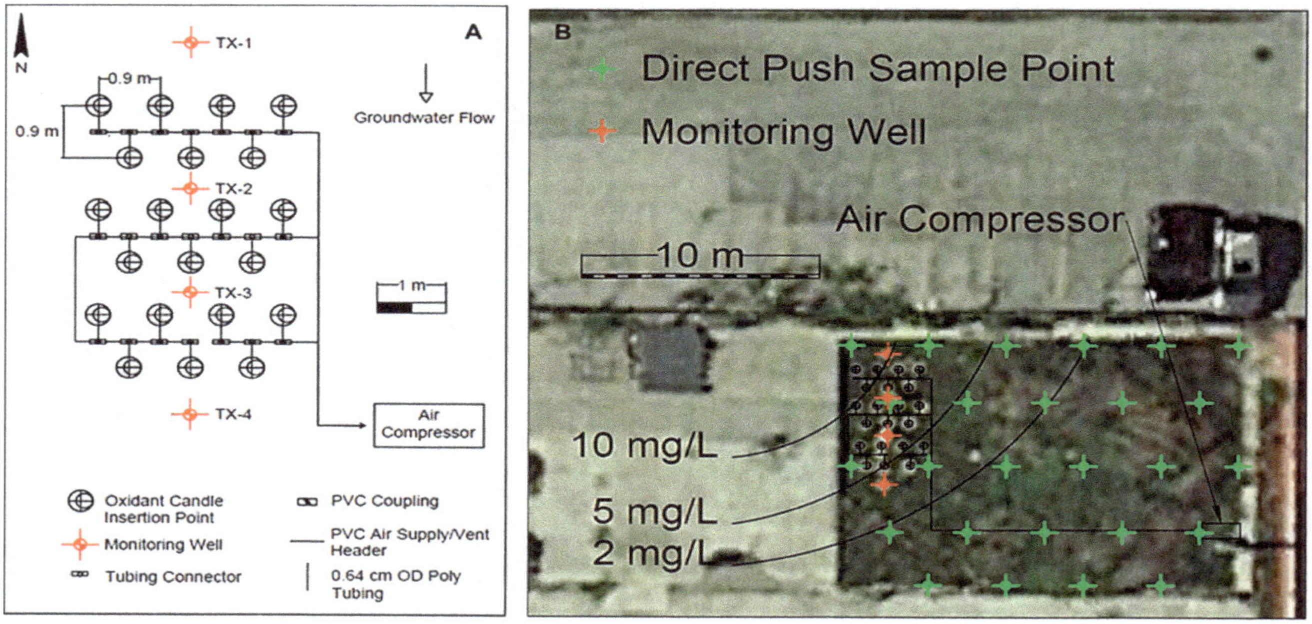

Figure 6. (**A**) Plot plan of the Textron site showing the location of monitoring wells and direct-push oxidant candles with aeration and (**B**) plan view of the Textron site showing BTEX contours and oxidant drive points (circled).

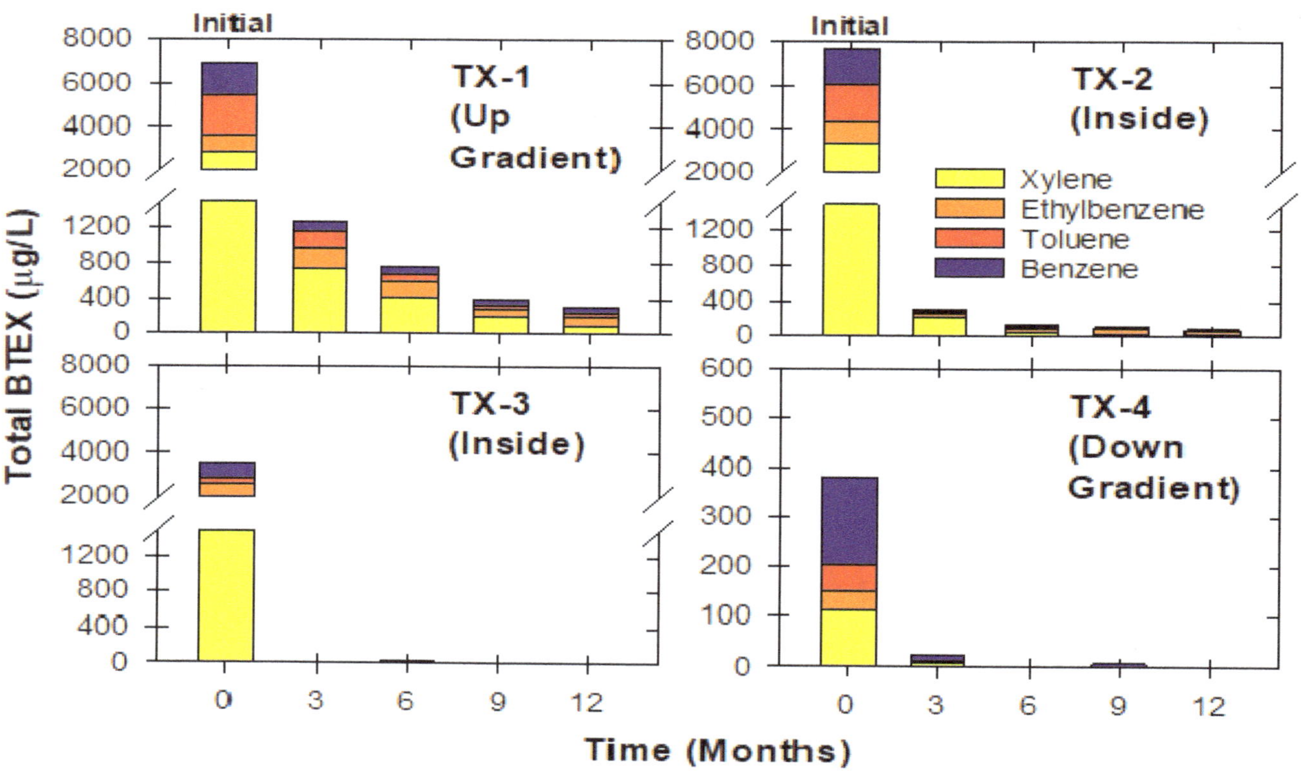

Figure 7. Changes in BTEX concentrations (μg L^{-1}) following treatment with reloadable direct-push aerated oxidant candles at the Textron site.

We also used this site to measure oxidant candle longevity in the field. Preweighed sodium persulfate candles were first installed (Figure 8). Specific candles placed in the drive points were marked and measured over time through sacrificial sampling. Results showed that under field conditions, the sodium persulfate candles would likely last 4–6 months. Once spent, we removed the sodium candles and replaced them with preweighed potassium persulfate candles. These results showed that the potassium persulfate candles would last approximately 8–9 months (Figure 8).

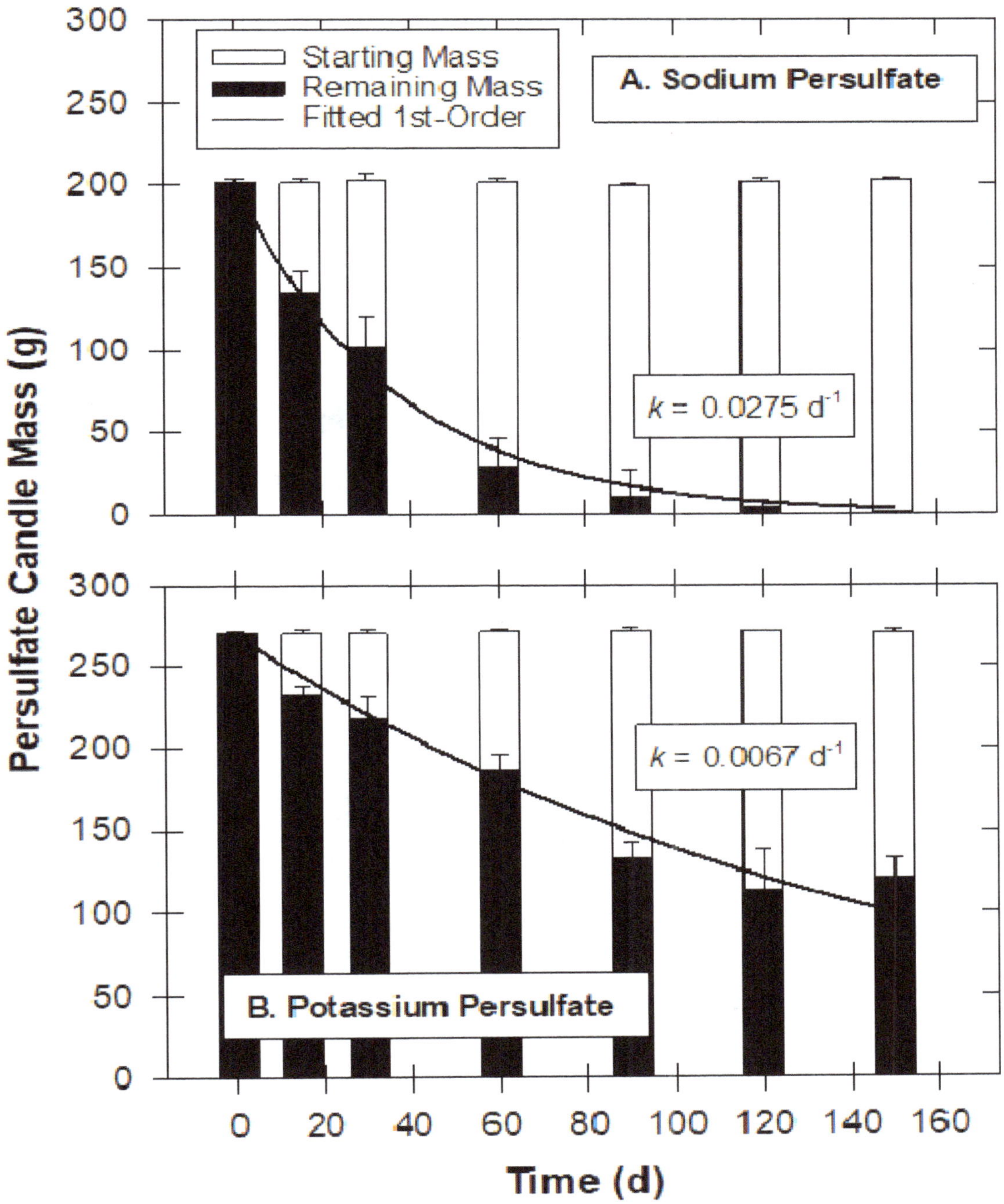

Figure 8. Temporal changes in persulfate mass in field-deployed oxidant candles at the Textron site; (**A**) sodium persulfate candles and (**B**) potassium persulfate candles. Open bars show candle weights at T = 0 day; solid bars show candle weights once removed from the aquifer and dried.

3.6. Hoover Site

Previous subsurface assessments of Hoover's property indicated detectable levels of metals, VOCs, SVOCs and hydrocarbons. The contaminant source areas include a former manufactured gas plant (MGP) site and two underground storage tanks, one containing gasoline and the other containing many different chemicals. Seepage from these source zones and the reducing conditions of the aquifer likely contributed to the suite of chlorinated degradation products present, including some reduced chlorinated degradation products (i.e., DCE and VC).

Initially, we performed a treatability test by incubating groundwater from the Hoover site with miniature permanganate candles (Table S2). Based on an initial treatability study, we installed permanganate candles into 14 reloadable drive points (Figure 9). One additional challenge this site posed was that groundwater was only 1 m deep above bedrock. The shallow depth reduced the strength of the airlift pump created in the drive points and thus likely provided less circulation. Moreover, a shallow water depth limits the number of candles that can be exposed, which results in less oxidant mass being released.

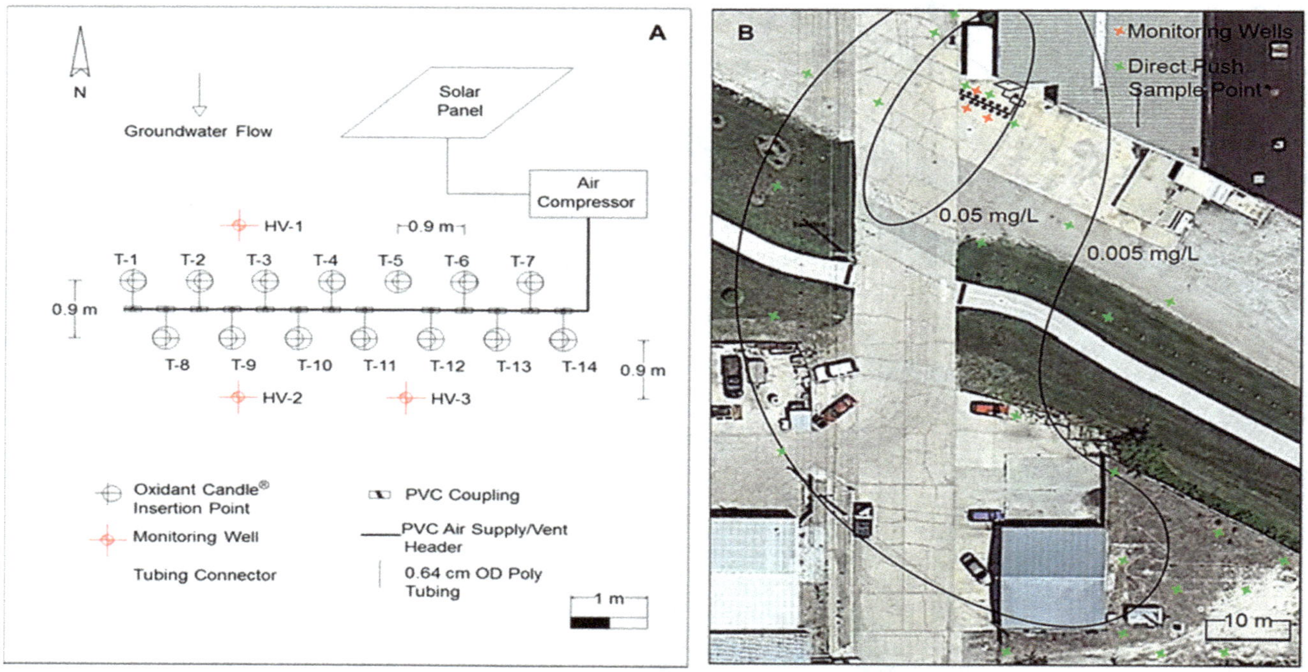

Figure 9. (**A**) Plot plan of the Hoover site showing the location of direct-push oxidant candles with aeration and (**B**) plan view of the Hoover site showing locations of TCE, BTEX and 1,4-dioxane plume and oxidant drive points (circled).

Monitoring the two downgradient wells after installation (HV-2 and HV-3, Figure 9) showed no consistent downward trends in contaminant concentrations (Figure 10). To verify that the permanganate candles were providing ample concentrations of permanganate to the treatment zone, we sampled inside the individual drive points (Figure 9) for permanganate and found concentrations in excess of 6000 mg/L. This proved the candle barrier was delivering adequate permanganate concentrations to the treatment zone.

We present two possible reasons for the initial lack of efficacy. The first is that this site was using a solar-powered compressor. Since aeration fluctuated diurnally, outlet ends of the aeration tubes became filled with permanganate when the aeration was off (nighttime hours) and some became coated with manganese dioxide (MnO_2), a product of permanganate oxidation. This likely clogged some of the aeration tubes and hindered oxidant spreading.

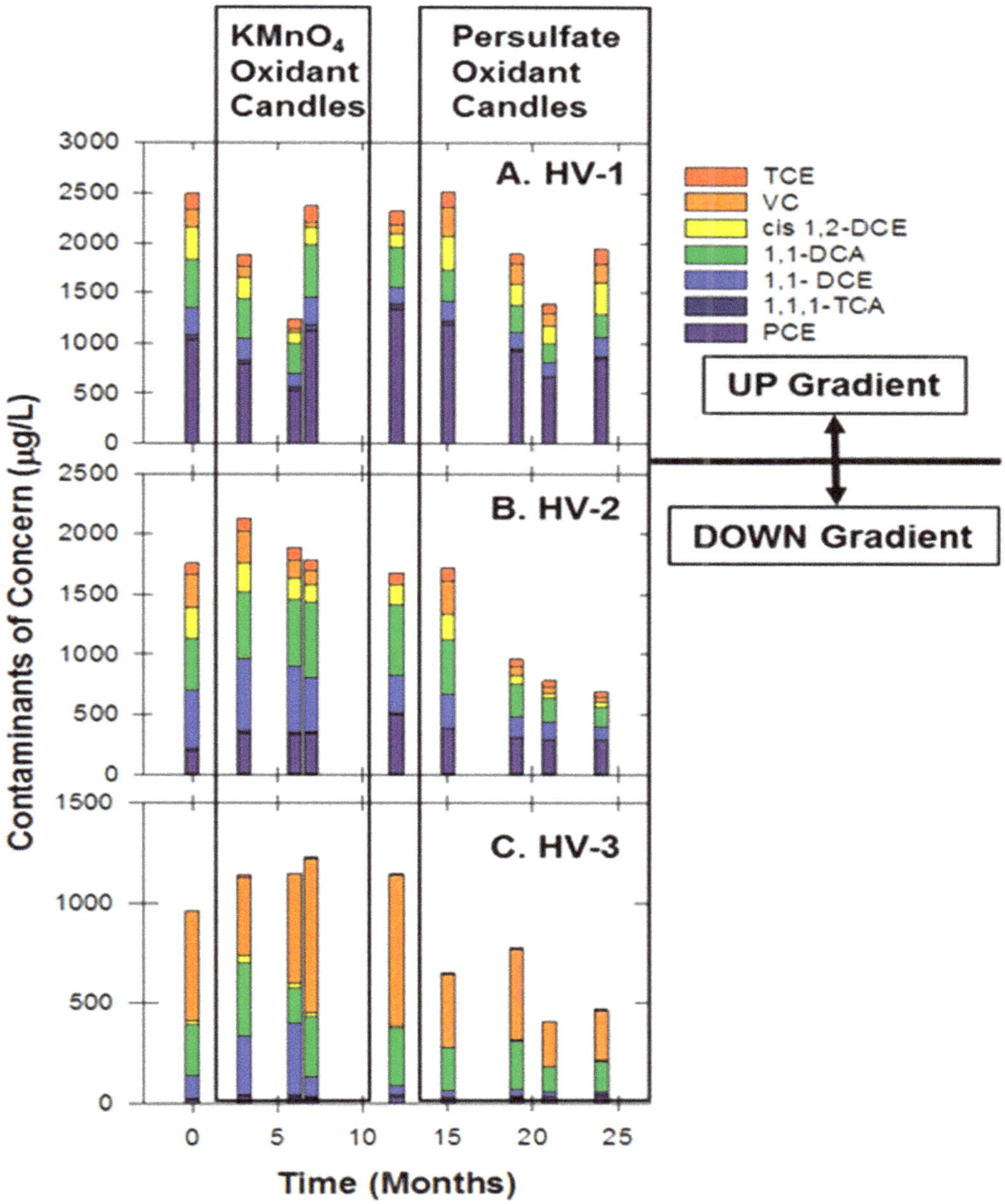

Figure 10. Temporal changes in contaminants of concern concentrations following installation of permanganate oxidant candles followed by persulfate oxidant candles at the Hoover site.

The second possible reason for the lack of efficacy by the permanganate candles was the observed production of acetone. Acetone and 2-butanone were not initially present in the Hoover groundwater but were observed in our treatability study (Table S2) and in the upgradient and downgradient wells after the permanganate candles were installed, with acetone concentrations ranging from 14 to 500 µg/L within the first 12 months. While acetone could have been an oxidation product of some of the other contaminants present, we performed additional testing and determined that both permanganate and persulfate could react with the wax and release acetone (Table S3). While the saturated paraffin structure would generally be inert to an oxidant attack, the presence of any unsaturated bonds near the end of a carbon chain could lead to formation of ketones (i.e., acetone and 2-butanone).

While acetone is known to biodegrade [46] and there currently is no federally enforceable maximum contaminant level (MCL) for acetone in drinking water, its production is problematic for other reasons. Treatment of acetone with permanganate can form oxalic acid [47] and oxalic acid treated with permanganate can form Mn(II) [48]. Manganese (II) is known to react with permanganate and cause autocatalysis [49]. These three reactions are summarized in Figure 11. Chokejaroenrat et al. [50] similarly found that small volumes of acetone in an aqueous matrix (1 mL/150 mL, *v*/*v*) caused permanganate reactions with the contaminant RDX to stall or plateau and correspondently showed that the loss of permanganate from solution (when acetone was present) was responsible for the stalled reaction. Chokejaroenrat [51] further showed that by keeping the pH acidic, the autocatalysis of permanganate induced by acetone could be avoided. Given that acetone was showing up in the downgradient wells without permanganate and that the contaminant concentrations were not decreasing, we believe that acetone was causing the permanganate to decrease and thus limiting permanganate effectiveness to degrade the contaminants.

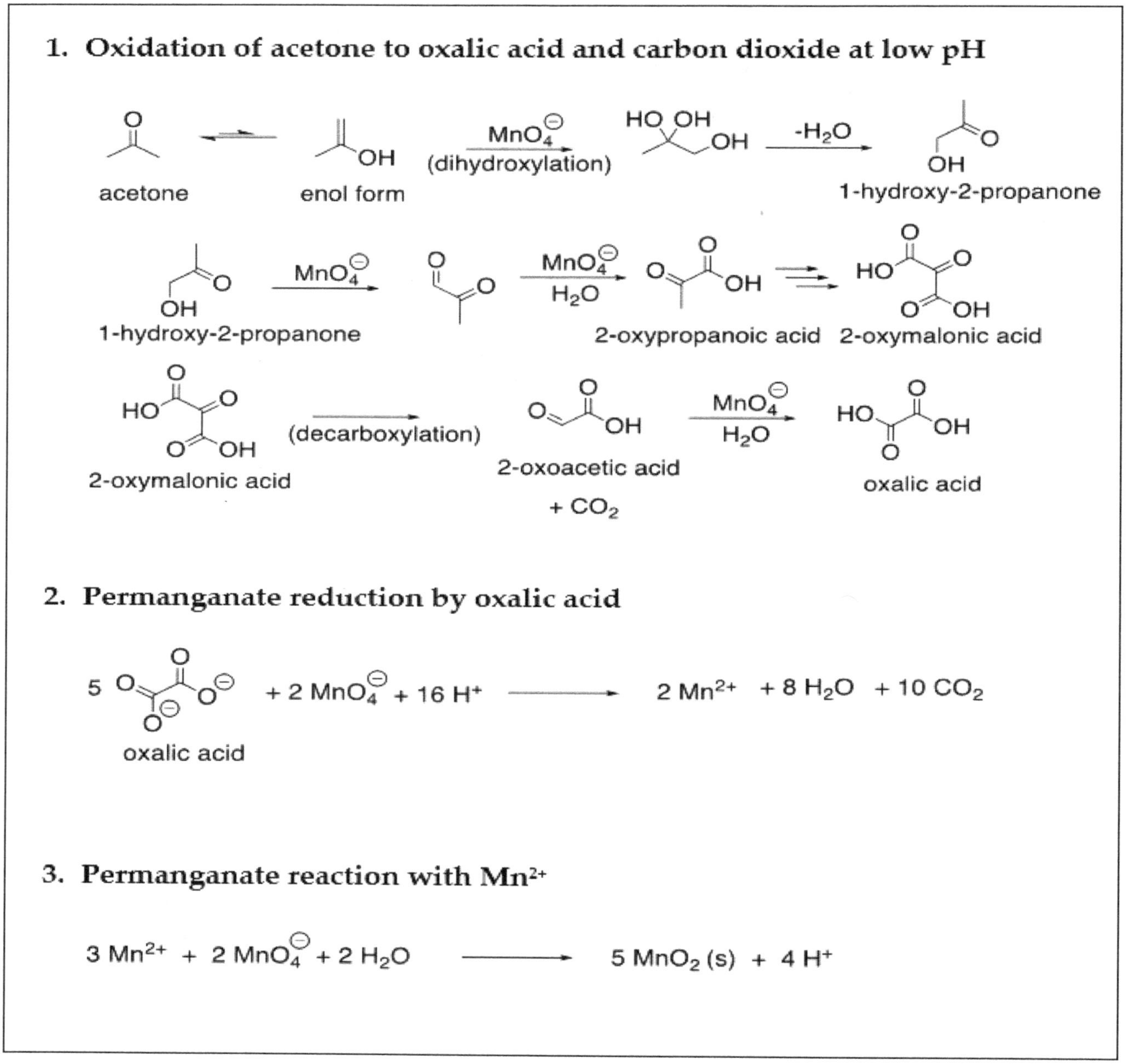

Figure 11. The reaction of acetone with permanganate leading to autocatalysis of permanganate.

To correct the situation, we performed a second treatability study, which included persulfate candles. The second treatability study showed that persulfate alone and persulfate + zerovalent iron combined [21] could both readily degrade the majority of contaminants present in the Hoover groundwater. The persulfate candles however, would provide the advantage of not being influenced by the production of acetone. For this reason, we removed the solid permanganate candles from the Hoover site and replaced them with persulfate and zerovalent iron candles. The flexibility of the reloadable design allowed for this change with minimal effort (i.e., the inside screen containing the oxidant was removed and replaced). In addition to changing the chemical oxidant, an electrically powered compressor replaced the solar-powered compressor.

Sampling results from the downgradient wells (HV-2 and HV-3, Figure 10) following installation of the persulfate candles showed that contaminant concentrations decreased by 52% (HV-3) and 61% (HV-2) from the initial concentrations with a clear downward trend observable in monitoring well HV-2 (Figure 10). As observed in the monitoring wells at the Textron site, persulfate was also observed in the upgradient well, HV-1 (<0.7–3.5 mg/L), and downgradient wells (HV-2, 0.7–2.1 mg/L and HV-3, 7–14 mg/L). This site is still being treated but sampling of monitoring wells has been infrequent since 2018. Still, the 2020 sampling showed continued marked decreases in contaminant concentrations, indicating the aerated, slow-release persulfate treatment is effective.

Finally, by isolating the effects of the oxidant candles on the contaminant 1,4-dioxane at the Hoover site, we similarly found the permanganate candles were not effective. However, decreases in 1,4-dioxane concentrations in the upgradient and downgradient wells were observed after the persulfate candles were installed (Figure S4). These field results corroborate Kambhu et al. [21] research, which showed that provided Fe^{2+} is not in excess, slow-release persulfate and zerovalent iron candles could effectively degrade 1,4-dioxane. Using unactivated persulfate, Evans et al. [52,53] also confirmed that slow release persulfate cylinders could degrade 1,4-dioxane under field conditions.

4. Conclusions

In O'Connor et al. [24] review of previously published papers on controlled-release remedial substances, they noted that in order for slow-release oxidants to become viable, it is imperative that new technologies be developed that increase the radius of influence under various geological settings. While the use of extraction wells to create gradients around oxidant cylinders has been proposed as one means of increasing oxidant coverage [13,14], Evans et al. [52,53] reported density-driven flow was still observed from persulfate-wax cylinders under an induced hydraulic gradient in the field.

We report herein, the design and installation of an aerated, modular oxidant delivery system that can be installed by direct-push equipment [28]. This technology avoids some of the burdens and inconveniences associated with liquid oxidant injections by eliminating the need for chemical mixing tanks, water source, pumps or pressurized injections. Likewise, the reloadable drive points allow for months to years of oxidant exposure to the treatment zone. Longer exposure times can facilitate oxidant penetration into lower permeable zones and reduce the potential for rebound [54]. By conducting various flow tank studies, we confirmed that a circulation pattern of oxidant and water is created around the aerated drive points and that this circulation pattern increases lateral dispersion, minimizes density-flow and thus eliminates the need to induce a gradient with extraction wells [26]. Field results confirm that the technology is effective in remediating contaminated groundwater either through the treatment of source zones or by creating a barrier.

5. Patents

Christenson M., Comfort S.D. Modular Oxidant Delivery System. United States Patent. Patent No. US 9,925,574 B2. 27 March 2018.

Supplementary Materials: The following are available online at http://www.mdpi.com/2073-4441/12/12/3383/s1, Table S1. Laboratory results from treatability experiments using groundwater from SG-10 (Freeman site) and oxidant candles. Table S2. Treatability results showing changes in contaminant concentrations of Hoover Site groundwater following treatment with miniature permanganate oxidant candles for 14 days. Table S3. Acetone and 2-butanone analysis following wax-oxidant 14 days incubations. Figure S1. Examples of observed and simulated flow around aerated oxidant candles in narrow and wide flow boxes along with simulated oxidant flow. Figure S2. (A) Schematic of non-reloadable aerated, slow-release oxidant deployed by direct-push equipment for Freemont Site; (C) Schematic of reloadable aerated, slow-release oxidant deployed by direct-push equipment for Textron and Hoover Sites. Figure S3. Schematic of wide flow box used to compare liquid oxidant versus aerated oxidant candles during transport. Figure S4. Temporal changes in 1,4-dioxane concentrations following installation of permanganate oxidant candles and persulfate oxidant candles at Hoover site.

Author Contributions: J.R.: Investigation, Methodology, Writing—original draft, M.C.: Funding acquisition, Investigation, Methodology, Writing—Review and Editing, A.K.: Software modeling, Writing—Review and Editing, Y.L.: Software modeling, Writing—Review and Editing, C.E.H.: Chemistry, Writing—Review and Editing, S.C.: Conceptualization, Writing—Review and Editing, Funding acquisition. All authors have read and agreed to the published version of the manuscript.

Acknowledgments: Appreciation is expressed to project managers Phil Hargis, Daniel Ross and Laurie Brunner (Nebraska Department of Environment and Energy), project engineer, Mike Reif (Terracon) and drill team leader, Matt Marxsen (Conservation and Survey Division, UNL) for site assistance and drive-point installations. The authors also acknowledge the undergraduate students, Wei Tang, Laura LeCuyer and Katie Spencer for assistance with data collection.

References

1. Huling, S.G.; Ross, R.R.; Prestbo, K.M. In situ chemical oxidation: Permanganate oxidant volume design considerations. *Groundw. Monit. Remediat.* **2017**, *37*, 78–86. [CrossRef]
2. Suthersan, S.; McDonough, J.; Schnobrich, M.; Divine, C. In situ chemical treatment: A love-the relationship. *Groundw. Monit. Remediat.* **2017**, *37*, 17–26. [CrossRef]
3. Krembs, F.J.; Siegrist, R.L.; Crimi, M.L.; Furrer, R.F.; Petri, B.G. ISCO for groundwater remediation: Analysis of field applications and performance. *Groundw. Monit. Remediat.* **2010**, *30*, 42–53. [CrossRef]
4. PERF. *Performance Evaluation of In situ Chemical Oxidation of Petroleum Impacts in Soil and Groundwater*; Petroleum Environmental Research Forum (PERF) Project 2009-01; ExxonMobil: Irving, TX, USA, 2013.
5. Liu, C.X.; Ball, W.P. Back diffusion of chlorinated solvent contaminants from a natural aquitard to a remediated aquifer under well-controlled field conditions: Predictions and measurements. *Groundwater* **2002**, *40*, 175–184. [CrossRef]
6. Mundle, K.; Reynolds, D.A.; West, M.R.; Kueper, B.H. Concentration rebound following in situ chemical oxidation in fractured clay. *Groundwater* **2007**, *45*, 692–702. [CrossRef]
7. McGuire, T.M.; McDade, J.M.; Newell, C.J. Performance of DNAPL source depletion technologies at 59 chlorinated solvent-impacted sites. *Groundw. Monit. Remediat.* **2006**, *26*, 73–84. [CrossRef]
8. Kang, N.; Hua, I.; Rao, P.S. Production and characterization of encapsulated potassium permanganate for sustained release as an in situ oxidant. *Ind. Eng. Chem. Res.* **2004**, *43*, 5187–5193. [CrossRef]
9. Ross, C.; Murdoch, L.C.; Freedman, D.L.; Siegrist, R.L. Characteristics of potassium permanganate encapsulated in polymer. *J. Environ. Eng.* **2005**, *131*, 1203–1211. [CrossRef]
10. Schwartz, F.W. *Semi-Passive, Chemical Oxidation Schemes for the Long-Term Treatment of Contaminants*; The Ohio State University: Columbus, OH, USA, 2005.
11. Lee, E.S.; Schwartz, F.W. Characteristics and applications of controlled-release $KMnO_4$ for groundwater remediation. *Chemosphere* **2007**, *66*, 2058–2066. [CrossRef]
12. Lee, E.S.; Schwartz, F.W. Characterization and optimization of long-term controlled release system for groundwater remediation: A generalized modeling approach. *Chemosphere* **2007**, *69*, 247–253. [CrossRef]
13. Lee, E.S.; Liu, G.M.; Schwartz, F.W.; Kim, Y.J.; Ibaraki, M. Model-based evaluation of controlled-release systems in the remediation of dissolved plumes in groundwater. *Chemosphere* **2008**, *72*, 165–173. [CrossRef] [PubMed]
14. Lee, E.S.; Woo, N.C.; Schwartz, F.W.; Lee, B.S.; Lee, K.C.; Woo, M.H.; Kim, J.H.; Kim, H.K. Characterization of controlled release $KMnO_4$ (CRP) barrier system for groundwater remediation: A pilot-scale flow-tank study. *Chemosphere* **2008**, *71*, 902–910. [CrossRef] [PubMed]
15. Lee, B.S.; Kim, J.H.; Lee, K.C.; Bin Kim, Y.; Schwartz, F.W.; Lee, E.S.; Woo, N.C.; Lee, M.K. Efficacy of controlled-release $KMnO_4$ (CRP) for controlling dissolved TCE plume in groundwater: A large flow-tank study. *Chemosphere* **2009**, *74*, 745–750. [CrossRef]

16. Swearingen, J.; Swearingen, L. Encapsulated Reactant and Process. U.S. Patent No. US 7,431,849 B1, 7 October 2008.
17. Kambhu, A.; Comfort, S.; Chokejaroenrat, C.; Chainarong, C. Developing slow-release persulfate candles to treat BTEX contaminated water. *Chemosphere* **2012**, *89*, 656–664. [CrossRef] [PubMed]
18. Rauscher, L.; Sakulthaew, C.; Comfort, S. Using slow-release permanganate candles to remediate PAH-contaminated water. *J. Hazard. Mater.* **2012**, *241–242*, 441–449. [CrossRef] [PubMed]
19. Yuan, B.L.; Li, F.; Chen, Y.; Fu, M.-L. Laboratory-scale column study for remediation of TCE-contaminated aquifers using three-section controlled-release potassium permanganate barriers. *J. Environ. Sci.* **2013**, *25*, 971–977. [CrossRef]
20. Liang, S.H.; Chen, K.F.; Wu, C.S.; Lin, Y.H.; Kao, C.M. Development of $KMnO_4$-releasing composites for in situ chemical oxidation of TCE-contaminated groundwater. *Water Res.* **2014**, *54*, 149–159. [CrossRef] [PubMed]
21. Kambhu, A.; Gren, M.; Tang, W.; Comfort, S.; Harris, C. Remediating 1,4-dioxane-contaminated water with slow-release persulfate and zerovalent iron. *Chemosphere* **2017**, *175*, 170–177. [CrossRef]
22. Angkaew, A.; Sakulthaew, C.; Santapanajaur, T.; Poapolathep, A.; Chokejaroenrat, C. UV-activated persulfate oxidation of 17β-estradiol: Implications for discharge water remediation. *J. Environ. Chem. Eng.* **2019**, *7*, 102858. [CrossRef]
23. Ma, Y.; Feng, Y.; Feng, Y.; Liao, G.; Sun, Y.; Ma, J. Characteristics and mechanisms of controlled-release KMnO4 for groundwater remediation: Experimental and modeling investigation. *Water Res.* **2020**, *171*, 115385. [CrossRef]
24. O'Connor, D.; Hou, D.; Ok, Y.S.; Song, Y.; Sarmah, A.K.; Li, X. Sustainable in situ remediation of recalcitrant organic pollutants in groundwater with controlled release materials: A review. *J. Control. Release* **2018**, *283*, 200–213. [CrossRef]
25. Christenson, M.; Kambhu, A.; Comfort, S.D. Using slow-release permanganate candles to remove TCE from a low permeable aquifer at a former landfill. *Chemosphere* **2012**, *89*, 680–687. [CrossRef]
26. Kambhu, A.; Li, Y.; Gilmore, T.; Comfort, S. Modeling the Release and Spreading of Permanganate from Aerated Slow-Release Oxidants in a Laboratory Flow Tank. *J. Hazard. Mater.* **2021**, *403*, 123719. [CrossRef]
27. Christenson, M.; Kambhu, A.; Reece, J.; Comfort, S.D.; Brunner, L. A five-year performance review of field-scale, slow-release permanganate candles with recommendations for second-generation improvements. *Chemosphere* **2016**, *150*, 239–247. [CrossRef]
28. Christenson, M.; Comfort, S.D. Modular Oxidant Delivery System. U.S. Patent No. US 9,925,574 B2, 27 March 2018.
29. Yoshinaga, T.; Sato, Y. Performance of an air-lift pump for conveying coarse particles. *Int. J. Multiph. Flow* **1996**, *22*, 223–238. [CrossRef]
30. Hanafizadeh, P.; Ghorbani, B. Review study on airlift pumping systems. *Multiph. Sci. Technol.* **2012**, *24*, 323–362. [CrossRef]
31. Kambhu, A. Treating 1,4-Dioxane with Slow-Release Persulfate and Zerovalent Iron and Modeling the Radius of Influence of Aerated Oxidant Candles. Ph.D. Thesis, University of Nebraska, Lincoln, NE, USA, 2019.
32. Kolthoff, I.M.; Miller, I.K. The chemistry of persulfate: I. The kinetics and mechanism of the decomposition of the persulfate ion in aqueous medium. *J. Am. Chem. Soc.* **1951**, *73*, 3055–3059. [CrossRef]
33. Dash, S.; Patel, S.; Mishra, B.K. Oxidation by permanganate: Synthetic and mechanistic aspects. *Tetrahedron* **2009**, *65*, 707–739. [CrossRef]
34. Huang, K.-C.; Hoag, G.E.; Chheda, P.; Woody, B.A.; Dobbs, G.M. Kinetics and mechanism of oxidation of tetrachloroethylene with permanganate. *Chemosphere* **2002**, *46*, 815–825. [CrossRef]
35. Huang, K.-C.; Zhao, Z.; Hoag, G.E.; Dahmani, A.; Block, P.A. Degradation of volatile organic compounds with thermally activated persulfate oxidation. *Chemosphere* **2005**, *61*, 551–560. [CrossRef]
36. Mahmoodlu, M.G.; Hassanizadeh, S.M.; Hartog, N. Evaluation of the kinetic oxidation of aqueous volatile organic compounds by permanganate. *Sci. Total Environ.* **2014**, *485–486*, 755–763. [CrossRef]
37. Lee, D.G.; Chen, T. Oxidation of hydrocarbons. 18. Mechanism of the reaction between permanganate and carbon-carbon double bonds. *J. Am. Chem. Soc.* **1989**, *111*, 7534–7538. [CrossRef]
38. Yan, Y.E.; Schwartz, F.W. Oxidative degradation and kinetics of chlorinated ethylenes by potassium permanganate. *J. Contam. Hydrol.* **1999**, *37*, 343–365. [CrossRef]

39. Yan, Y.E.; Schwartz, F.W. Kinetics and mechanism for TCE oxidation by permanganate. *Environ. Sci. Technol.* **2000**, *34*, 2535–2541. [CrossRef]
40. Woo, N.C.; Hyun, S.G.; Park, W.W.; Lee, E.S.; Schwartz, F.W. Characteristics of permanganate oxidation of TCE at low reagent concentrations. *Environ. Technol.* **2009**, *30*, 1337–1342. [CrossRef]
41. Crimi, M.L.; Taylor, J. Experimental evaluation of catalyzed hydrogen peroxide and sodium persulfate for destruction of BTEX contaminants. *Soil Sediment Contam.* **2007**, *16*, 29–45. [CrossRef]
42. Killian, P.F.; Bruell, C.J.; Liang, C.; Marley, M.C. Iron (II) activated persulfate oxidation of MGP contaminated soil. *Soil Sediment Contam.* **2007**, *16*, 523–537. [CrossRef]
43. Liang, C.; Huang, C.F.; Chen, Y.J. Potential for activated persulfate degradation of BTEX contamination. *Water Res.* **2008**, *42*, 4091–4100. [CrossRef]
44. Achugasim, D.; Osuji, L.C.; Ojinnaka, C.M. Use of activated persulfate in the removal of petroleum hydrocarbons from crude oil polluted soils. *Res. J. Chem. Sci.* **2011**, *1*, 57–67.
45. Lemaire, J.; Croze, V.; Maier, J.; Simonnot, M. Is it possible to remediate a BTEX contaminated chalky aquifer by in situ chemical oxidation? *Chemosphere* **2011**, *84*, 1181–1187. [CrossRef]
46. Agency for Toxic Substances and Disease Registry (ATSDR). Toxicological Profile for Acetone. U.S. Department of Health and Human Services. 1994. Available online: https://www.atsdr.cdc.gov/substances/toxsubstance.asp?toxid=1 (accessed on 8 January 2020).
47. Wiberg, K.B.; Geer, R.D. The kinetics of the permanganate oxidation of acetone. *J. Am. Chem. Soc.* **1965**, *87*, 5202–5209. [CrossRef]
48. Harcourt, A.V.; Esson, W. On the laws of connexion between the Conditions of a Chemical Change and Its Amount. *Philos. Trans. R. Soc. Lond.* **1866**, *156*, 193–221. [CrossRef]
49. Gates-Anderson, D.D.; Siegrist, R.L.; Cline, S.R. Comparison of potassium permanganate and hydrogen peroxide as chemical oxidants for organically contaminated soils. *J. Environ. Eng.* **2001**, *127*, 337–347. [CrossRef]
50. Chokejaroenrat, C.; Comfort, S.D.; Harris, C.; Snow, D.; Cassada, D.; Sakulthaew, C.; Satapanajaru, T. Transformation of Hexahydro-1,3,5-trinitro-1,3,5-triazine (RDX) by permanganate. *Environ. Sci. Technol.* **2011**, *45*, 3643–3649. [CrossRef]
51. Chokejaroenrat, C. Laboratory and Pilot-Scale Investigations of RDX Treatment by Permanganate. Master's Thesis, University of Nebraska, Lincoln, NE, USA, 2008.
52. Evans, P.J.; Dugan, P.; Nguyen, D.; Lamar, M.; Crimi, M. Slow-release permanganate versus unactivated persulfate for long-term in situ oxidation of 1,4-dioxane and chlorinated solvents. *Chemosphere* **2019**, *221*, 802–811. [CrossRef]
53. Evans, P.; Hooper, J.; Larmar, M.; Nguyen, D.; Dugan, P.; Crimi, M.; Ruiz, N. *Sustained In Situ Chemical Oxidation (ISCO) of 1,4-Dioxane and Chlorinated VOCs Using Slow-Release Chemical Oxidant Cylinders*; Environmental Security Technology Certification Program (ESTCP) Cost and Performance Report (ER-201324); Environmental Security Technology Certification Program (ESTCP): Alexandria, VA, USA, July 2018; p. 73.
54. Chokejaroenrat, C.; Negin, K.; Sakulthaew, C.; Comfort, S.; Li, Y. Improving the sweeping efficiency of permanganate into low permeable zones to treat TCE: Experimental results and model development. *Environ. Sci. Technol.* **2013**, *47*, 13031–13038. [CrossRef]

2

Transport of Contamination under the Influence of Sea Level Rise in Coastal Heterogeneous Aquifer

Qiaona Guo *, Yahui Zhang, Zhifang Zhou and Zili Hu

School of Earth Sciences and Engineering, Hohai University, No.8 Focheng West Road, Nanjing 211100, China; yahui@hhu.edu.cn (Y.Z.); zhouzf@hhu.edu.cn (Z.Z.); CoweixongYytiupteud@gmail.com (Z.H.)
* Correspondence: guoqiaona2010@hhu.edu.cn

Abstract: This paper provided for the first time an experimental study on the influence of sea level rise on transport of contamination in the heterogeneous unconfined aquifer of the coastal zone. The experiments were conducted using the tank, considering the difference between sea level and inland head 1 cm for Case 1 and 2 cm for Case 2. Observed data were validated using the numerical model, which matched well with the toe length of seawater wedge and the shape of the contaminant plume. The results showed that the observed and simulated values of Cl^- concentration at the sampling points increased sharply at the initial time, and then they increased slowly and tended to be stable. The seawater wedge migrated inland with time under the effects of the hydraulic gradient toward the inland and the density difference between saltwater and freshwater. The steady state length of the 50% isoline of the seawater wedge was 167 cm in Case 2, which was larger than that of Case 1. The maximum area of plume in Case 2 was 0.13 m^2, larger than that in Case 1, which indicated that the velocity of diffusion of the contaminant plume increased as the sea level increased. As the velocity of diffusion increased, the time for pollutant migration to the intersection between seawater and freshwater became shorter. The maximum area and vertical depth of pollutant plume were sensitive to the hydraulic conductivity, dispersivity, and contamination concentration. The infiltration depth and range of the contaminant plume in the heterogeneous aquifer were greater than those in the homogeneous aquifer of the actual beach.

Keywords: seawater level rising; contaminant transport; inland heterogeneous unconfined aquifer; coastal zone

---

1. Introduction

The coastal area is densely populated, due to the rapid development of the economy there. Therefore, a large amount of groundwater is pumped to provide people with water for living and industrial production. Nevertheless, the problems of water resources, ecology, and environment pollution in coastal areas are complex and serious, due to the influences of interaction between land and sea, global warming, and sea level rising [1–5]. Among them, environmental pollution is a very serious problem. For example, the spilled oil deposits within the beach, caused by the accident of offshore oil tankers, which causes the long-term pollution of groundwater in the aquifer [6–8]. In addition, the wastewater infiltrates into the aquifer due to seepage from facilities, with the rapid development of the aquaculture in recent years, which results in the deterioration of groundwater quality and water balance of coastal aquifers [9–11]. Therefore, understanding the contaminate transport process and remediation in coastal groundwater is crucial to the economic, social, and environmental sustainability of coastal areas.

The spatial and temporal distribution of salinity in coastal zones have been focused on, because the seawater encroaches into the coastal aquifer e.g., [2,12–16]. In the recent decade, there has been increasing interests in the problem of seawater intrusion due to sea level rising, which is considered to be one of the most serious threats to the world's freshwater resources [17–21]. The analytical e.g., [16,22,23], numerical e.g., [4,21,24], and experimental e.g., [18,25,26] methods have been used to study seawater intrusion caused by sea level rise. The seawater toe location and circulation rates of seawater under the condition of seawater level rising in the aquifers are investigated in previous studies.

The studies on coastal contaminant transport are relatively less, compared with those of seawater intrusion, because the movement of contaminant is inherently complex. The contaminant transport in the coastal aquifer is affected by many factors, such as the tide fluctuation, inland freshwater recharge, variable density flow, concentration of contaminant, and aquifer heterogeneity [27–32]. Among them, seawater intrusion has an important impact on pollutant transport, because the variable density effect can change the flow pattern and eventually affect the contaminant transport processes [33].

In previous studies, the methods of laboratory experiments, analytical solution, and numerical simulation were used to investigate the shape and movement of the solute plume in the coastal aquifer e.g., [7,33–43]. The laboratory experiment was applied to study the contaminant transport in the coastal aquifer earlier. For example, Zhang et al. [34] observed the landward plume transport during the experiment, which was affected by the tide fluctuation. Chang and Clement [38] investigated the contaminant transport patterns above the wedge and within it, and they found that the transport rate active was influenced by dispersion. Moreover, some analytical solutions were derived to explain the transport process directly. For instance, Koohbor et al. [33] derived a semi-analytical solution of pollutant transport in the coastal confined aquifer under the condition of variable velocity field. However, the analytical solution could be used to solve the pollutant transport problem under ideal conditions. The numerical simulation was applied to solve the problem of contaminant transport under the aforementioned complex conditions. Boufadel et al. [43] simulated the nutrient transport in the beach, and found that the solute transport was sensitive to tidal boundary condition and unsaturated parameters. Geng et al. [44] developed a numerical reactive transport model to simulate the fate and transport of the contaminant plume, which was influenced by tide amplitude, capillarity, and hydraulic conductivity. Shen et al. [32] considered the solute transport in beach aquifers, which was influenced by the unstable flow.

However, most of the previous studies focused on the contaminant transport in homogeneous aquifer. Up to now, several research studies on the contaminant transport in the heterogeneous aquifer were explored. In reality, the distribution of the plume was affected by the degree of heterogeneity of aquifer dramatically [7,45]. For example, Guo et al. [46] investigated the lithium tracer migration in a gravel beach with two-layered structure. Jang et al. [47] identified that the factor of soil heterogeneity affected the nitrate reduction process. Therefore, the heterogeneity of aquifers affecting the movement of pollutants in coastal aquifers should receive more attention. In addition, it was not clear how the rising sea level affects the contaminant transport in the heterogeneous aquifer, although there were many studies on the seawater intrusion caused by the rising sea level.

Fish ponds were built in the freshwater areas for aquaculture using the topography in coastal areas [10,48,49]. The seawater at the interface between saltwater and freshwater was pumped into the fish pond for aquaculture, and then the waste water was discharged to the sea. The infiltrated seawater caused by the leakage of aquaculture wastewater in the beach was an environmental pollution problem, which was different from the seawater intrusion caused by tide and artificial pumping. How does the seawater flow from the leakage of the fishpond in the aquifer? What is the relation between the infiltrated seawater from the fishpond and seawater intrusion from the sea? There was no research in this area so far. The aim of this paper is to explore the influence of sea level rise on contaminant transport in heterogeneous aquifers. The saltwater tracer is used to evaluate the contaminate transport in the heterogeneous aquifer, from the seepage of surface reservoirs. The complicated chemical and microbiological processes are not considered. The variation of the area and depth of contaminant under the condition of the rising sea level in the heterogeneous aquifer is examined, based on the laboratory experiments and numerical simulation. The effects of boundary condition, aquifer heterogeneity, and seawater intrusion on contaminant transport are discussed. The main hydrogeological factors affecting the contaminant transport in the coastal aquifer are determined.

2. Materials and Method

2.1. Laboratory Experiment

The dimension of the tank for experiment was 260 cm × 30 cm × 100 cm, which was shown in Figure 1. The tank was made of plexiglass and protected by a steel frame. It was composed of three parts: a freshwater chamber, a central chamber, and a seawater chamber. The left and right side chambers were separated from the central chamber by plates, which were made of PVC materials. In order to avoid the sand flowing from the central chamber to the two side chambers, the geotextile was pasted on the surface of the PVC board drilled with small holes. The bottom of the seawater chamber and freshwater chamber were respectively connected with a water reservoir, which was made of organic plastic plates with the length of 84 cm, width of 84 cm, and height of 84 cm. The seawater and freshwater were respectively pumped into the seawater chamber and freshwater chamber by peristaltic pumps. The left and right side chambers represented the freshwater and seawater boundary conditions, respectively. Three different particle sizes were used during the experiment, namely fine sand, medium sand, and coarse sand from up to down. The material was homogeneous in each layer. The particle sizes of fine sand, medium sand, and coarse sand were ranging from 0.1 to 0.25 mm, from 0.25 to 0.5 mm, and from 0.5 to 1 mm, respectively. The material in each layer was compacted. In reality, the coastal aquifer inclines to the sea boundary. However, the slope of the aquifer was generally small. Therefore, the slope of the aquifer for the experiment was set horizontal. The thickness of the surface sandy aquifer was generally larger, considering the actual beach condition. Therefore, the fine sand layer, medium sand layer, and coarse sand layer were laid horizontally in the sand tank, with the thickness of 23 cm, 16 cm, and 16 cm, respectively. The thickness of the whole aquifer was 55 cm. First of all, a certain amount of freshwater was put into the tank before filling the sand. Then, the coarse sand with the thickness of 2 cm was filled. The sand was left for 30 min to make it fully saturated. Subsequently, the above steps were repeated in turn until the coarse sand layer reach 16 cm. The method of laying sand in the medium sand layer and fine sand layer was similar to that of coarse sand. During the process of laying sand, each layer was compacted as much as possible to remove the air. The hydraulic conductivities of the fine sand, medium sand, and coarse sand were 2.3×10^{-5} m/s, 2.9×10^{-4} m/s, and 6.9×10^{-4} m/s, respectively, which were obtained from the variable head permeameters in the laboratory.

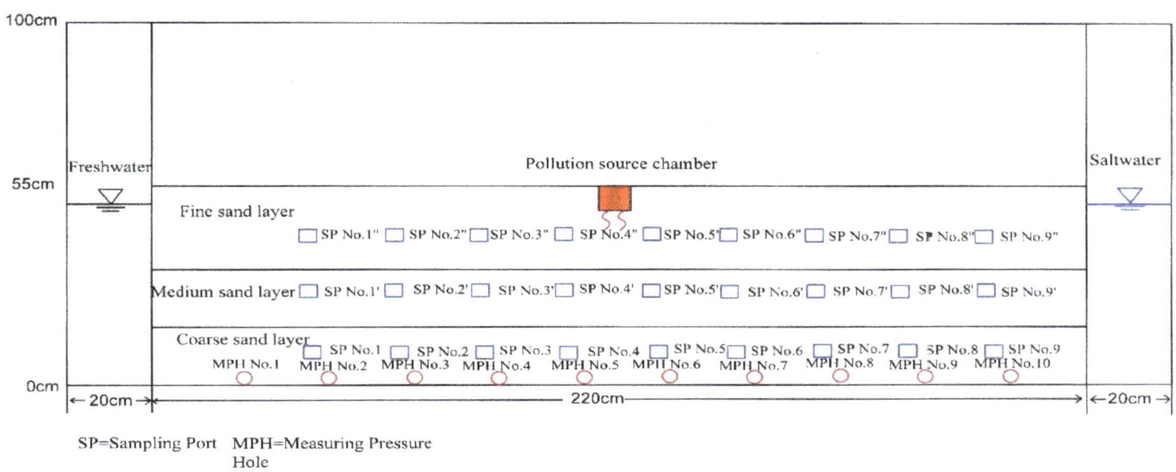

Figure 1. Schematic diagram of the test model.

The pollution source chamber was installed on the top of the aquifer, which was 106 cm away from the left freshwater chamber. It was made by the transparent organic plastic plates, with the length of 30 cm, width of 5 cm, and height of 20 cm. A layer of geotextile was pasted at the bottom of the pollution source chamber drilled with holes, in order to prevent the sand from entering into the pollution source chamber. A peristaltic pump was used to pump the contaminant with constant concentration into the pollution source chamber at the constant speed. Ten pressure taps were arranged at the bottom of the sand box, and each tap was connected with a transparent polyurethane pipe to measure the variation of the groundwater level. A 0.2 mm filter screen was set at the pressure tap, in order to prevent the sand from entering the pressure pipe and blocking the pipe. At the back of the sand tank, there were nine sampling ports in each layer. The distance between each two sampling ports was 20 cm. During the experiment, the sample was taken from the sampling port using a syringe. Then, the water sample was diluted and the electrical conductivity of the sample was measured by the DDS-307 conductivity meter. The concentration of Cl^- can be calculated, because there is a linear relationship between the concentration of NaCl and electrical conductivity. Two sets of tests were considered for sea level rise. In the first experiment (Case 1), the left inland head and right sea level were fixed at 45 and 46 cm, respectively. In the second experiment (Case 2), the constant inland head and sea water level were maintained at 45 and 47 cm, respectively. The water level at different positions was observed at the measuring pressure holes during the experiment. The deionized water was used to make freshwater, and the density of freshwater was 1.0×10^3 kg/m^3. The concentration and density of saltwater were 35.0 g/L and 1.02×10^3 kg/m^3, respectively. The bright blue was added into the right seawater chamber, in order to show the seawater intrusion process. The saltwater with the concentration of 35.0 g/L was injected into the pollution source chamber. The carmine tracer with the concentration of 1.0 g/L was added into the saltwater to trace the movement of contaminant.

Each layer of the aquifer was filled with freshwater at the beginning of experiment. Then, the seawater flowed into the seawater chamber from the reservoir, and overflow was discharged through the overflow of the seawater chamber. Subsequently, the seawater gradually entered the water tank from the seawater chamber. The steady flow was formed on both sides. When the length of seawater wedge did not change in 5 min, the interface between seawater and freshwater reached steady state. Once the steady state was established, the peristaltic pump was open and the contaminant was injected into the pollution source chamber. The flow was set to 0.44 m/h, by controlling the peristaltic pump. At this time, the contaminant moved, influenced by the hydraulic gradient and density difference between seawater and freshwater. Finally, the experiment was finished when the contaminant plume intersected with the interface of seawater and freshwater in the coarse sand layer. The experimental process of Case 1 was similar to that of Case 2. During the experiment, the Canon digital camera (IXUS 175) was used to record the process of seawater wedge and contaminant migration. The salinity at each sampling port was monitored every 5 min. At the end of the experiment, the contaminant plume and saltwater wedge were analyzed by Matlab software.

2.2. Numerical Model and Procedure

Based on the finite difference model MODFLOW, the influence of density on groundwater flow was considered. A numerical simulation model of solute transport in groundwater flow was established by using SEAWAT [50,51]. The softwater was widely used in seawater intrusion and groundwater discharge e.g., [2,28,52–54]. The coupling controlling equations of groundwater flow and solute transport with variable density were as follows:

$$\begin{aligned} &\frac{\partial}{\partial x}\left[\rho K_{fx}\left(\frac{\partial h_f}{\partial x} + \frac{\rho-\rho_f}{\rho_f}\frac{\partial Z}{\partial x}\right)\right] + \frac{\partial}{\partial y}\left[\rho K_{fy}\left(\frac{\partial h_f}{\partial y} + \frac{\rho-\rho_f}{\rho_f}\frac{\partial Z}{\partial y}\right)\right] + \\ &\frac{\partial}{\partial z}\left[\rho K_{fz}\left(\frac{\partial h_f}{\partial z} + \frac{\rho-\rho_f}{]\rho_f}\frac{\partial Z}{\partial z}\right)\right] = \rho S_f\frac{\partial h_f}{\partial t} + \theta\frac{\partial \rho}{\partial C}\frac{\partial C}{\partial t} - \rho_s q_s \end{aligned} \quad (1)$$

$$\frac{\partial \theta C}{\partial t} = \nabla(\theta D \nabla C) - \nabla(\theta V C) + q_s C_s + R_n \,, \tag{2}$$

where ρ is the density of saline groundwater at a point in aquifer [ML^{-3}]; ρ_f is the density of fresh water [kg/m^3]; ρ_s is density of water entering from a source or leaving through a sink [ML^{-3}]; q_s is the volumetric flow rate per unit volume of porous medium representing sources and sinks [T^{-1}]; K_{fx}, K_{fy}, K_{fz} are the hydraulic conductivities in different directions [LT^{-1}]; Z is the position head [L]; S_f is the specific storage [L^{-1}]; h_f is the freshwater head [L]; θ is the effective porosity of porous medium [–]; C is the solute concentration [ML^{-3}], and t is the time [T]; D is the hydrodynamic dispersion tensor [L^2T^{-1}]; C_s is the solute concentration of water entering or leaving from sources or sinks [$M.L^{-3}$]; V is the fluid velocity vector [LT^{-1}]; and R_n is reaction term of chemical substance.

The numerical simulation area was a heterogeneous and isotropic unconfined aquifer. During the setup of numerical simulation, the laboratory settings and parameters were followed as closely as possible. A two-dimensional vertical cross section was built, with the length of 220 cm and height of 55 cm. The domain of model was discretized into 24 layers and 5000 columns. A free surface boundary was defined on the upper part. A no-flow boundary condition was set on the bottom of the numerical model, because the bottom of the aquifer was bedrock in reality. For Case 1, the inland head and seawater level were set to 45 and 46 cm on the left and right boundaries, respectively. The left inland head and right seawater level were fixed at 45 and 47 cm for Case 2. The concentrations of saltwater and freshwater were set to 35 and 0 g/L, respectively. It was assumed that the contamination on the water table of the model was uniformly distributed at the distance between 1.04 and 1.08 m. The constant concentration of contamination was set to 35 g/L for Case 1 and Case 2. The injection time was at 55 and 50 min for Case 1 and Case 2, respectively. The simulated period was 110 min for Case 1 and 100 min for Case 2. The time step was set to 60 s for both cases.

In order to fit the observed head and salinity, the parameters of aquifers were calibrated repeatedly by a trial-and-error method. The process of calibration was to adjust the values of the parameters, including the hydraulic conductivities, specific yield, specific storage, effective porosity, and dispersivity, until the mean error between the calculated salinity values at the sampling ports was less than 10% of the maximum observed value variations. The horizontal hydraulic conductivity was assumed to be the same to its vertical value of the aquifer system. The estimated hydraulic conductivities were 4.6×10^{-5} m/s, 2.9×10^{-4} m/s, and 6.9×10^{-4} m/s, respectively, for the fine sand layer, medium sand layer, and coarse sand layer, respectively. The other parameters used in the simulations were listed in Table 1.

Table 1. Model parameters and their values used in the numerical simulations.

Parameter	Definition	Unit	Value
K	Hydraulic Conductivity	m/s	4.6×10^{-5} (fine sand layer) 2.9×10^{-4} (medium sand layer) 6.9×10^{-4} (coarse sand layer)
μ	Specific Yield	-	0.2 (fine sand layer) 0.25 (medium sand layer) 0.30 (coarse sand layer)
S_s	Specific Storage	1/m	10^{-5}
ρ_f	Density of the Freshwater	kg/m^3	1.0×10^3
ρ	Density of the Seawater	kg/m^3	1.02×10^3
n	Porosity	-	0.3
α_L	Longitudinal Dispersivity	m	0.1
α_T	Transverse Dispersivity	m	0.01
τD_m	Molecular Diffusion Coefficient in Porous Media	m^2/s	10^{-9}
t	Time Period	min	110 for Case 1 100 for Case 2

3. Results and Discussion

3.1. Numerical Validation

The numerical simulations of variable density groundwater flow and the solute transport model reproduced the observed water, Cl^- concentration, and contamination concentration in the transect. Here, the focus was on the Cl^- concentration. Figure 2 shows the experimental versus the simulated Cl^- concentrations variation with time at the four sampling points for Case 1 and Case 2. From the figure, one can see that the simulated and observed Cl^- exhibited good correlation generally. The calculated correlation coefficient (R^2) ranged from 0.96 to 0.98, which indicated that the numerical model can successfully simulate the contaminant movement in the aquifer. From Figure 2, it can be seen that the observed and simulated values of Cl^- concentration increased sharply at the initial time, and then they increased slowly and tended to be stable.

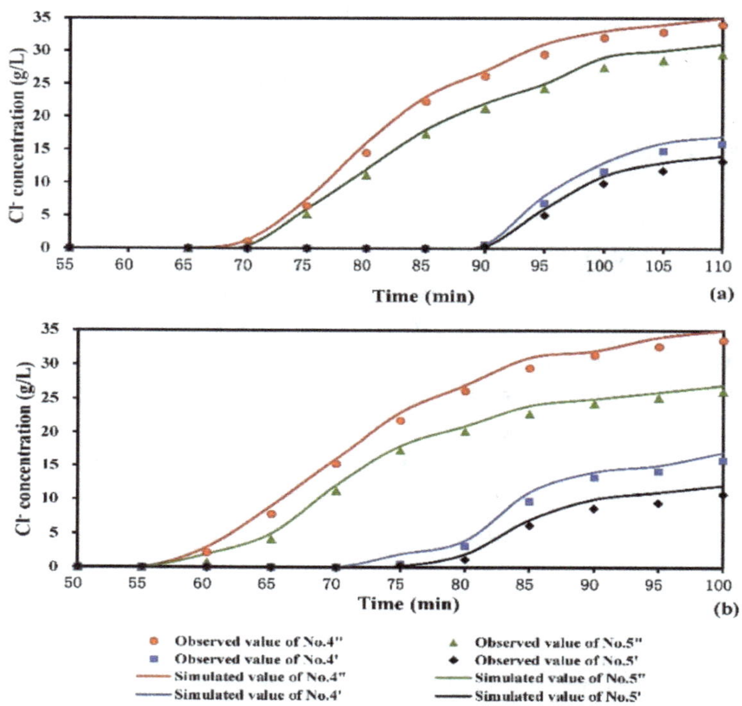

Figure 2. Comparison of simulated and observed Cl^- concentrations over time at selected points for (**a**) Case 1 and (**b**) Case 2.

The Cl^- concentration values at the sampling ports No. 4″ and No. 5″ in the upper layer were higher than those at No. 4′ and No. 5′ in the middle layer. It indicated that the concentration of contaminant decreased gradually when the contaminant plume migrated downward in the aquifer. The migration of plume in the medium sand layer lagged behind that in the fine sand layer. From Figure 2a, one can see that the maximum value of Cl^- concentration at left No. 4″ was close to 35 g/L, which was higher than that at right No. 5″ in the upper layer. The value of Cl^- concentration at left No. 4′ was higher than that at right No. 5′ in the middle layer, which was similar to that in the upper layer. It was evident that the distribution of contaminant was affected by the flow from the sea to inland. Compared with Figure 2a, Figure 2b showed that the values of concentration at the sampling points (No. 4″, No. 5″, No. 4′, and No. 5′) in Case 2 were little smaller than those in Case 1. It demonstrated that the contaminant was diluted due to the increase of water flow velocity influenced by the sea level rising.

3.2. *Seawater Intrusion Effected by Seawater Level Rise*

In order to investigate the groundwater flow and salt transport under the influence of seawater level rising, simulations of seawater intrusion processes for different cases were conducted. Figure 3 reported the variation of observed data and simulated concentration color maps of the seawater wedge with time for Case 1. The comparison of Figure 3a,b reflected that the simulated toe lengths of the seawater wedge agreed with the observed ones well. The observed and simulated seawater wedge migrated inland with time, under the effects of the hydraulic gradient toward the inland and the density difference between saltwater and freshwater. The shape of the seawater wedge was parabolic. At the intial time, the migration velocity of the saltwater wedge was fast in the aquifer. The area increased in each layer of the aquifer. As for the hydraulic conductivity, the area and migration velocity of the seawater wedge increased from the top layer to bottom layer. Then, the migration velocity of the seawater wedge slowed down gradually, and it was close to zero when the time was 55 min. The interface of saltwater and freshwater reached equilibrium state, when the saltwater wedge did not move in the next five min. The toe position of 50% isoline of the seawater wedge represented the length of seawater wedge, which was wildly applied in depicting the saltwater wedge in the coastal aquifer, in which the dispersivity was small. As the invasion of seawater reaches the equilibrium status at the time of 55 min, the horizontal length of the 50% isoline of the observed seawater wedge can reach 142 cm. When the intruded seawater wedge was stable, the simulated length of the 50% isoline of seawater wedge was 144 cm, which was consistent with that of the observed seawater wedge.

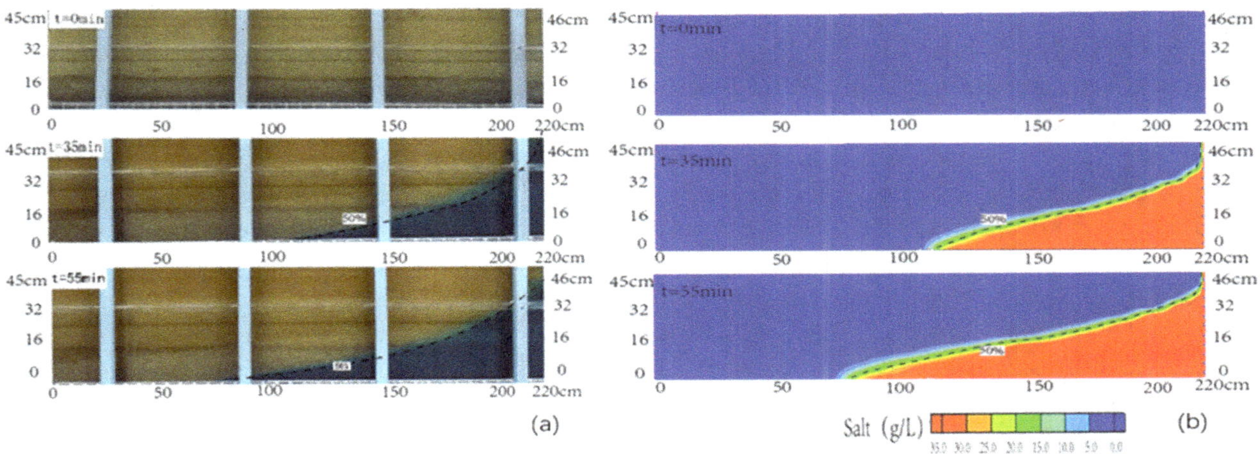

Figure 3. Comparison between (**a**) transient experimental and (**b**) numerical toe position of the saltwater wedge for Case 1, the dashed line represents the 50% isoline of the seawater wedge.

Figure 4 showed the variation of observed data and simulated results of the toe position of seawater wedge at various time for Case 2. The seawater wedge encroached into the inland with time, which was influenced by the density difference between saltwater and freshwater and the hydraulic gradient slope toward the sea. Similar to Case 1, the shape of the seawater wedge was parabolic. The migration velocity of seawater intrusion was fast initially, and then it slowed down gradually in Case 2. The migration velocity of the seawater wedge was close to zero at the time of 50 min, and the steady state of seawater intrusion was formed at this time in the three layers. The observed length of the 50% isoline of the seawater wedge was 167 cm, which was larger than that of Case 1. It was because the head at the seaside boundary and the flow velocity for Case 2 were higher than those for Case 1. It reflected that the rising sea level had a positively impact on the seawater intrusion process in the layered aquifer, which was consistent with the observation in Guo et al. [18]. At the time of 50 min, the simulated length of 50% isoline of the seawater wedge was 167 cm (Figure 4b), which was consistent with the observed value (Figure 4a). In general, the numerical simulation model can describe the variation of the seawater wedge well.

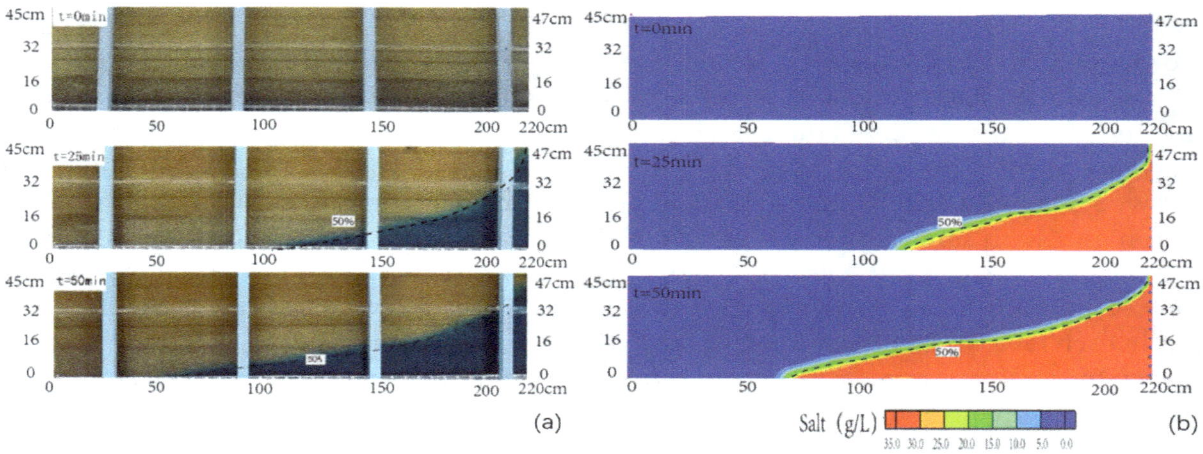

Figure 4. Comparison between (**a**) transient experimental and (**b**) numerical toe position of the saltwater wedge for Case 2, the dashed line represents the 50% isoline of the seawater wedge.

3.3. *Contamination Transport Effected by Seawater Level Rise*

The contaminate transport was simulated for both Case 1 and Case 2, when the seawater wedge was stable. Figure 5 showed the comparison between the observed and simulated contamination concentration at different times in the physical and numerical model for Case 1. The results showed that the numerical model could reproduce relatively well the overall shape of the plume, albeit the finger shape in the front of the plume observed in the laboratory was not captured in the simulation. The observed and simulated plume diffused mainly in the lateral and downward directions in the fine sand layer during the period of 55 and 70 min, and 70 and 85 min, respectively. The area of the contaminant plume increased linearly, and the shape of the plume was semi-elliptical. The maximum area of the contaminant plume was 0.02 m^2, and the maximum depth of the contaminant plume was 0.13 m. At the time of 85 min, the contaminant entered into the medium sand layer and moved downward. The pollution plume became concave at the interface between the upper layer and middle layer, because the velocity of the contaminant plume migration increased gradually. The contaminant plume migrated in the medium sand layer at the time between 85 and 100 min. The variation rate of area in the medium sand layer was twice as much as that of the fine sand layer. The maximum area of the contaminant plume was 0.067 m^2, and the maximum depth of the plume was 0.29 m. During the time of 100 and 110 min, the contaminant plume moved downward and inland in the coarse sand layer. The plume became concave at the interface between the middle layer and lower layer, due to the increasing of velocity of contaminant plume migration. The front of the pollutant flowed in the form of dominant flow, due to the heterogeneity in the coarse sand layer. The variation rate of the area in the coarse sand layer was 1.5 times of that in the middle sand layer, and the maximum plume area was 0.11 m^2 at the time of 110 min.

Figure 6 showed the comparison between the transient experimental and numerical contaminant transport for Case 2. In all, the results showed that the transient movement of the plume was reasonably well simulated by the numerical model, although the finger shape in the front of the plume observed in the laboratory was not captured in the simulation. During the period of 50 and 70 min, the observed and simulated plume diffused laterally in the fine sand layer. The area of the contaminant plume increased linearly, and the shape of the plume was semi-elliptical, which was similar to Case 1. The maximum area and depth of the contaminant plume were 0.025 m^2 and 0.13 m, respectively. At the time of 70 min, the contaminant entered into the medium sand layer and moved downward. The pollution plume became concave at the interface between the upper layer and middle layer, because the velocity of the contaminant plume migration increased gradually. The contaminant plume migrated in the medium sand layer at the time between 70 and 90 min. The maximum area of the contaminant plume was 0.069 m^2, and the maximum depth of plume was 0.29 m. During the time of 90 and 100 min, the contaminant plume moved downward and inland in the coarse sand layer. The plume became concave at the interface between the middle layer and lower layer, due to the increasing of velocity of contaminant plume migration. At the time of 100 min, the observed front of the pollutant was lower than that of the simulated one. Due to the heterogeneity in the coarse sand layer, the front of the pollutant flowed in the form of dominant flow. The variation rate of the area in the coarse sand layer was 2.0 times of that in the middle sand layer, and the maximum plume area was 0.13 m^2 at the time of 100 min. The end of the front of the contaminant plume migrated towards inland. Compared with the observed one, the numerical plume has a deviation toward the inland, because of the effect of seawater intrusion. Compared with Case 1, the area of the plume in Case 2 was larger than that in Case 1 in the same layer. It indicated that the velocity of diffusion of the contaminant plume increased as the sea level increased. As the velocity of diffusion increases, the time for the observed pollutant migration to the intersection between seawater and freshwater in Case 2 was shorter than that in Case 1.

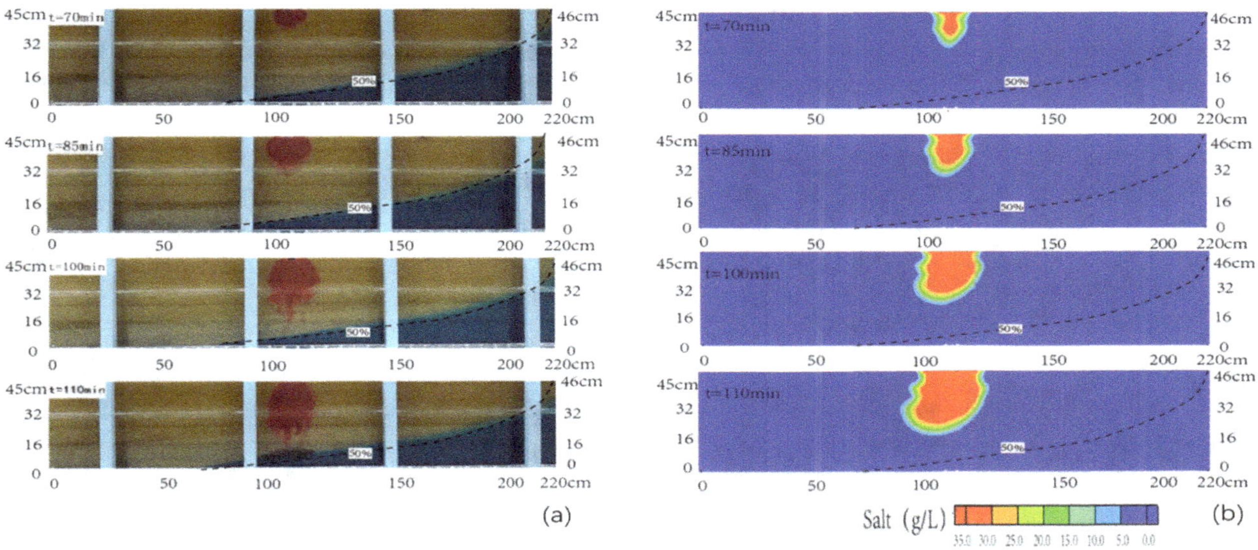

Figure 5. Observed (**a**) and simulated (**b**) contamination concentration distributions in the cross section for Case 1.

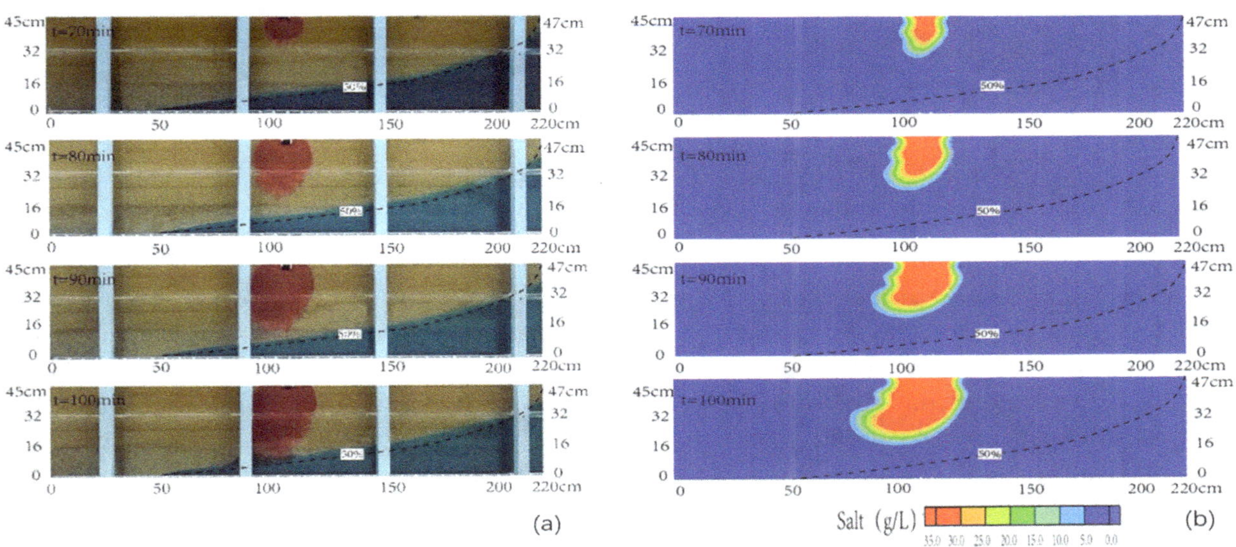

Figure 6. Observed (**a**) and simulated (**b**) contamination concentration distributions in the cross section for Case 2.

3.4. Sensitivity Analysis

The hydraulic conductivity, dispersivity, contaminate concentration, and heterogeneity were important parameters for solute transport in the coastal aquifer. Therefore, the sensitivities of saltwater contamination transported in the aquifer to the model parameters were analyzed, taking the Case 2 as an example. The area and velocity of the plume diffusion were discussed by changing the above parameters. It should be noted that the parameter values for analyzing the sensitivity were increased or decreased, whereas the values of the other parameters were not changed, which was shown in Table 1.

3.4.1. Effect of Hydraulic Conductivity

Figure 7a showed the simulated spatial contaminant plume distributions with time by the new model, as the values of K were raised to 8.0 m/d for the upper layer, 50.0 m/d for the middle layer, and 120.0 m/d for the bottom layer. Compared with Figure 6b, Figure 7a showed that the match to the observed contours of the saltwater contamination in the new model was worse, compared with the simulated ones in the Case 2. It was found that the larger length of the seawater wedge than in reality moved into the aquifer of the new model. The region of the transition zone in the front edge of the plume in the new model became larger, compared with that in the basic model. This was because the velocities were larger with higher hydraulic conductivities. The plume migrated more downward and landward, especially in the coarse sand layer. Increasing the hydraulic conductivity increased remarkably the area of the plume in the heterogeneous aquifer. The maximum area of the pollution plume simulated by the new model reached 0.143 m^2, which was larger than that simulated in Case 2.

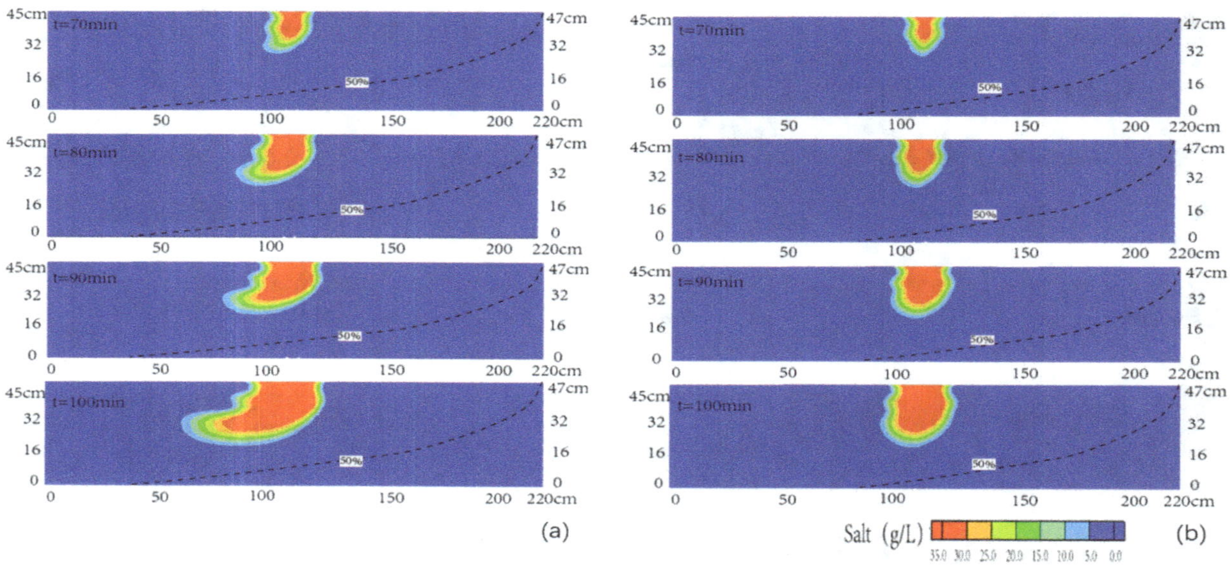

Figure 7. Simulated spatial saltwater contamination concentration distributions for (**a**) increasing hydraulic conductivities and (**b**) decreasing hydraulic conductivities at different times.

As the values of K were reduced to 2.0 m/d for the upper layer, 12.5 m/d for the middle layer, and 30.0 m/d for the bottom layer, Figure 7b showed that the match to the observed contours of the saltwater contamination was considerably worse, compared with the simulated ones in Figure 6b of Case 2. One can see that the smaller length of the seawater wedge than in reality migrated into the aquifer of the new model. Obviously, the maximum vertical pollution depth decreased when the hydraulic conductivities of the heterogeneous aquifer decreased, which indicated that the hydraulic conductivities for sensitivity analysis were too low. Additionally, with the lower hydraulic conductivities in each layer, the velocity of plume migration became slowly. However, the migration direction of the pollution plume remained downward and landward. The area of the pollution plume simulated by the new model increased slowly, and the maximum value of it reached 0.083 m^2, which was smaller than that of Case 2. Therefore, one can conclude that increasing or decreasing the hydraulic conductivities by the new models were not able to reproduce the spatial distributions of saltwater contamination concentration.

3.4.2. Effect of Dispersion

The sensitivity of the plume transport to the dispersivity was explored for two new models, due to the importance of dispersivity for solute transport. The variation of contours of the saltwater contamination was used for analyzing sensitivity with respect to the dispersivity. When the value of α_L increased to 0.5 m and the ratio α_T/α_L was 0.1, Figure 8a showed that more saltwater intruded into the aquifer of the new model than that of the basic case. The length of seawater in the new model was 0.13 m longer than that in Case 2. The simulated maximum vertical pollution depth was larger, compared with that in the basic model, which could be because high dispersivity induced widening of the transition zone. The area of the pollution plume simulated by the new model was 0.163 m^2, which was larger than that of Case 2. It indicated that higher dispersion has great influence on plume transport. However, there was no change in the direction of pollutant migration.

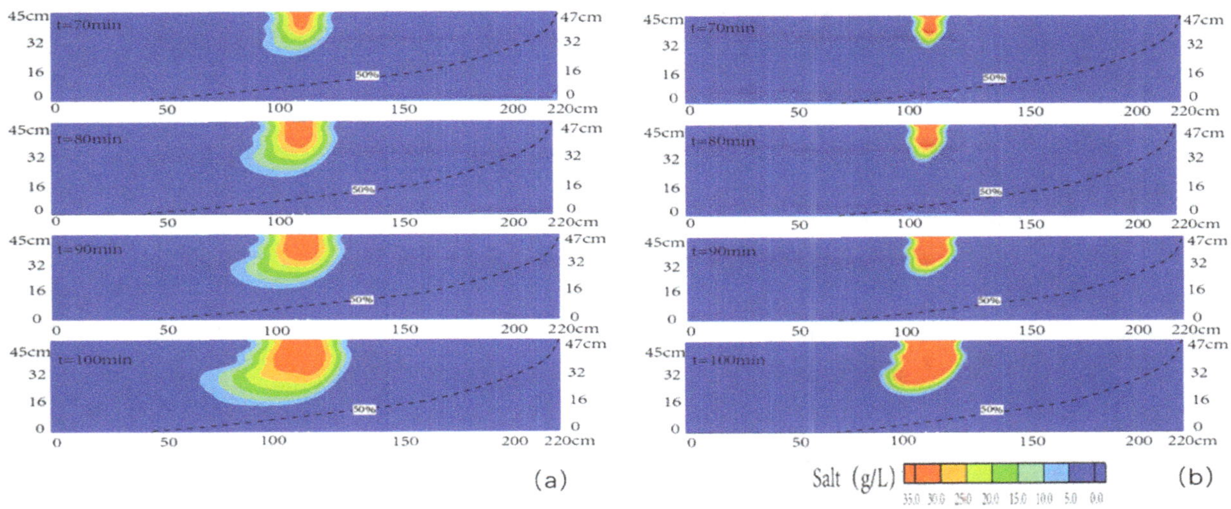

Figure 8. Simulated spatial saltwater contamination concentration distributions for (**a**) increasing dispersivity and (**b**) decreasing dispersivity at different times.

When the value of α_L decreased to 0.01 m and the ratio α_T/α_L was 0.1, Figure 8b showed that the area simulated by the new model could not match the observed ones at different times, compared with Figure 6b. The length of seawater in the new model was 0.093 m smaller than that in Case 2. It depicted that the area of pollution plume simulated by the new model was 0.086 m^2, which was much smaller than that of Case 2. It indicated that decreasing the dispersion would lead to the slow migration velocity of plume. Overall, changes in dispersivity have effects on plume transport.

3.4.3. Effect of Contaminant Concentration

The saltwater concentration was reduced from 35 to 20 g/L in the new model. Figure 9 shows that the fitting to the observed plume was worse, compared with that of the basic case in Figure 6b. One can see that the area of the pollutant plume and maximum vertical pollution depth decreased sharply, whereas the movement of the plume decreased slightly in the horizontal direction of the new model, compared with the simulated ones in Case 2. It was because the dispersion was too small. At the time of 100 min, the simulated area of the plume was 0.063 m^2, which was smaller than that of the basic case. However, the direction of the pollutant plume migration was downstream and landward in the aquifer. Therefore, the range of the plume was sensitive to the concentration of contamination.

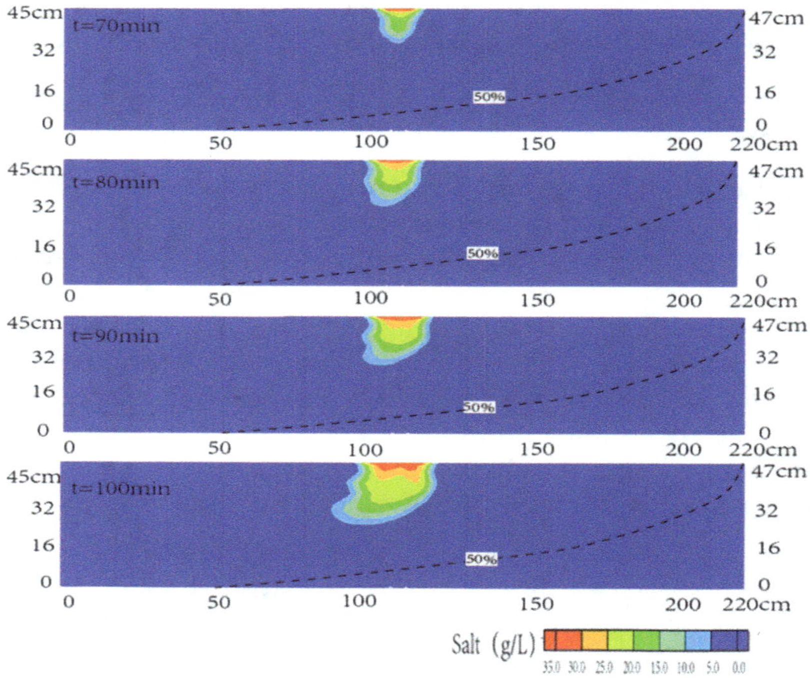

Figure 9. Simulated spatial saltwater contamination concentration distributions at different times when the injected saltwater concentration decreased to 20 g/L.

3.4.4. Effect of Homogeneous Aquifer

In order to discuss the influence of heterogeneity on the contamination transport, numerical contamination simulation in the homogeneous aquifer was considered. The aquifer was assumed to be filled with fine sand based on Case 2. The value of K of the whole aquifer was taken as 4.0 m/d, and the other parameters were fixed as shown in Table 1. Figure 10 showed the simulated contamination concentration distributions of the cross section in the homogeneous aquifer for the new model. The steady state length of the seawater in the new model was smaller than that in Case 2. At the time of 70 min, the shape of the pollution plume was semi-elliptic, which was the same to that of the basic case. Then, the plume migrated downward and laterally during the period between 70 to 100 min, compared to that of the basic case. The vertical migration of the plume decreased, nevertheless, the movement in the horizontal direction increased slightly for the new model. This was because the vertical velocity of the new model was smaller than that of the basic case. There was no concave observed in the plume during the process of pollutant transport, compared to Figure 6b. At the time of 100 min, the maximum area and vertical pollution depth of the pollutant plume reached 0.11 m^2 and 0.25 m in the new model, which were smaller than those simulated by the basic case. Therefore, one can conclude that the infiltration depth and range of the contaminant plume in the heterogeneous aquifer were greater than those in the homogeneous aquifer of the actual beach.

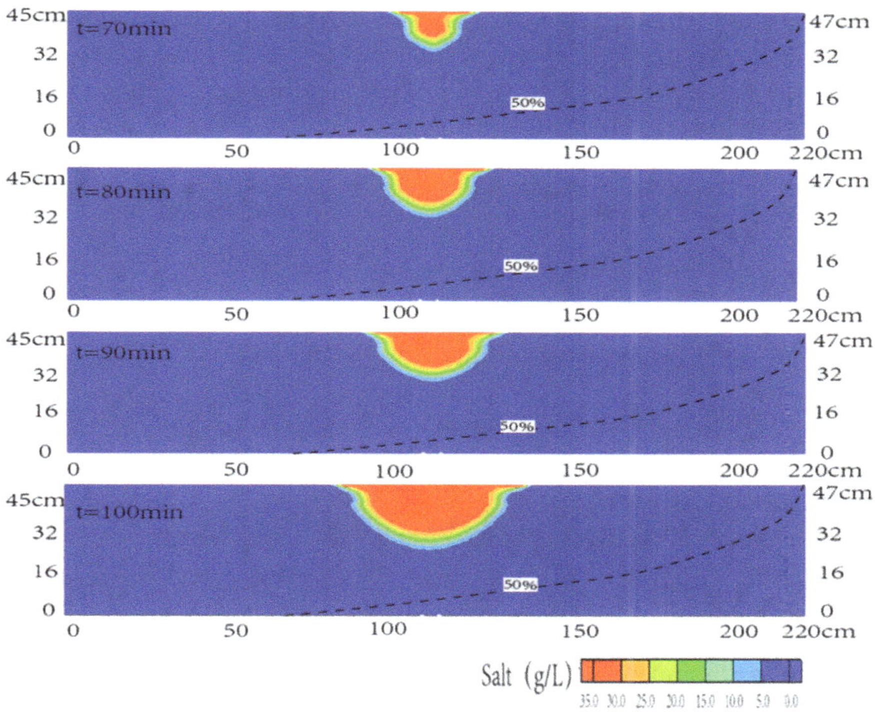

Figure 10. Simulated contamination concentration distributions of the cross section in the homogeneous aquifer for new model 1.

4. Conclusions

This paper presented for the first time analyzing the influence of sea level rise on contaminate transport in the heterogeneous unconfined aquifer of the coastal zone. Two sets of laboratory experiments were tested considering the rising sea level (the difference between sea level and inland head was 1 cm for Case 1 and 2 cm for Case 2). The variation of seawater intrusion and contaminant transport with time were observed during both sets of experiments. The numerical model SEAWAT was used for validation. The numerical results of the model provided matched well with the experimental data for both the toe length of seawater wedge and the shape of the contaminant plume for the rising sea level in the heterogeneous unconfined aquifer.

The observed and simulated values of Cl^- concentration at the sampling points increased sharply at the initial time, and then they increased slowly and tended to be stable. The Cl^- concentration in the medium sand layer was smaller than that in the fine sand layer, which indicated that the migration of plume in the middle layer lagged behind that in the upper layer. The values of concentration at the sampling points in Case 2 were a little smaller than those in Case 1, because the contaminant was diluted due to the increase of water flow velocity influenced by the rising sea level. The observed and simulated seawater wedge migrated inland with time, under the effects of the hydraulic gradient toward the inland and the density difference between saltwater and freshwater. The steady state length of the 50% isoline of the seawater wedge was 167 cm in Case 2, which was larger than that of Case 1. It was because the head at the seaside boundary and the flow velocity for Case 2 were higher than those for Case 1.

In both cases, the observed and simulated plume diffused laterally and downward in the fine sand layer and medium sand layer. The contaminant plume moved downward and inland in the coarse sand layer. The front of the pollutant flowed in the form of the dominant flow, due to the heterogeneity of the aquifer. At the interface between each two layers, the plume became concave due to the increasing velocity of contaminant plume migration. The maximum area of plume in Case 2 was 0.13 m^2, larger than that in Case 1, which indicated that the velocity of diffusion of the contaminant plume increased as the sea level increased. As the velocity of diffusion increased, the time for pollutant migration to the intersection between seawater and freshwater became shorter. The maximum area and vertical depth of the pollutant plume were sensitive to the hydraulic conductivity, dispersivity, and contamination concentration. The infiltration depth and range of the contaminant plume in the heterogeneous aquifer were greater than those in the homogeneous aquifer of the actual beach.

Author Contributions: Q.G. conceived and designed the study; Y.Z. and Z.H. analyzed laboratory data; Q.G. wrote the paper with the assistance of Z.Z. All authors have read and agreed to the published version of the manuscript.

References

1. Ferguson, G.; Gleeson, T. Vulnerability of coastal aquifers to groundwater use and climate change. *Nat. Clim. Chang.* **2012**, *2*, 342–345. [CrossRef]
2. Guo, Q.N.; Huang, J.W.; Zhou, Z.F.; Wang, J.G. Experiment and numerical simulation of seawater intrusion under the influences of tidal fluctuation and groundwater exploitation in coastal multilayered aquifers. *Geofluids* **2019**, *2019*, 2316271. [CrossRef]
3. Post, V.E.A. Fresh and saline groundwater interaction in coastal aquifers: Is our technology ready for the problems ahead. *Hydrogeol. J.* **2005**, *13*, 120–123. [CrossRef]
4. Sefelnasr, A.; Sherif, M. Impacts of seawater rise on seawater intrusion in the Nile Delta aquifer, Egypt. *Ground Water* **2014**, *52*, 264–376. [CrossRef] [PubMed]
5. Werner, A.D.; Bakker, M.; Post, V.E.A.; Vandenbohede, A.; Lu, C.H.; Ataie-Ashtianiab, B.; Simmons, C.T.; Barry, D.A. Seawater intrusion processes, investigation and management: Recent advances and future challenges. *Adv. Water Resour.* **2013**, *51*, 3–26. [CrossRef]
6. Guo, Q.N.; Li, H.L.; Boufadel, M.C.; Sharifi, Y. Hydrodynamics in a gravel beach and its impact on the *Exxon Valdez* oil. *J. Geophys. Res.* **2010**, *115*. [CrossRef]
7. Li, H.L.; Boufadel, M.C. Long-term persistence of oil from the Exxon Valdez spill in two-layer beaches. *Nat. Geosci.* **2010**, *3*, 96–99. [CrossRef]
8. Boufadel, M.; Geng, X.L.; An, C.J.; Owens, E.; Chen, Z.; Lee, K.N.; Taylor, E.; Prince, R.C. A Review on the factors affecting the deposition, retention, and biodegradation of oil stranded on beaches and guidelines for designing laboratory experiments. *Curr. Pollut. Rep.* **2019**, *5*, 407–423. [CrossRef]
9. Herbeck, S.L.; Unger, D.; Wu, Y.; Jennerjahn, C.T. Effluent, nutrient and organic matter export from shrimp and fish ponds causing eutrophication in coastal and back-reef waters of NE Hainan, tropical China. *Cont. Shelf Res.* **2013**, *57*, 92–104. [CrossRef]
10. Tal, A.; Weinstein, Y.; Yechieli, Y.; Borisover, M. The influence of fish ponds and salinization on groundwater quality in the multi-layer coastal aquifer system in Israel. *J. Hydrol.* **2017**, *551*, 768–783. [CrossRef]
11. Tal, A.; Weinstein, Y.; Walman, S.; Goldman, M.; Yechieli, Y. The interrelations between a multi-layered coastal aquifer, a surface reservoir (fish ponds) and the sea. *Water* **2018**, *10*, 1426. [CrossRef]
12. Abd-Elhamid, H.F.; Javadi, A.A. Impact of sea level rise and over-pumping on seawater intrusion in coastal aquifers. *J. Water Clim. Chang.* **2011**, *2*, 19–28. [CrossRef]
13. Anwar, N.; Robinson, C.; Barry, D.A. Influence of tides and waves on the fate of nutrients in a nearshore aquifer: Numerical simulations. *Adv. Water Resour.* **2014**, *73*, 203–213. [CrossRef]
14. Ataie-Ashtiani, B.; Volker, R.E.; Lockington, D.A. Tidal effects on seawater intrusion in unconfined aquifers. *J. Hydrol.* **1999**, *216*, 17–31. [CrossRef]
15. Ataie-Ashtiani, B.; Werner, A.D.; Simmons, C.T.; Morgan, L.K.; Lu, C. How important is the impact of land-surface inundation on seawater intrusion caused by sea-level rise? *Hydrogeol. J.* **2013**, *21*, 1673–1677. [CrossRef]
16. Lu, C.; Xin, P.; Li, L.; Luo, J. Seawater intrusion in response to sea-level rise in a coastal aquifer with a general-head inland boundary. *J. Hydrol.* **2015**, *522*, 135–140. [CrossRef]
17. Green, N.R.; MacQuarrie, K.T.B. An evaluation of the relative importance of the effects of climate change and groundwater extraction on seawater intrusion in coastal aquifers in Atlantic Canada. *Hydrogeol. J.* **2014**, *22*, 609–623. [CrossRef]
18. Guo, Q.N.; Zhang, Y.H.; Zhou, Z.; Zhao, Y. Saltwater transport under the influence of sea-level rise in coastal multilayered aquifers. *J. Coast. Res.* **2020**, *36*, 1040–1049. [CrossRef]

19. Ketabchi, H.; Mahmoodzadeh, D.; Ataie-Ashtiani, B.; Simmons, C.T. Sea-level rise impacts on seawater intrusion in coastal aquifers: Review and integration. *J. Hydrol.* **2016**, *535*, 235–255. [CrossRef]
20. Lemieux, J.M.; Hassaoui, J.; Molson, J.; Therrien, R.; Therrien, P.; Chouteau, M.; Ouellet, M. Simulating the impact of climate change on the groundwater resources of the Magdalen Islands, Quebec, Canada. *J. Hydrol. Reg. Stud.* **2015**, *3*, 400–423. [CrossRef]
21. Mehdizadeh, S.S.; Karamalipour, S.E.; Asoodeh, R. Sea level rise effect on seawater intrusion into layered coastal aquifers (simulation using dispersive and sharp-interface approaches). *Ocean Coast. Manag.* **2017**, *138*, 11–18. [CrossRef]
22. Ketabchi, H.; Mahmoodzadeh, D.; Ataie-Ashtiani, B.; Werner, A.D.; Simmons, C.T. Sea-level rise impact on fresh groundwater lenses in two-layer small islands. *Hydrol. Process.* **2014**, *28*, 5938–5953. [CrossRef]
23. Shi, W.L.; Lu, C.H.; Ye, Y.; Wu, J.C.; Li, L.; Luo, J. Assessment of the impact of sea-level rise on steady-state seawater intrusion in a layered coastal aquifer. *J. Hydrol.* **2018**, *563*, 851–862. [CrossRef]
24. Webb, M.D.; Howard, W.F. Modeling the transient response of saline intrusion to rising sea levels. *Groundwater* **2011**, *49*, 560–569. [CrossRef]
25. Abdoulhalik, A.; Ahmed, A.A. Transience of seawater intrusion and retreat in response to incremental water-level variations. *Hydrol. Process.* **2018**, *32*, 2721–2733. [CrossRef]
26. Morgan, L.K.; Stoeckl, L.; Werner, A.D.; Post, V.E.A. An assessment of seawater intrusion overshoot using physical and numerical modeling. *Water Resour. Res.* **2013**, *49*, 6522–6526. [CrossRef]
27. Ataie-Ashtiani, B.; Volker, R.E.; Lockington, D.A. Contaminant transport in the aquifers influenced by tide. *Aust. Civil Eng. Trans. Institut. Eng. Aust.* **2002**, *43*, 1–11.
28. Mao, X.; Enot, P.; Barry, D.A.; Li, L.; Binley, A.; Jeng, D.S. Tidal influence on behavior of a coastal aquifer adjacent to a low-relief estuary. *J. Hydrol.* **2006**, *327*, 110–127. [CrossRef]
29. Bakhtyar, R.; Brovelli, A.; Barry, D.A.; Robinson, C.; Li, L. Transport of variable-density solute plumes in beach aquifers in response to oceanic forcing. *Adv. Water Resour.* **2013**, *53*, 208–224. [CrossRef]
30. Brovelli, A.; Mao, X.; Barry, D.A. Numerical modeling of tidal influence on density-dependent contaminant transport. *Water Resour. Res.* **2007**, *43*, W10426. [CrossRef]
31. Liu, Y.; Jiao, J.J.; Luo, X. Effects of inland water level oscillation on groundwater dynamics and land-sourced solute transport in a coastal aquifer. *Coast. Eng.* **2016**, *114*, 347–360. [CrossRef]
32. Shen, C.; Zhang, C.; Kong, J.; Xin, P.; Lu, C.; Zhao, Z.; Li, L. Solute transport influenced by unstable flow in beach aquifers. *Adv. Water Resour.* **2019**, *125*, 68–81. [CrossRef]
33. Koohbor, B.; Fahs, M.; Ataie-Ashtiani, B.; Simmons, C.T. Semi-analytical solutions for contaminant transport under variable velocity field in a coastal aquifer. *J. Hydrol.* **2018**, *560*, 434–450. [CrossRef]
34. Zhang, Q.; Volker, R.E.; Lockington, D.A. Experimental investigation of contaminant transport in coastal groundwater. *Adv. Environ. Res.* **2002**, *6*, 229–237. [CrossRef]
35. Ataie-Ashtiani, B. MODSharp: Regional-scale numerical model for quantifying groundwater flux and contaminant discharge into the coastal zone. *Environ. Modell Softw.* **2007**, *22*, 1307–1315. [CrossRef]
36. Chen, J.S.; Lai, K.H.; Liu, C.W.; Ni, C.F. A novel method for analytically solving multi-species advective–dispersive transport equations sequentially coupled with first-order decay reactions. *J. Hydrol.* **2012**, *420*, 191–204. [CrossRef]
37. Hayek, M.; Kosakowski, G.; Jakob, A.; Churakov, S.V. A class of analytical solutions for multidimensional multispecies diffusive transport coupled with precipitation-dissolution reactions and porosity changes. *Water Resour. Res.* **2012**, *48*, WR3525. [CrossRef]
38. Chang, S.W.; Clement, T.P. Laboratory and numerical investigation of transport processes occurring above and within a saltwater wedge. *J. Contam. Hydrol.* **2013**, *147*, 14–24. [CrossRef]
39. Miller, C.T.; Dawson, C.N.; Farthing, M.W.; Hou, T.Y.; Huang, J.F.; Kees, C.E.; Kelley, C.T.; Langtangen, H.P. Numerical simulation of water resources problems: Models, methods, and trends. *Adv. Water Resour.* **2013**, *51*, 405–437. [CrossRef]
40. Liu, Y.; Mao, X.; Chen, J.; Barry, D.A. Influence of a coarse interlayer on seawater intrusion and contaminant migration in coastal aquifers. *Hydrol. Process.* **2014**, *28*, 5162–5175. [CrossRef]
41. Parker, J.C.; Kim, U. An upscaled approach for transport in media with extended tailing due to back-diffusion using analytical and numerical solutions of the advection dispersion equation. *J. Contam. Hydrol.* **2015**, *182*, 157–172. [CrossRef] [PubMed]

42. Shahkarami, P.; Liu, L.; Moreno, L.; Neretnieks, I. Radionuclide migration through fractured rock for arbitrary-length decay chain: Analytical solution and global sensitivity analysis. *J. Hydrol.* **2015**, *520*, 448–460. [CrossRef]
43. Boufadel, M.C.; Xia, Y.Q.; Li, H.L. Modeling solute transport and transient seepage in a laboratory beach under tidal influence. *Environ. Modell Softw.* **2011**, *26*, 899–912. [CrossRef]
44. Geng, X.; Boufadel, M.C.; Cui, F. Numerical modeling of subsurface release and fate of benzene and toluene in coastal aquifers subjected to tides. *J. Hydrol.* **2017**, *551*, 793–803. [CrossRef]
45. Li, H.L.; Boufadel, M.C. A tracer study in an Alaskan gravel beach and its implications on the persistence of the Exxon Valdez oil. *Mar. Pollut. Bull.* **2011**, *62*, 1261–1269. [CrossRef]
46. Guo, Q.N.; Li, H.L.; Boufadel, M.C.; Liu, J. A field experiment and numerical modeling of a tracer at a gravel beach in Prince William Sound, Alaska. *Hydrogeol. J.* **2014**, *22*, 1795–1805. [CrossRef]
47. Jang, E.; He, W.; Savoy, H.; Dietrich, P.; Kolditz, O.; Rubin, Y.; Schüth, C.; Kalbacher, T. Identifying the influential aquifer heterogeneity factor on nitrate reduction processes by numerical simulation. *Adv. Water Resour.* **2017**, *99*, 38–52. [CrossRef]
48. Lalumera, G.M.; Calamari, D.; Galli, P.; Castgilioni, S.; Crosa, G.; Fanelli, R. Preliminary investigation on the environmental occurrence and effects of antibiotics used in aquaculture in Italy. *Chemosphere* **2004**, *54*, 661–668. [CrossRef]
49. Scribner, E.A.; Dietze, J.E.; Meyer, M.T.; Kolpin, D.W. *Occurrence of Antibiotics in Water from Fish Hatcheries*; U.S. Geological Survey: Lawrence, KS, USA, 2002; p. 68.
50. Guo, W.; Langevin, C.D. User's guide to SEAWAT: A computer program for simulation of three-dimensional variabledensity groundwater flow. In *Techniques of Water-Resources Investigations*; U.S. Geological Survey: Reston, VA, USA, 2002; p. 434.
51. Langevin, C.D.; Shoemaker, W.B.; Guo, W. *MODFLOW-2000, the U.S. Geological Survey Modular Ground-Water Model: Documentation of the SEAWAT-2000 Version with Variable Density Flow Process (VDF) and the Integrated MT3DMS Transport Process (IMT). U.S. Geological Survey Water-Resources Investigations Report*; U.S. Geological Survey: Reston, VA, USA, 2003; pp. 3–426.
52. Bakker, M.A. Dupuit formulation for modeling seawater intrusion in regional aquifer system. *Water Resour. Res.* **2003**, *39*, 1131. [CrossRef]
53. Li, X.Y.; Bill, X.H.; Burnett, W.C.; Santos, I.R.; Chanton, J.P. Submarine groundwater discharge driven by tidal pumping in a heterogeneous aquifer. *Ground Water.* **2009**, *47*, 558–568. [CrossRef]
54. Chang, Y.W.; Hu, B.X.; Xu, Z.X.; Li, X.; Tong, J.X.; Chen, L.; Zhang, H.X.; Miao, J.J.; Liu, H.W.; Ma, Z. Numerical simulation of seawater intrusion to coastal aquifers and brine water/freshwater interaction in south coast of Laizhou Bay, China. *J. Contam. Hydrol.* **2018**, *215*, 1–10. [CrossRef] [PubMed]

3

Radon as a Natural Tracer for Monitoring NAPL Groundwater Contamination

Martina Mattia [1], Paola Tuccimei [1,*], Michele Soligo [1] and Claudio Carusi [2]

[1] Dipartimento di Scienze, Università "Roma Tre", 00154 Rome, Italy; martina.mattia@uniroma3.it (M.M.); michele.soligo@uniroma3.it (M.S.)

[2] Protezione Ambiente, Mares S.r.l., 00144 Rome, Italy; claudiocarusi@maresitalia.it

* Correspondence: paola.tuccimei@uniroma3.it

Abstract: In this research, the radioactive noble gas radon was used as a tracer for Non-Aqueous Phase Liquids (NAPLs) contamination, since it is much more soluble in these substances than in air or water. Soil radon remains trapped within the NAPLs, resulting in a local reduction in the radon concentration within close proximity to the contaminated area. This technique was applied to a contaminated site in Roma (Italy). The main residual NAPLs are total hydrocarbons and methyl-tertiary-butyl ether (MTBE), a water-soluble additive. The monitoring activities included two sampling campaigns of groundwater from 18 wells in February and May 2020. Concentration maps were produced using radon data. The results show that the radon deficit traces the location of NAPLs in the fuelling station very well, with a residual source zone extending in a NNW-SSE direction. A good correspondence between a low amount of radon and a higher concentration of NAPLs was found. A reduction in the average amount of radon in the May 2020 survey indicated a stronger remobilization of NAPLs compared to that of the February 2020 monitoring campaign. The peaks of Volatile Organic Compounds (VOCs) detected between 8–9 and 11–12 m depths indicate the presence of residual blobs of NAPLs in the vadose zone of the aquifer.

Keywords: NAPLs; radon; groundwater contamination; monitoring; MTBE; geochemical tracer

---

1. Introduction

The monitoring and detection of organic contaminants—e.g., Non-Aqueous Phase Liquids (NAPLs)—in the subsoil using geochemical and non-invasive methods are crucial due to the direct and strong impact that these substances have on groundwater resources. Indeed, NAPLs may persist in the subsurface for a long period, thus becoming long-term sources of groundwater contamination. Therefore, effective remediation planning requires the identification and assessment of residual NAPLs in the subsoil or contaminants dissolved in groundwater to form a NAPL plume. Geochemical tracers such as radon gas are often used to study the spatial distribution of residual contaminants in aquifers and to deduce information on subsoil contamination [1,2].

Radon is a naturally occurring radioactive noble gas. Due to its ubiquitous presence in nature, its chemical and physical properties, and its straightforward detectability, radon fulfils all the requirements for use as an environmental tracer [2,3]. The equilibrium radon concentration in the pore space of an aquifer or soil depends on the aquifer emanation coefficient, the aquifer porosity, the radium activity concentration, and the bulk density of the mineral matrix [4].

Radon is usually employed to trace NAPL contamination in soil and groundwater because of its remarkable solubility in these substances. Radon gas is much more soluble in a wide range of these contaminants than in air or water. Consequently, as it rises towards the interface with the atmosphere it remains trapped within the pure residual NAPLs, but not necessarily in an aqueous solution of BTEX

(Benzene, Toluene, Ethylbenzene and Xylene) or MTBE, for example, and produces a concentration deficit compared to nearby unpolluted sites. The mapping of this process, known as the radon-deficit technique [1,2,5–16], allows observers to identify the contamination affecting either the vadose or the saturated portions of an aquifer. This topic has been addressed using a steady-state radon partitioning model and a one-dimensional model for radon transport with NAPL partitioning [1,17,18]. Both approaches are based on some assumptions, such as a homogeneous ^{226}Ra distribution in the aquifer rock, a constant porosity, no ^{222}Rn loss, equilibrium ^{222}Rn partitioning between the NAPLs and the water phase, and a negligible ^{222}Rn sorption from the matrix. These need to be demonstrated and accounted for in any single situation [6].

Other works have dealt with modification [19] by the aging and biodegradation of NAPLs in aquifers [20,21]. An excellent review of these, with an extensive list of related references, has recently been published [2].

In this context, the present work is mainly a case study, and its purpose is to verify and support the potential of the radon-deficit technique to assess contamination by NAPLs in a study area located in Roma (Italy).

2. Study Area

2.1. Geological Setting

At the study site, located in the south-east portion of Roma (Italy), there is a fuelling station where, more than twenty years ago, a first oil spill occurred, and other NAPLs spills have occurred over time. This area is characterized by the presence of volcanic deposits from the Alban Hills complex [22–24], belonging to the formation of Villa Senni, which consists of two main ignimbrites: "Tufo Lionato" (VSN1) at the bottom and "Pozzolanelle" (VSN2) at the top. The area belongs to the "Alban Hills Hydrogeological Unit" (Figure 1), as shown in the recent hydrogeological map of Rome [25]. The degree of relative permeability is high for both porosity and fracturing.

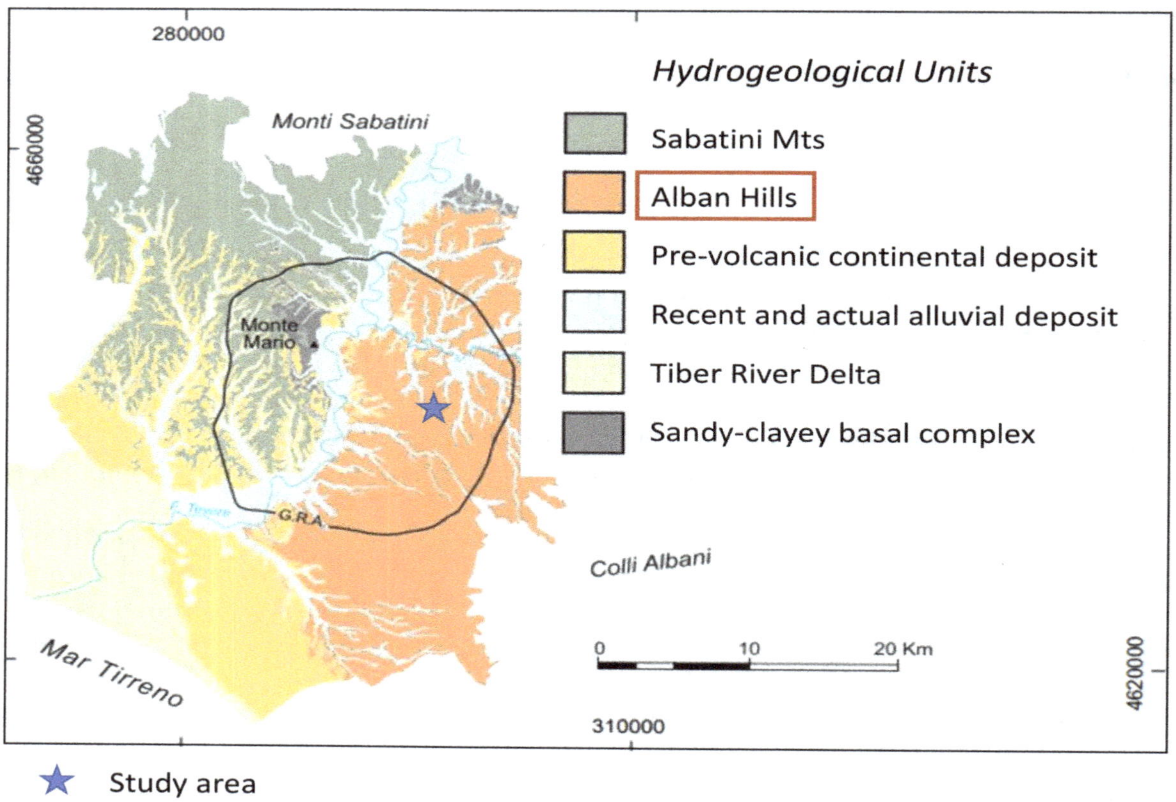

Figure 1. Map of hydrogeological units (modified from [26]).

The simplified local stratigraphy, obtained from a drilling carried out in May 2020 to check for the presence of residual Volatile Organic Compounds (VOCs) in a vadose aquifer (Figure 2), generally consists of 4–5 m of backfill soils and at least 14–15 m of ignimbrites. Specifically, there is 2 m of backfill, about 3 m of reworked pozzolan, 5 m of pozzolan belonging to the "Pozzolanelle" unit, and about 10 m of "Tufo Lionato" ignimbrite that is more granular on top and more sandy at the log bottom. The groundwater table can be found at a depth of approximately 18 m, with a seasonal variation of about 0.5 m.

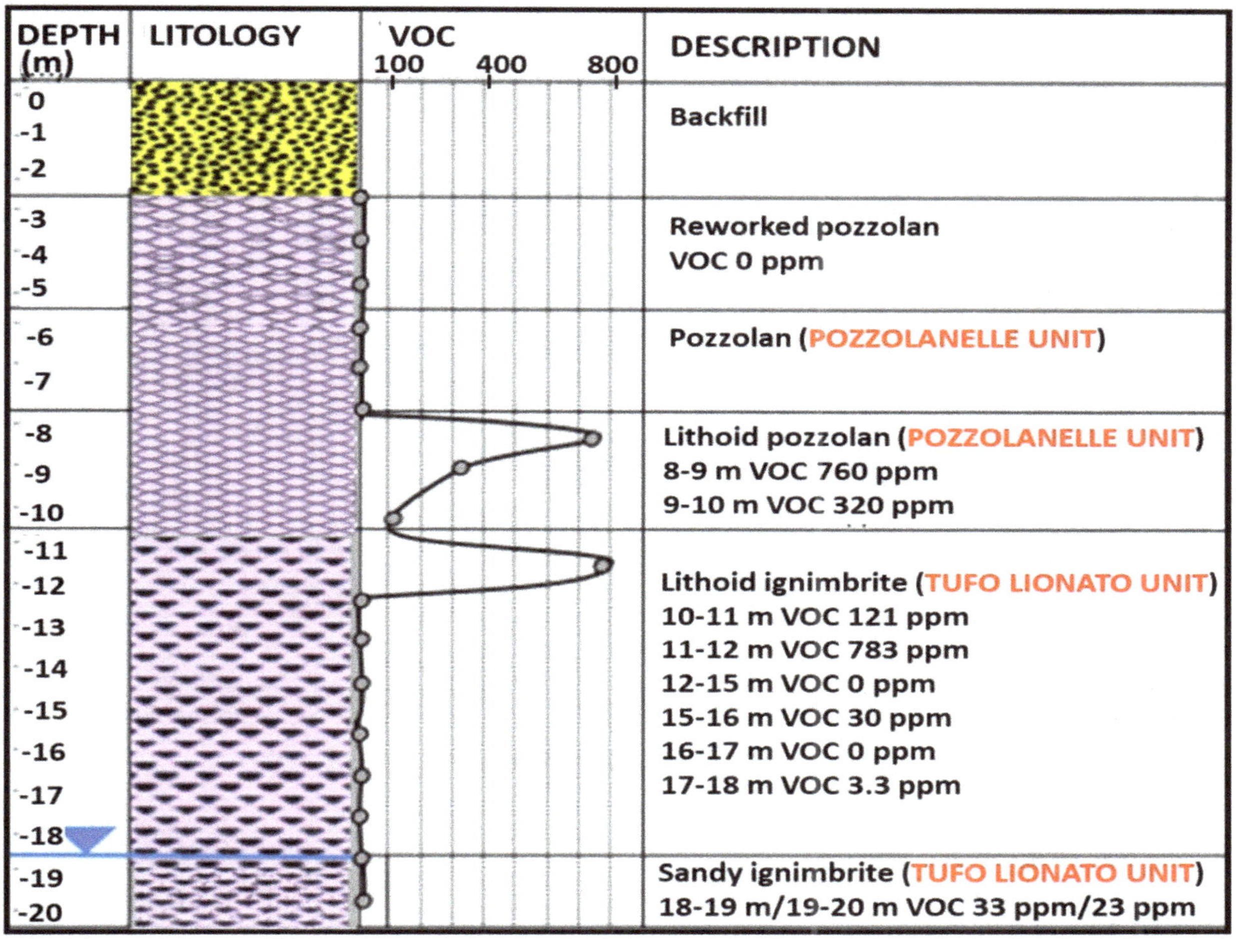

Figure 2. Stratigraphic log (SN10) of the study area.

It is worth noting that the groundwater table frequently rises to a depth of about −9 m below ground level in two wells (PZ10 and PZ4) located downstream and far from the recharge wells of the "Pump and treat" system, probably due to leaks from the city underground pipeline network, as evidenced by an increasing concentration of chlorinated organic compounds. In particular, Mares S.r.l. found up to 1240 mg/L of 1,2-dichloroethane in July 2016, demonstrating that, even if low values of substances nowadays represent the urban background, they may reach concentration peaks through fractures and cavities locally intersecting some wells.

2.2. Remediation System

The reclamation of the area began in 2016, when a network of 18 sampling wells was set up to screen the extent and evolution of the contamination; of these wells, 14 are located inside the study area, while the other four are placed outside. These outermost wells are used to provide the background values for comparison with those obtained from wells located inside the area (Figure 3). The remediation scheme consists of four abstraction wells of the "pump & treat" system, and the other five are recharge wells for the reintroduction of treated water. "Pump and treat" is a common method for cleaning up groundwater contaminated with dissolved chemicals, including industrial solvents, metals, and fuel oil [26]. In the study area, groundwater is pumped from the abstraction wells to a treatment system consisting of activated carbon filters. Treatment occurs above ground, and the cleaned groundwater is re-injected into the subsurface. This reclamation system has been active (and is still active) for about four years. A soil vapor extraction system completes the remediation plant.

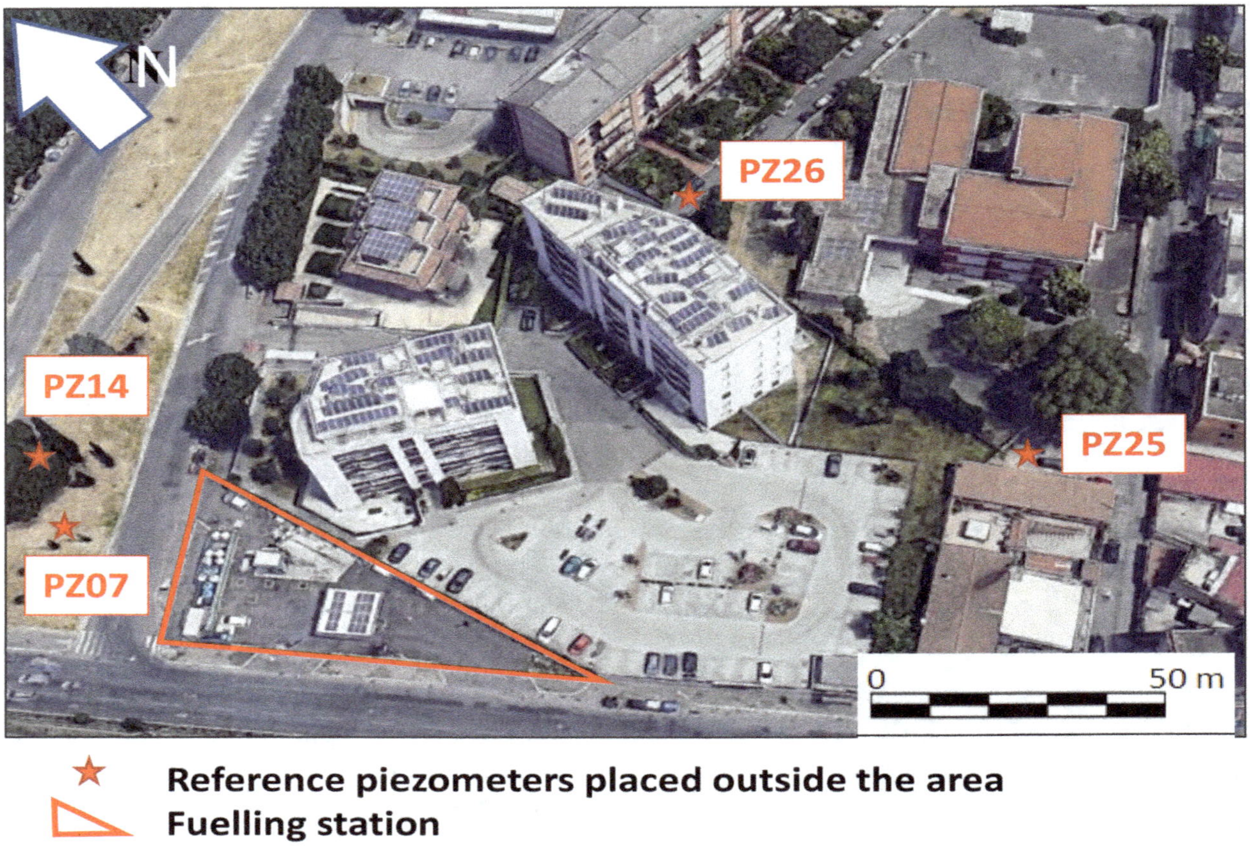

Figure 3. Study area with the location of reference wells placed outside the fuelling station.

3. Materials and Methods

3.1. Field Work

Two groundwater sampling surveys were carried out in February and May 2020. In both campaigns, samples were taken from 14 wells in the study area and from the other 4 outside the site. No residual NAPLs were detected during the two surveys. Groundwater level was always recorded before extracting from 3 to 5 times the water volume in the wells. This action was necessary to sample groundwater that was really representative of the aquifer conditions in terms of its composition and radon concentration. During the sampling of the water, degassing was avoided by using a flow reducer valve. Polyethylene terephthalate (PET) bottles were used to store water samples for the laboratory determination of radon.

During the field survey (May 2020), continuous core drilling was carried out to check for residual Volatile Organic Compounds (VOCs) in the vadose aquifer and to collect soil samples. This drilling gave us the opportunity to check the local stratigraphy. The VOC levels in ppm were measured using a photoionization detector (PID). PID detectors employ an ultraviolet source to bombard gas samples, allowing them to detect a broad range of VOCs.

3.2. Laboratory Methods

The radon activity concentration in the groundwater was measured using a RAD7 monitor (Durridge Company Inc., Billerica, MA, USA) with a Big Bottle RAD H2O accessory.

Radon Measurements and Data Processing

The radon activity concentration in the water samples collected in 500- or 660-mL PET bottles was measured using the RAD7 equipped with the Big Bottle RAD H2O accessory. This experimental set-up has been extensively and thoroughly described in earlier studies [6–9,11,27–29]. It consists of several components: the PET bottle, the bubble trap, the aeration system, the temperature data logger, vinyl tubing with an aeration stone fixed at its lower end and a cap at the upper end, the laboratory dryer, and the RAD7. The experimental apparatus is shown in Figure 4. Two cycles of 15 min each have been set for the analysis.

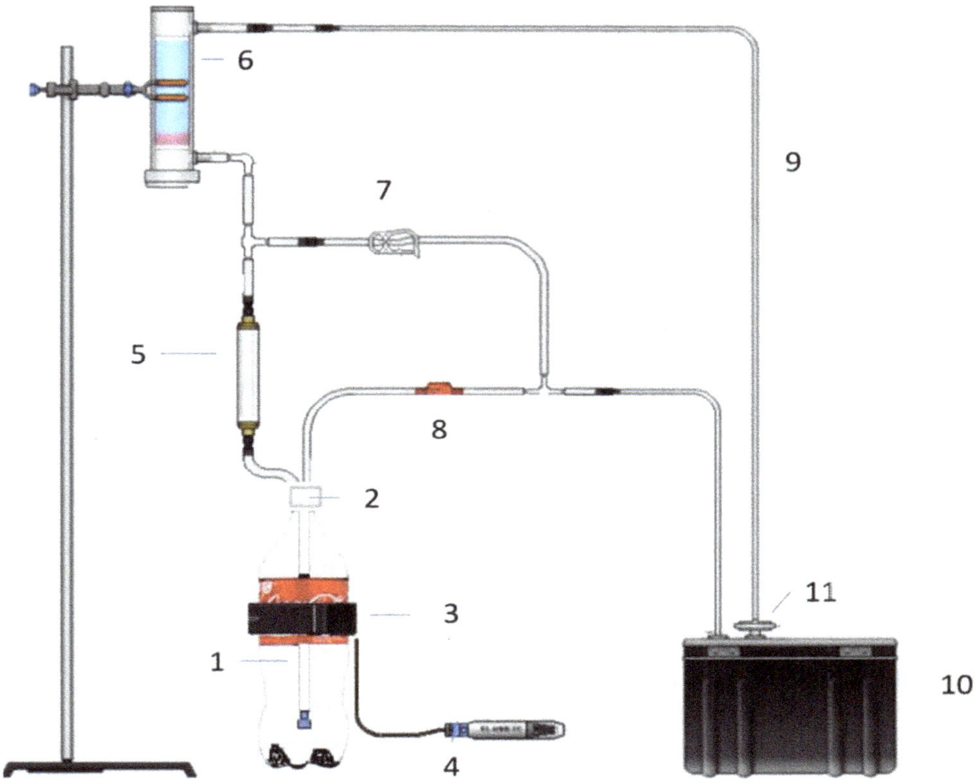

Figure 4. RAD7 monitor with the Big Bottle RAD H2O accessory: (**1**) plastic soda bottle; (**2**) screw-on Teflon aerator with a single air stone; (**3**) elastic clinching strap; (**4**) temperature data logger; (**5**) bubble trap; (**6**) laboratory dryer; (**7**) clip; (**8**) check valve; (**9**) vinyl tubing; (**10**) RAD7 radon detector; (**11**) inlet filter (after [6]).

Radon data obtained from the water analyses were used to produce a dot map of the average radon activity concentration in the groundwater during the study period.

4. Results

The radon activity concentrations of the groundwater from the 18 wells are reported in the Supplementary Material (Table S1). The results of the radon determination carried out in February and May 2020 show the concentration ranging between 40 (well 13 in May 2020) and 180 Bq/L (well 14 in May 2020). Higher values were found in the external sampling wells 7, 14, 25, and 26 (see Figure 3 for locations), as they were located away from the contamination, while the lowest radon values were measured in some wells within the study area (wells 2, 4, 8, 9, 10, 13, and 19; see Figure 5).

Radon data higher than 100 Bq/L placed upgradient or to the side of PZ13 (adjacent to the removed leaking underground tanks) were used to calculate the average radon in the clean wells (112 Bq/L in February 2020 and 128 Bq/L in May 2020). This concentration (the background radon) was then employed as a reference value to estimate the radon deficit of "dirty" wells (see Table S1) for the ratio between the radon level of single "contaminated" wells and the background [6].

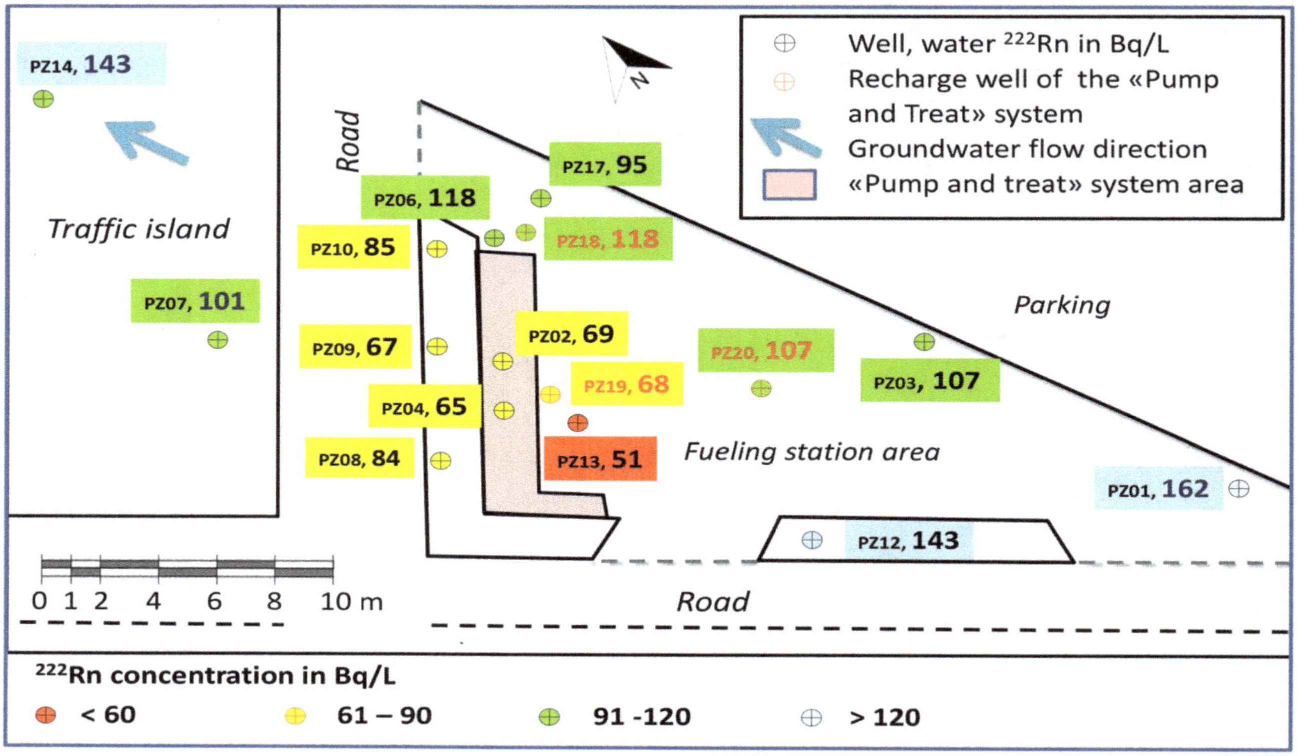

Figure 5. Dot map of the average radon concentrations in the study area obtained from the two sampling campaigns (February and May 2020).

The chemical analyses of the groundwater show that the main residual NAPLs still present in the site are total hydrocarbons, expressed as n-hexane and methyl-tertiary-butyl ether (MTBE), an additive introduced in gasoline in place of lead. MTBE is a hydrophilic ether with a solubility of 50 g/L in water, and it is generally 30 times more soluble in water than the hydrocarbon constituents in gasoline [21]. MTBE and other components, commonly known as "oxygenates", are added to gasoline to increase the octane number and reduce the carbon monoxide emissions. MTBE is the main contaminant at the site because it is biologically and chemically stable and thus very persistent in the environment. It is absorbed onto soil grains, such as other residual NAPLs, but its high water solubility and persistence cause it to travel faster and farther than many other components of gasoline when released into an aquifer [6]. It is worth stressing that ^{222}Rn is used here as a natural tracer for monitoring the groundwater contamination by highly soluble MTBE and BTEX because it is highly soluble in a wide range of residual NAPLs, but it is not necessarily soluble in the aqueous solutions of these.

MTBE reached average maximum concentrations of about 2200 and 1700 μg/L in wells 2 and 13, respectively. Furthermore, high mean concentration values of total hydrocarbons were also detected in wells 2 and 13 (values of about 46,000 and 72,000 μg/L, respectively). Finally, moderate concentrations of benzene and toluene were found, which was in agreement with the MTBE peaks in the same wells.

5. Discussion

The groundwater radon concentrations obtained for the February and May 2020 surveys were used to calculate the average values. With these data, a concentration dot map was drawn (Figure 5). The map shows the radon distribution in the sampled wells.

The average radon concentration is plotted using different colors according to changing values. First of all, it is important to remember that the equilibrium radon concentration in the pore space of an aquifer or soil depends on the aquifer emanation coefficient, the aquifer porosity, the radium activity concentration, and the bulk density of the mineral matrix [4], and that the natural aquifer could not always fulfill these requirements [6,30–32].

The map shows that the residual source zone of contamination is presumably located around well 13 (lowest radon concentration, 51 Bq/L), and that the residual NAPLs source zone, identified by the further very low radon values in wells 2, 4, 8, 9, 10, 19, follows the NNW-SSE flow direction of the aquifer. This is consistent because when light non-aqueous phase liquids (LNAPLs), such as total hydrocarbons or MTBE, infiltrate the soil, being lighter than water they tend to float on top of the water table and therefore follow its flow direction. Nonetheless, it is worth noting that a residual source zone, with low volumes of residual NAPLs occupying the pore space, is confined in the fueling station area and does not extend beyond it, as the higher radon data in wells 7 and 14 suggest.

In addition, if we analyze the relationship between the radon values and the locations of recharge wells, the distribution of the data does not always appear to be simple and clear. There are actually low radon levels downstream of the recharge wells, where you would expect a higher radon concentration, suggesting that an equilibrium between the ^{222}Rn in treated water and ^{226}Ra in the host rock has been reached. Probably, this equilibrium was not achieved because groundwater treated with activated carbon filters and depleted in ^{222}Rn did not have the time necessary to accomplish the full equilibrium (low residence time).

These apparent inconsistencies could also depend on a heterogeneous distribution of ^{226}Ra in the aquifer, the degree of weathering and fracturing, and the emanation coefficient of the rock [30–32]. Finally, it is worth adding that some pumps did not work properly between February and May 2020, and this could have affected the local groundwater level and the flow direction. Generally speaking, the "pump and treat system" worked properly, confining residual NAPLs in the fueling station area.

This finding is supported by the average values (mean of the NAPL levels from February and May 2020 surveys) calculated for the MTBE and total hydrocarbon concentration. The average values exceeding reference limits established by Italian legislation (40 and 350 μg/L limits for MTBE and total hydrocarbons, respectively) are reported in a new map (Figure 6). Wells 2, 4, 10, and 13 are the most polluted and are coherently characterized by a very low radon concentration and low radon deficits (Table S1). This scenario highlights that the radon partitioning in NAPLs accurately traces them in the subsoil and is still visible after years.

In order to strengthen the potential of radon-deficits to trace NAPLs distribution, we calculated the equilibrium radon concentration in the Alban Hills Hydrogeological Unit [25] and then compared it to the measured background radon reference. We obtained a calculated value ($^{222}Rn_C$) of equilibrium radon concentration in the Alban Hills Hydrogeological Unit [25], characterized by a ^{226}Ra activity concentration of the aquifer matrix (A_{Ra}, Bq/kg) of 80, a mineral density (ρ, kg/L) of 2.65, a porosity (n) of 0.35, and an emanation radon efficiency (E) of 0.2, using the following equation [15]:

$$^{222}Rn_C = A_{Ra} \times \rho \times E/n. \quad (1)$$

The result of the calculation is 121 Bq/L, very close to the experimental value (120 Bq/L), with averages of 112 and 128 Bq/L, which are the average radon in the monitoring campaigns of February and May 2020, respectively.

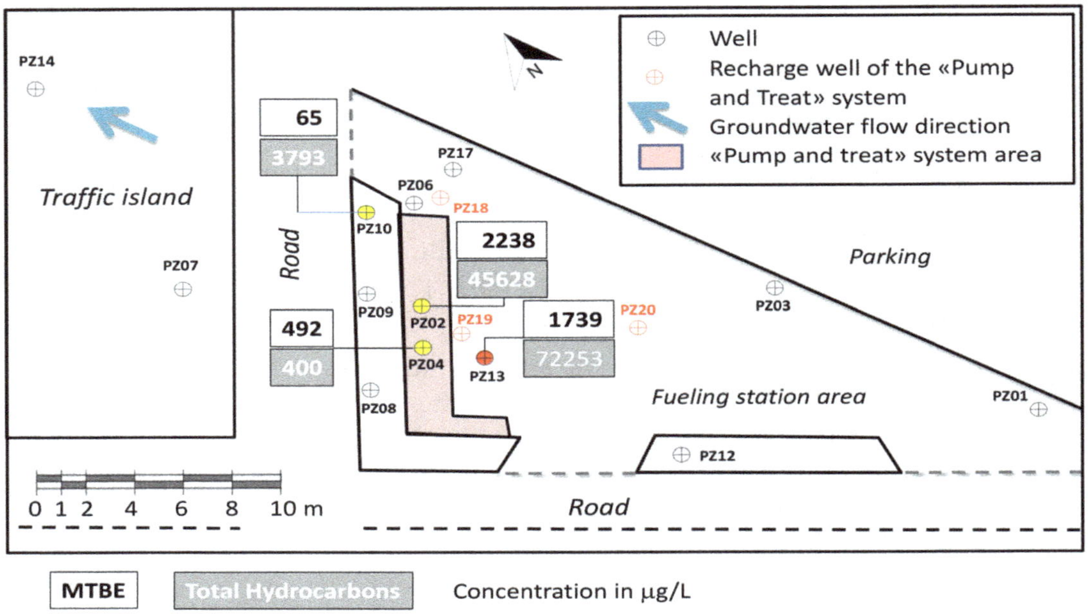

Figure 6. Dot map of the average MTBE and the total hydrocarbon concentrations in the groundwater during the field survey (February–May 2020). We reported only the values of wells where the reference limit was exceeded. Italian law quotes 40 and 350 μg/L as the reference limits for MTBE and total hydrocarbons, respectively.

Another point in favor of the ability of radon-deficits to trace NAPLs contamination arises from the comparison between the average radon deficits in "dirty" wells (wells 2, 4, 8, 9, 10, 13, and 19) of the fueling station in February 2020 (0.63) and the corresponding value of May 2020 (0.54). It is evident that a lower amount of radon was detected in May 2020 in those aquifer domains polluted with residual NAPLs when the sum of the total hydrocarbon and MTBE concentrations reached a higher level (183,640 μg/L), whereas a lower groundwater NAPL concentration (142,425 μg/L) was measured in February 2020.

In situ analyses of VOCs performed on soil samples extracted from the above-mentioned drilling showed that there are concentration peaks both inside the "Pozzolanelle" Unit (between 8 and 9 m depth) and on the top of the "Tufo Lionato" Unit (between 11 and 12 m depth). This indicates that, at these depths, there are still hypothetical blobs of residual NAPLs that can be removed by groundwater, as evidenced by chemical analyses of the water samples.

Based on the available information, the following conceptual site model can be proposed (Figure 7). About five years ago, a remediation plant based on a "pump and treat" system, soil vapor extraction, and water re-emission, started in the site where gasoline leakage from underground tanks had occurred, with infiltration throughout the aquifer. Dilution, retardation, degradation, and volatilization processes took place at the same time, causing contaminant attenuation. Regardless, residual NAPLs blobs are still present at depth. They are probably absorbed onto largely abundant zeolite minerals characterizing the ignimbrites [6]. Occasionally, rainfalls wash the terrain and cause the water table to rise, removing hydrocarbons and MTBE to different extents.

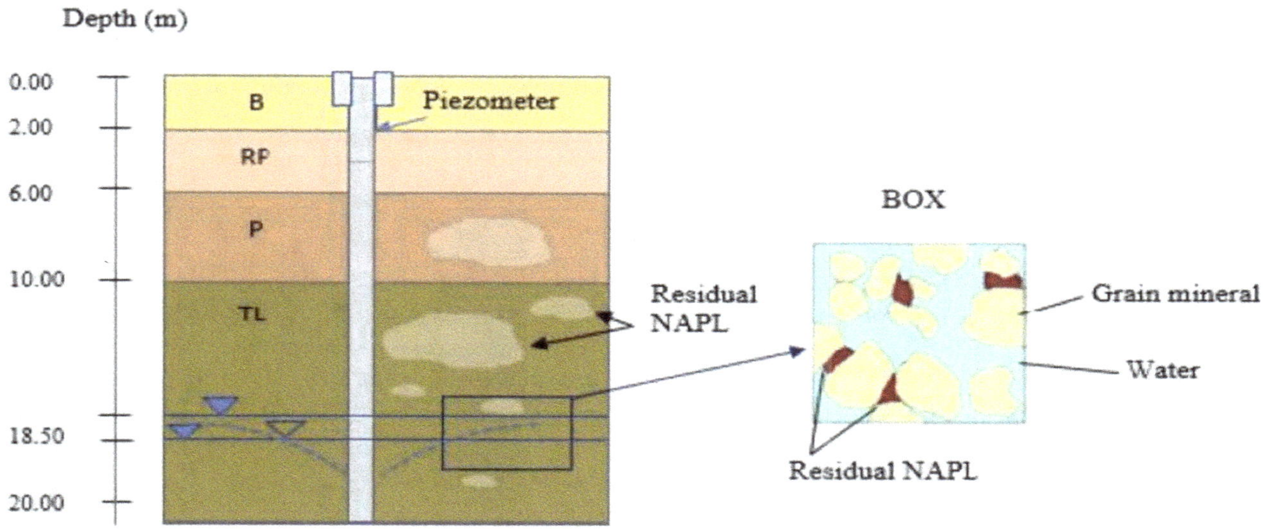

Figure 7. Conceptual site model. B is backfill, RP is reworked pozzolan, P is pozzolan (pozzolanelle unit), and TL is "Tufo Lionato" unit. Solid lines in TL are the groundwater static levels, while the dashed line is the dynamic level. Box: red blobs are residual NAPLs in the aquifer which are leached by groundwater either when the water table rises or during water pumping.

6. Conclusions

More than twenty years ago, the first NAPL spill occurred at the study site. Other spills took place until 2016, when a monitoring and remediation system was activated in the area. In the year 2020, dissolved radon measurements were started in order to use this gas as a natural tracer for NAPL spills. Radon is far more soluble in these substances than in water or air, producing a concentration deficit in those aquifer domains polluted with residual NAPLs compared to the background values in a close and similar uncontaminated area.

The VOC measurements performed on soil samples collected from the drilling highlighted the presence of residual NAPL pockets in the vadose zone at different depths, occasionally removed from groundwater. The presence of these pockets is compatible with the chemical analyses, which show high MTBE and total hydrocarbon concentrations in the water.

A dot map of the average radon concentration in groundwater was produced, suggesting a residual source zone close to the mapped radon minimum zone, well confined in the fueling station area by the "pump and threat" system. This portion of low radon values and low radon deficits corresponds to the places where the dissolved total hydrocarbons and MTBE reach their highest values, also exceeding the reference limits of Italian legislation. The lower average radon deficit allowed us to discriminate the stronger remobilization of NAPLs, which was recorded in the May 2020 survey, compared to the previous one (February 2020).

The residual source zone, although interfered by the pump and treat system, tends to follow the flow direction of the aquifer due to the low density of NAPLs that, once they have infiltrated the subsoil and reached the groundwater table, float on it, following its flow. In places, the low radon levels downstream of the recharge wells are possibly due to the injection of groundwater treated with activated carbon and depleted in radon which did not reach full equilibrium with the ^{226}Ra in the host rock.

Based on that, it is possible to conclude that radon is a powerful and cheap natural tracer of NAPL contamination, providing useful information to study the distribution and evolution of residual blobs of NAPLs in the subsoil and related mobilization processes. Further data are necessary in order to implement the conceptual model of the study site and further explore the potential of radon as a tracer of soil NAPL contamination, as well as as a proxy of groundwater pollution.

Author Contributions: Conceptualization, M.M., P.T., M.S., and C.C.; methodology, M.M. and P.T; investigation, M.M., P.T., and M.S.; resources, M.M. and P.T.; data curation, M.M. and P.T.; writing—original draft preparation, M.M.; writing—review and editing, P.T.; visualization, M.M. and P.T.; supervision, P.T., M.S., C.S. project administration, P.T., M.S., C.S. All authors have read and agreed to the published version of the manuscript.

Acknowledgments: Authors wish to thank Mares S.r.l. for cooperation in the field work and for providing us with data on the contaminants in groundwater.

References

1. Semprini, L.; Hopkins, O.S.; Tasker, B.R. Laboratory, field, and modeling studies of Radon-222 as a natural tracer for monitoring NAPL contamination. *Transp. Porous Med.* **2000**, *38*, 223. [CrossRef]
2. Schubert, M. Using radon as environmental tracer for the assessment of subsurface non-aqueous phase liquid (NAPL) contaminate on a review. *Eur. Phys. J. Special Topics* **2015**, *224*, 717–730. [CrossRef]
3. Krishnaswami, S.; Graustein, W.S.; Turekian, K.K.; Dowd, J.F. Radium, thorium and radioactive lead isotopes in groundwaters: Application to the in-situ determination of adsorption-desorption rate constants and retardation factors. *Water Resour. Res.* **1982**, *18*, 1633–1675. [CrossRef]
4. Andrews, J.N.; Wood, D.F. Mechanism of radon release in rock matrices and entry into groundwaters. *Inst. Mining Met. Trans. Sect. B* **1972**, *81*, 198–209.
5. Briganti, A.; Voltaggio, M.; Soligo, M.; Tuccimei, P.; Carusi, C.; Lucchetti, C.; Castelluccio, M. Using radon as tracer of an old NAPL contamination in groundwater (Roma, Italy). In Proceedings of the XIV GARRM International Workshop on the Geological Aspects of Radon Risk Mapping, Prague, Czech Republic, 18–20 September 2018.
6. Briganti, A.; Tuccimei, P.; Voltaggio, M.; Carusi, C.; Galli, G.; Lucchetti, C. Assessing methyl tertiary butyl ether residual contamination in groundwater using radon. *J. Appl. Geochem.* **2020**, 116. [CrossRef]
7. Briganti, A.; Voltaggio, M.; Soligo, M.; Tuccimei, P. Assessing the age of a NAPL-spill by Radium isotopes. In Proceedings of the 7th International RaRn Workshop, Delmenhorst, Germany, 2–6 July 2018.
8. Castelluccio, M.; Agrahari, S.; De Simone, G.; Pompilj, F.; Lucchetti, C.; Sengupta, D.; Galli, G.; Friello, P.; Curatolo, P.; Giorgi, R.; et al. Using a multi-method approach based on soil radon deficit, resistivity and induced polarization measurements to monitor NAPL contamination in two study areas in Italy and India. *Environ. Sci. Pollut. Res.* **2018**, *25*, 12515–12527. [CrossRef] [PubMed]
9. De Simone, G.; Lucchetti, C.; Pompilj, F.; Galli, G.; Tuccimei, P.; Curatolo, P.; Giorgi, R. Soil radon survey to assess NAPL contamination from an ancient spill. Do kerosene vapors affect radon partition? *J. Environ. Radioact.* **2017**, *171*, 138–147. [CrossRef]
10. Cohen, G.; Jousse, F.; Luze, N.; Hohener, P.; Atteia, O. LNAPL source zone delineation using soil gases in a heterogeneous silty-sand aquifer. *J. Contam. Hydrol.* **2016**, *192*, 20–34. [CrossRef]
11. De Simone, G.; Galli, G.; Lucchetti, C.; Tuccimei, P. Calibration of Big Bottle RAD H2O set-up for radon in water using HDPE bottles. *Radiat. Meas.* **2015**, *76*, 1–7. [CrossRef]
12. Schubert, M.; Freyer, K.; Treutler, H.; Weiss, H. Using the soil gas radon as an indicator for ground contamination by non-aqueous phase-liquids. *J. Solids Sediments* **2001**, *1*, 217–222. [CrossRef]
13. Schubert, M.; Freyer, K.; Treutler, H.; Weiss, H. Using radon-222 in soil gas as an indicator of subsurface contamination by non- aqueous phase-liquids (NAPLs). *Geofis. Int.* **2002**, *41*, 433–437.
14. Schubert, M.; Pena, P.; Balcazar, M.; Meissner, M.; Lopez, A.; Flores, J.H. Determination of radon distribution patterns in the upper soil as a tool for the localization of subsurface NAPL contamination. *Radiat. Meas.* **2005**, *40*, 633–637. [CrossRef]
15. Schubert, M.; Paschke, A.; Lau, S.; Geyer, W.; Knoller, K. Radon as a naturally occurring tracer for the assessment residual NAPL contamination of aquifers. *Environ. Pollut.* **2007**, *145*, 920–927. [CrossRef] [PubMed]
16. De Miguel, E.; Barrio-Parra, F.; Izquierdo-Diaz, M.; Fernandez, J.; Garzia-Gonzalez, J.E.; Alvarez, R. Applicability and limitations of the radon-deficit technique for the preliminary assessment of sites contaminated with complex mixture organic chemicals: A blind filed-test. *Environ. Int.* **2020**, *138*, 105591. [CrossRef] [PubMed]
17. Tasker, B.R. Radon-222 as an Indicator for a Nonaqueous Phase Liquids in the Saturated Zone: A Numerical Methods Analysis, Engineering Report. Master's Thesis, Department of Civil Engineering, Oregon State University, Corvallis, OR, USA, 1995.

18. Hunkeler, D.; Hoehn, E.; Hohener, P.; Zeyer, J. ^{222}Rn as a partitioning tracer to detect diesel fuel contamination in aquifer: Laboratory study and field observations. *Environ. Sci. Technol.* **1997**, *31*, 3180–3187. [CrossRef]
19. Johansson, S.; Fiandaca, G.; Dahlin, T. Influence of non-aqueous phase liquid configuration on induced polarization parameters: Conceptual models applied to a time-domain field case study. *J. Appl. Geophys.* **2015**, *123*, 295–309. [CrossRef]
20. Kinner, N.E. *Fate, Transport and Remediation of MTBE*; University of New Hampshire: Durham, NH, USA, 2001.
21. Kamath, R.; Connor, J.A.; McHugh, T.E.; Nemir, A.; Le, M.P.; Ryan, A.J. Use of long-term monitoring data to evaluate benzene, MTBE, and TBA plume behavior in groundwater at retail gasoline sites. *J. Environ. Eng.* **2012**, *138*, 458–469. [CrossRef]
22. Funiciello, R.; Giordano, G. La nuova Carta Geologica di Roma: Litostratigrafia e organizzazione stratigrafica. In *La Geologia di Roma: Dal Centro Storico alla Periferia. Memorie Descrittive della Carta Geologica d'Italia*; B.E.L.C.A.: Firence, Italy, 2008.
23. Ventriglia, U. *Geologia del Territorio del Comune di Roma*; Amministrazione Provinciale di Roma: Rome, Italy, 2002; pp. 1–767.
24. Giordano, G.; De Benedetti, A.A.G.; Diana, A.; Diano, G.; Gaudioso, F.; Marasco, F.; Miceli, M.; Mollo, S.; Cas, R.A.F.; Funiciello, R. The Colli Albani mafic caldera (Roma, Italy): Stratigraphy, structure and petrology. *J. Volcanol. Geoth. Res.* **2006**, *155*, 49–80. [CrossRef]
25. La Vigna, F.; Mazza, R.; Amanti, M.; Di Salvo, C.; Petitta, M.; Pizzino, L.; Pietrosante, A.; Martarelli, L.; Bonfa, I.; Capelli, G.; et al. Groundwater of Rome. *J. Maps* **2016**, *12*, 88–93. [CrossRef]
26. Speight, J.G. Remediation technologies. In *Natural Water Remediation: Chemistry and Technology*; Butterworth-Heinemann: Oxford, UK, 2020; pp. 263–303.
27. De Simone, G.; Lucchetti, C.; Galli, G.; Tuccimei, P. Correcting for H2O interference using a RAD7 electrostatic collection-based silicon detector. *J. Environ. Radioact.* **2016**, *162–163*, 146–153. [CrossRef]
28. Lucchetti, C.; De Simone, G.; Galli, G.; Tuccimei, P. Evaluating radon loss from water during storage in standard PET, bio-based PET, and PLA bottles. *Radiat. Meas.* **2016**, *84*, 1–8. [CrossRef]
29. Tuccimei, P.; Lane-Smith, D.; Galli, G.; Simko, J.; Cook, I.; Bond, C.E.; Lucchetti, C.; De Simone, G. Our PET project: An unlimited supply of big and small water sample vials for the assay of radon in water. *J. Radioanal. Nucl. Chem.* **2016**, *307*, 2277–2280. [CrossRef]
30. Davis, B.M.; Istok, J.D.; Semprini, L. Static and push-pull methods using radon-222 to characterize dense nonaqueous phase liquid saturations. *Ground Water* **2003**, *4*, 470–481. [CrossRef] [PubMed]
31. Cranswick, R.H.; Cook, P.G.; Lamontagne, S. Hyporheic zone exchange fluxes and residence times inferred from riverbed temperature and radon data. *J. Hydrol.* **2014**, *519*, 1870–1881. [CrossRef]
32. Sukanya, S.; Noble, J.; Joseph, S. Factors controlling the distribution of radon (^{222}Rn) in groundwater of a tropical mountainous river basin in southwest India. *Chemosphere* **2021**, *263*, 128096. [CrossRef]

4

Spatial Assessment of Groundwater Quality and Health Risk of Nitrogen Pollution for Shallow Groundwater Aquifer around Fuyang City, China

Nigus Kebede Wegahita [1], Lei Ma [1,*], Jiankui Liu [2], Tingwei Huang [1], Qiankun Luo [1] and Jiazhong Qian [1,*]

1 School of Resources and Environmental Engineering, Hefei University of Technology, Hefei 230009, China; nigusk19@hotmail.com (N.K.W.); 2018170635@mail.hfut.edu.cn (T.H.); QKLuo@hfut.edu.cn (Q.L.)
2 Anhui Geological Environment Monitoring Station, Hefei 230001, China; liujiankui696@163.com
* Correspondence: Lei8505@hfut.edu.cn (L.M.); qianjiazhong@hfut.edu.cn (J.Q.)

Abstract: Prolonged exposure to intensive and extensive agricultural and industrial activities is leading to an increased deterioration of groundwater quality, especially nitrogen pollution in shallow groundwater aquifers. This study was carried out using the fuzzy comprehensive method to assess the overall groundwater quality, and the noncarcinogenic risks were estimated using the human health risk assessment method recommended by the United States Environmental Protection Agency (USEPA) via drinking water intake pathways around Fuyang City, China. A total of 34 samples were collected from shallow groundwater private wells, and 16 parameters were analyzed for each groundwater sample. The evaluation results of groundwater quality show 14.7% of groundwater samples classified as poor and very poor quality, and NO_3-N, TH, TDS, Fe^{3+}, and Mn are of high potential to affect the quality of potable drinking water. These are mainly derived from anthropogenic pollutants, predominantly due to uncontrolled agricultural and industrial activities, as well as some natural processes. The noncarcinogenic risk of nitrate indicates that 8.82% of groundwater samples surpass the permissible limit recommended by the USEPA for both adults and children. This study may provide the local authority with insights into making scientific decisions for exploiting groundwater in a sustainable manner so as to protect public health.

Keywords: groundwater pollution; fuzzy comprehensive method; water quality assessment; nitrogen; health risk assessment

---

1. Introduction

Groundwater plays an indispensable role in healthy human life and ecosystem sustainability. It is also the predominant source for different uses (e.g., domestic, agriculture, and industrial) [1]. This is particularly true in arid and semi-arid regions, owing to insufficient rainfall and surface water availability [2,3]. An overwhelming number of people depend on groundwater for drinking purposes, either directly or indirectly [4,5]. The majority of the rural population (exceeding 70%) in China rely on groundwater for domestic purposes [6]. Recently, the demand for fresh groundwater is enhanced more than ever. However, groundwater quality is severely deteriorating, which is greatly attributed to huge population growth, rapid industrialization, increasing discharge of domestic sewage, overexploitation of groundwater, excessive application of agrochemicals, as well as the geogenic process that causes groundwater pollution through water–rock interaction, strong evaporation, and ion exchange in many countries and regions all over the world [7,8]. The World Health Organization (WHO) report states that, as significant as four-fifths of diseases are due to poor quality of drinking water in the world, as cited in [8,9]. Hence, the purpose of groundwater quality assessment is not only to rate groundwater quality, but it is also closely related to human health and to save a lot of people from suffering from various diseases. Therefore, it is necessary to evaluate groundwater quality correctly. Various scientific approaches that contributed to significant quality improvements in water quality have been reported [10–12]. Some of these methods include set pair analysis [13,14], matter element extension analysis [15,16], water quality index [17], and others that were mainly studied before. All these methods mentioned above have their own merits and demerits. For example, in groundwater quality assessment, set pair analysis and matter element extension analysis can be more important to deal with the uncertainties, but their calculation processes are complex, so they are not easy for common engineers to understand their basic principles and concepts. Water quality index is easy to calculate, but unable to deal with uncertainties associated with the water quality standards and water quality parameters [12]. Fuzzy comprehensive assessment method was used to overcome the shortcomings associated with the above methods in this study. Fuzzy comprehensive assessment method removes the influence of some uncertain factors, as compared to water quality indices [18,19]. Furthermore, the method treats main pollutants based on weights of evaluation factors in each sample site, which is another crucial reason to opt for the fuzzy comprehensive assessment method over other methods. Overdose and deficiency ingestion of hydrochemical effluents of groundwater can pose negative impacts on human health [20]. In the past two decades, intensive agricultural input and management have greatly increased production despite a high population increase [6]. However, this improvement has brought unexpected negative impacts on groundwater quality as well, because with the supply of rainwater or irrigation water, nitrogen can easily penetrate into groundwater through the soil [21]. With the continued rapid expansion of irrigation and intensive use of agrochemicals, groundwater nitrate pollution is becoming a serious problem that needs a great deal of attention [22–24]. In groundwater, Hudak reviewed that agricultural activities (especially fertilization) were a principal source of high nitrate levels throughout the country [25]. Rahmati et al. showed that the amount of applied nitrogenous fertilizer is closely related to high concentration nitrate in groundwater [26].

Owing to the rapid development of industry and the widespread use of agricultural fertilizers, groundwater nitrogen pollution has become more serious than ever, leading to multiple human health problems [27,28]. Several studies have revealed that nitrogen pollution (particularly nitrate) level in groundwater exceeds the standards set; this is further deteriorating water quality by creating eutrophication in surface water when groundwater discharges to surface water in their study areas [29,30]. Wu et al. [31] conducted a study on severe nitrate pollution and the health risks of a coastal aquifer and found that the groundwater nitrate concentrations exerted noncarcinogenic health risks for different age groups in the northern Shandong Peninsula of China. Furthermore, exposure to excessive nitrate concentration through drinking water is one of the reasons for methemoglobinemia in infants, miscarriage, and certain types of cancers, including gastric and stomach cancer [6,32]. Considering the importance of groundwater in arid and semi-arid regions, many interesting and comprehensive studies have been conducted on groundwater pollution and related health hazards around the world [33–35]. For example, in order to quantify the impact of different hazardous substances on human health, Batayneh [36], Chen et al. [6], and Ryu et al. [37] conducted a health risk assessment of hazardous substances in groundwater. All these studies indicated that the main route to expose carcinogenic and noncarcinogenic risks was the intake of contaminated groundwater, while a few also considered the dermal contact pathway [15,38]. Knowledge of human health risk assessments proposed by the USEPA is useful for evaluating potential health risks from water contaminants and provides scientific guidance for differential water supplies and cost-effective water treatments [5].

In Weining Plain, China, Li et al. reported that the pollution of nitrogen, Mn, and other pollutants in groundwater is predominantly due to intensive and extensive agricultural and industrial activities [27]. These factors help us to better understand the sources of nitrogen pollution and to investigate appropriate measures to control such pollution. WHO has set a permissible limit for nitrate in drinking water, which is 50 mg/L; the corresponding national standard for China is 20 mg/L, as N [39]. So far, many scholars not only notify about human health risk assessment of groundwater pollution around the world, but also provide useful insights for the protection of human health. Batheja et al. reported that a higher concentration of nitrate was found in the groundwater wells around the agricultural areas [40]. Su et al. suggested that health risk from nitrate-nitrogen in groundwater in an agricultural area was higher than that in urban areas of northeast China [38]. The study of [41] points out that a comprehensive understanding of the potential health risks of nitrate in drinking water is essential for making proper decisions to reduce the contamination and protect residents from health hazards. Researchers and organizations never stop worrying about the deterioration of groundwater quality and its potential human health risks.

A geographic information system (GIS) is one of the most powerful techniques commonly used for spatial information analysis and management. GIS can be used to delineate the spatial extent of contaminated sites both in natural and anthropogenic affected areas [42]. Moreover, it can be used to generate real-time figures of water-related problems in inaccessible areas [43]. The spatial distribution map of groundwater quality parameters, health risk, and water quality assessment can be outlined using spatial interpolation techniques in ArcGIS software. Inverse distance weight (IDW) is a widely used spatial interpolation method in assessing spatial distribution of groundwater quality [44]. Therefore, the IDW interpolation technique was used for analyzing spatial distribution of groundwater quality and health risk of nitrate around Fuyang City, Anhui Province, China.

The specific aims of this study were to (i) assess shallow groundwater quality around Fuyang City and (ii) assess noncarcinogenic human health risks of nitrogen for adults and children through drinking water intakes. The findings of the study could be helpful for local authorities to make sustainable decisions that imply mitigation measures to protect the groundwater quality in Fuyang City, in specific, and in China, in general. Finally, the research approach and findings might help global scholars to conduct further and future research in similar case studies around the world.

2. Description of the Study Area

Fuyang City is located in the northwest of Anhui Province, China (Figure 1). The study area is located in the southern part of the Huang-Huai-Hai and western Huaibei Plain. The topography of the area can be divided into two types of alluvial plain and denudation-alluvial plain based on their genetic morphology with 26 to 36 m elevation ranges above mean sea level (a.m.s.l.). Generally, the average land elevation is higher in the northwest and lower in the southeast, with an average ground surface slope of about 1/8000 [45].

The study area belongs to a warm temperature, semi-humid, monsoon climate, which is mild and humid, with adequate sunshine, moderate rainfall, and four distinct seasons. According to the meteorological data of the Fuyang Meteorological Bureau, the average annual temperature and annual precipitation are 14.9 °C and 901 mm, respectively. The maximum annual precipitation occurred in 1956, with a value of 1618.7 mm, and the minimum occurred in 1953 with a value of 440.8 mm. Generally, precipitation is subject to high temporal and spatial variation in the study area. The concentration of precipitation mostly occurs in June, July, and August in each annual flood season, accounting for about 48.3% of the annual precipitation. The rainfall was the highest in July, with an average monthly value of 217.9 mm. The annual average evaporation is 1604.2 mm.

The rivers around Fuyang City belong to the Huaihe River system. The main rivers flowing through the city include the Ying River, Quanhe River, and Cihuaixinhe River. All these rivers and ditches converge into the Huaihe River.

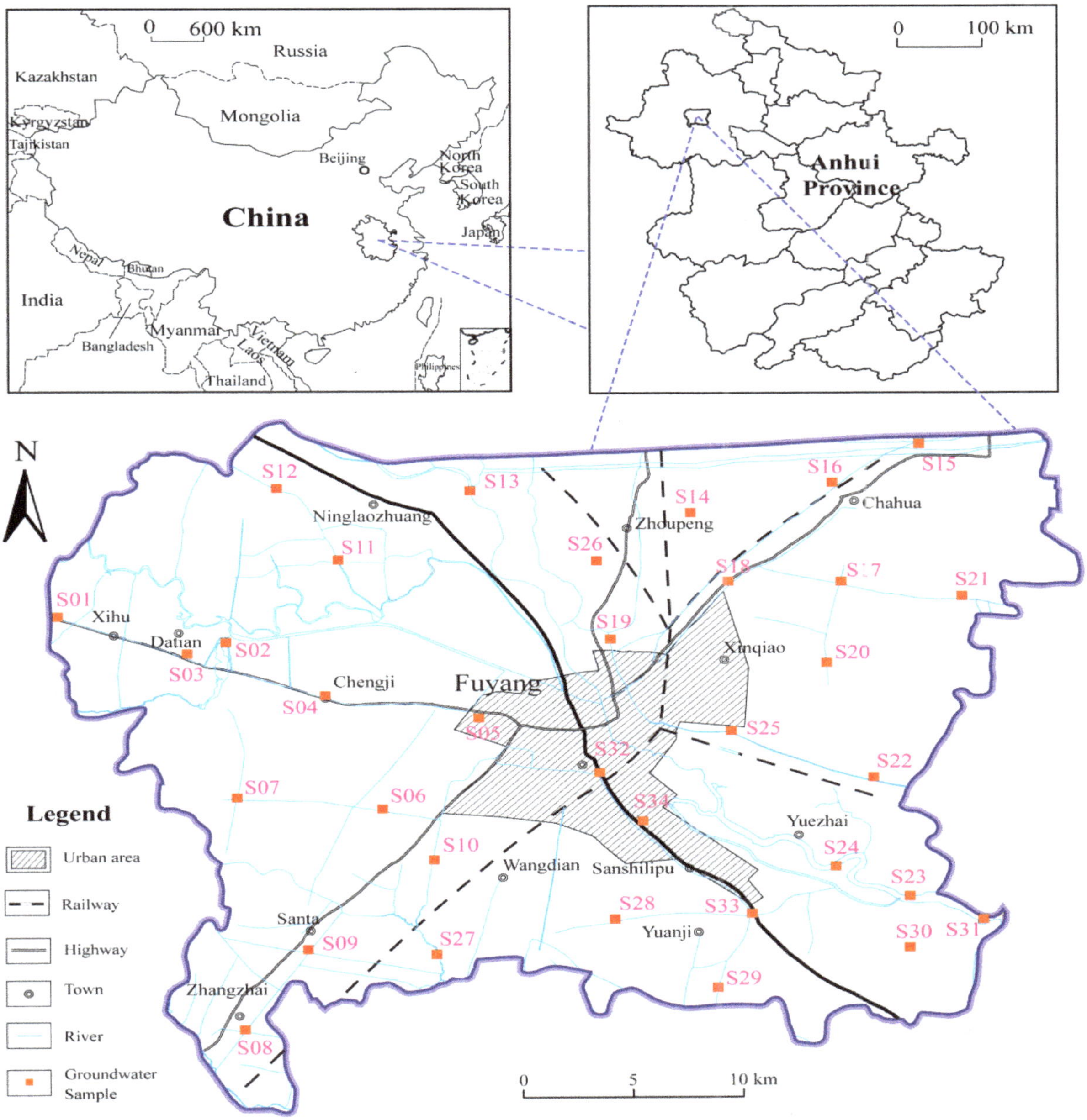

Figure 1. Map of the study area and location of shallow groundwater sampling sites.

3. Geological and Hydrogeological Settings

The stratigraphic lithology of the study area is mainly covered by Quaternary strata, with a thickness of about 130–150 m [45], and it varies from one location to another, which is thin in the northeast and thick in the northwest. Geologically, the strata of shallow groundwater are the upper Pleistocene and the Holocene. The upper Pleistocene is widely outcropped on the land surface, with a total thickness of 27–48 m, and it consists of silty clay, silty sand, and fine sand. The Holocene is exposed on both sides of the Yinghe River and the Laoquanhe River, the alluvial origin with a width of 2.6–6.5 km and an average thickness of 14.4 m [45]. The lithologies are mainly silty sand, sand, and silty clay, and there are weak interlayers composed of mud clay in some parts of the study area.

The aquifer formation in the study area is characterized by loose, sandy, and porous Cenozoic aquifers. Shallow groundwater occurs in shallow Holocene and late Pleistocene strata with a depth of about 50 m, which is closely related to atmospheric precipitation and surface water. The first aquifer group (shallow groundwater) is widely distributed throughout the region. The lithology of shallow aquifers is mainly loose structure and well-sorted silt sand.

The terrain in the area is flat, and the lithology of unsaturated zone is dominated by clay with micro-fissures and sub-sandy soil with a loose structure, which easily seeps into the atmospheric precipitation. Under the present conditions, atmospheric precipitation is the main source for shallow groundwater recharge. Farmland irrigation is more common in the area, and irrigation seepage and lateral runoff are other sources of recharge for shallow groundwater. Evaporation, water exploitation by rural residents, and lateral runoff are the ways of their discharge. According to the groundwater monitoring data of Fuyang City in 2014, the depth to groundwater is generally within the range of 2.5–4.7 m. The annual peak of water level occurred in the flood season from July to September, and the water level is lower from January to March, and the difference of water level in the other periods is small. The amplitude of the annual variation of the groundwater level ranges from 1 to 2 m. Generally, the flow of shallow groundwater in the study area is from northwest to southeast, with water level elevation varying from 25 to 28 m [45].

4. Materials and Methods

4.1. Sampling and Analysis

The sampling process was carried out with prescribed national technical guidelines, and recommended precautions were taken to avoid misinterpretation of data and induced contamination. Before sampling, the selected groundwater wells were pumped for ten minutes to remove the stagnant water stored at the well screen, and the water samples were stored in polyethylene containers, and each of the sampling containers was previously washed and rinsed thoroughly with the groundwater before sampling. In this sampling investigation, we have 10% duplicate samples, which were sent to another testing laboratory to test for quality control. Therefore, we had one sample at each of the 34 locations, in which we had one more parallel sample at each of 4 locations. Excluding the four parallel samples, a total of 34 samples (Figure 1) were collected from private groundwater wells, with depth less than 38 m, during January 2015, and their locations were obtained by using a portable GPS device. Part wells that are located in the urban areas and towns are not used for supplying drinking water now. Others are located in villages. Most of them were used for drinking purposes in our field investigation. All samples were sealed tightly to prevent contamination and timely sent to the Laboratory of Anhui Geological Environment Monitoring Station within a holding time of 24 h and stored in ice-box until analyzed for further processing. The samples were analyzed within 5 days. For this study, pH, total dissolved solids (TDS), total hardness (TH), potassium (K^+), sodium (Na^+), calcium (Ca^{2+}), magnesium (Mg^{2+}), chloride (Cl^-), sulfate ($SO_4{}^{2-}$), bicarbonate ($HCO_3{}^-$), ammonia nitrogen (NH_4-N), nitrate (NO_3-N), nitrite (NO_2-N), chlorides (Cl^-), fluoride (F^-), ferric iron (Fe^{3+}) and manganese (Mn), etc., were analyzed as the indicators of groundwater pollution.

4.2. Groundwater Quality Assessment

For groundwater quality assessment, several scientific approaches have been proposed, in which fuzzy comprehensive assessment method is extensively used in appraisal water quality assessment [10]. In this study, the overall groundwater quality assessment was carried out using fuzzy comprehensive assessment method to determine the suitability for human consumption [10,46,47]. With respect to national groundwater quality standards [39], groundwater quality can be classified into five grades: excellent (grade I), good (grade II), fair (grade III), poor (grade IV), and very poor (grade V), as displayed in Table 1.

Table 1. Classification of quality of groundwater according to national Chinese standard [39].

Variables	Grades				
	I	II	III	IV	V
NH_4-N	≤0.02	≤0.1	≤0.5	≤1.5	>1.5
NO_3-N	≤2	≤5	≤20	≤30	>30
NO_2-N	≤0.01	≤0.1	≤1	≤4.8	>4.8
Fe^{3+}	≤0.1	≤0.2	≤0.3	≤2	>2
Mn	≤0.05	≤0.05	≤0.1	≤1.5	>1.5
Cl^-	≤50	≤150	≤250	≤350	>350
F^-	≤0.2	≤0.5	≤1	≤2	>2
TDS	≤300	≤500	≤1000	≤2000	>2000
TH	≤150	≤300	≤450	≤550	>550
Water quality	Excellent	Good	Fair	Poor	Very poor

The principle procedure of fuzzy comprehensive assessment method is described as shown below. The membership function at the first level is expressed as:

$$r_{ij} = \begin{cases} 1, & x_i \le s_{ij} \\ \frac{s_{i(j+1)}-x_i}{s_{ij}-s_{i(j+1)}}, & s_{ij} < x_i \le s_{i(j+1)} \\ 0, & x_i > s_{i(j+1)} \end{cases}, \tag{1}$$

The membership function from level 2 to $m-1$ is:

$$r_{ij} = \begin{cases} 0, & x_i \le s_{i(j-1)} \text{ or } x_i > s_{i(j+1)} \\ \frac{x_i-s_{i(j-1)}}{s_{ij}-s_{i(j-1)}}, & s_{i(j-1)} < x_i < s_{ij} \\ 1, & x_i = s_{ij} \\ \frac{s_{i(j+1)}-x_i}{s_{i(j+1)}-s_{ij}}, & s_{ij} < x_i \le s_{i(j+1)}, \end{cases}, \tag{2}$$

The membership function at $j = m$ is:

$$r_{ij} = \begin{cases} 0, & x_i \le s_{i(j-1)} \\ \frac{x_i-s_{i(j-1)}}{s_{ij}-s_{i(j-1)}}, & s_{i(j-1)} < x_i \le s_{ij} \\ 1, & x_i > s_{ij} \end{cases}, \tag{3}$$

where r_{ij} is the fuzzy membership degree of parameter i ($i = 1, \ldots, n$) to grade j ($j = 1, \ldots, m$), x_i is the measured value of concentration for ith parameter, and s_{ij} is the standard value of the given parameter i at grade j. The evaluation of fuzzy membership matrix R obtains the water quality parameters and classes for each groundwater sample (k) and is expressed as follows:

$$R^{(k)}_{(n\times m)} = \begin{bmatrix} r_{11} & r_{12} & \cdots & r_{1m} \\ r_{21} & r_{22} & \cdots & r_{2m} \\ \vdots & \vdots & \vdots & \vdots \\ r_{n1} & r_{n2} & \cdots & r_{nm} \end{bmatrix}, \tag{4}$$

The weight vector of water quality indicator is defined as [10,48]:

$$w_i^k = \frac{\frac{x_i}{\sum_{j=1}^{m} s_{ij}}}{\sum_{i=1}^{n} \frac{x_i}{\sum_{j=1}^{m} s_{ij}}}, \tag{5}$$

where w_i^k is the weight vector of parameter i to the kth sample, x_i is the measured value of the ith parameter, and the variables of n, m, and k signify the number of parameters, water quality grades, and sample, respectively. Through the above Equation (5), the single weight vector can be constructed as $A^k = \left[w_1^k, w_2^k, \ldots, w_n^k\right]$.

After the fuzzy weight vector A and the fuzzy relation matrix R are obtained, the fuzzy membership subset B of evaluation of the groundwater quality can be obtained by the compound operation of A and R; finally, the comprehensive evaluation results of each sample $B_{k\times m}$ is formed as follows:

$$B_{n\times m} = \left[w_1^k, w_2^k, \ldots, w_n^k\right] \times \begin{bmatrix} r_{11} & r_{12} & \cdots & r_{1m} \\ r_{21} & r_{22} & \cdots & r_{2m} \\ \vdots & \vdots & \vdots & \vdots \\ r_{n1} & r_{n2} & \cdots & r_{nm} \end{bmatrix} = \{b_{11}, b_{12}, \ldots, b_{1m}\}, \tag{6}$$

$$B_{k\times m} = \begin{bmatrix} A^{(1)} \times R_{n\times m}^{(1)} \\ A^{(2)} \times R_{n\times m}^{(2)} \\ \vdots \\ A^{(k)} \times R_{n\times m}^{(k)} \end{bmatrix} = \begin{Bmatrix} b_{11} & b_{12} & \cdots & b_{1m} \\ b_{21} & b_{22} & \cdots & b_{2m} \\ \vdots & \vdots & \cdots & \vdots \\ b_{k1} & b_{k2} & \cdots & b_{km} \end{Bmatrix}, \tag{7}$$

The fuzzy $B_{k\times m}$ indicates the membership degree of the water quality of the evaluated water body (sample) to the standard water quality at all levels, which reflects the fuzziness of the determination of comprehensive water quality level based on the principle of maximum membership degree.

4.3. Health Risk Assessment

Health risk assessment is an important role to appraise the negative impacts of pollutants on human health. Hence, gaining a comprehensive knowledge of human health risk assessment is essential to protect human health hazards from various toxic chemical elements, to ascertain groundwater quality assessment and management [28,41]. Contaminated groundwater enters the human body through different pathways such as drinking water intake and dermal contact, and it may pose several risks to human health [15]. Numerous studies have recommended that the health risks of contaminated groundwater through dermal contact can be negligible as compared to drinking water intake [6,49,50]. Therefore, this study only focuses on the noncarcinogenicity risk assessment through drinking water intake on the health of local residents. The human health risk assessment models recommended by the Ministry of Environmental Protection of the P.R. China [51] are based on the model of the United States Environmental Protection Agency (USEPA) [52]. Considering USEPA models, many scholars broadly used these four steps including hazard identification, dose–response assessment, exposure assessment, and risk characterization during the assessment [28,53]. In the present study, nitrogen pollutants, mainly ammonia nitrogen (NH_4-N), nitrate (NO_3-N), and nitrite (NO_2-N), were selected as parameters for human health risk assessment. The calculated health risk assessment which is greater than one is unsafe for drinking purposes. Therefore, the noncarcinogenic risk through drinking water intake pathways can be calculated as follows [28,52]:

$$E = \frac{\mathrm{C} \times \mathrm{IR} \times \mathrm{EF} \times \mathrm{ED}}{\mathrm{BW} \times \mathrm{AT}}, \tag{8}$$

$$HQ = \frac{E}{\mathrm{RfD}}, \tag{9}$$

where E indicates the average daily dose through drinking water intake per unit weight (mg/(kg·d)), C is the concentration of the pollutant in water (mg/L). IR indicates the water ingestion rate (L/day). In this study, the ingestion rate with a value of 1.5 and 0.7 L/day was used respectively for adults and children [15,50]. EF is the exposure frequency (365 days/year), ED is the exposure duration, with a suggested value of 30 years for adults and 12 years for children. BW represents the body mass of a person (kg), with a proposed value of 56.8 kg for adults and 15.9 kg for children in this study [50]. AT represents the average exposure time for noncarcinogenic effect (days), which is 10,950 days for adults and 4380 days for children. *HQ* and RfD represent the hazard quotient and reference dosage, respectively, for noncarcinogenic pollutants through the drinking water intake pathway (mg/(kg·d)). In this study, the values of RfD are 0.97, 1.6, and 0.1 mg/(kg·d) for NH_4-N, NO_3-N, and NO_2-N, respectively [49,51,54].

To calculate the overall adverse effect of noncarcinogenic multiple pollutants through the drinking water intake exposure pathways on the human body, the total risk (HI_{total}) can be calculated as follows:

$$HI_{\text{total}} = \sum_{i=1}^{n} HQ_i \tag{10}$$

where HI_{total} indicates hazard index; it is the sum of each of the noncarcinogenic chemical pollutants of hazard quotients, i refers to target pollutants in groundwater. Additionally, when the value of HQ exceeds 1, which is vulnerable to health risks, the $HQ < 1$ indicates an acceptable level of noncarcinogenic risk. Alike the HQ, $HI_{\text{total}} > 1$ means that the noncarcinogenic risk of contaminant is beyond the acceptable limit, while $HI_{\text{total}} < 1$ indicates that the risk is within the acceptable limit.

5. Results and Discussion

5.1. Nitrogen Pollution in Groundwater and Other Parameters

The groundwater samples of different physicochemical parameters were analyzed, and the concentration of physicochemical parameters such as pH, TDS, TH, K^+, Na^+, Ca^{2+}, Mg^{2+}, Cl^-, $SO_4{}^{2-}$, $HCO_3{}^-$, NH_4-N, NO_3-N, NO_2-N, F^-, Fe^{3+}, and Mn were determined, and statistically described in terms of minimum (Min), maximum (Max), mean, standard deviation (SD), and coefficient of variation (CV), as illustrated in Table 2. The national standard prescribes that the limit for pH value of drinking water is 6.5–8.5, while if it is not within the acceptable limit, it causes impacts on human health. As shown in Table 2, the groundwater pH of the study area shows neutral to slightly alkaline ranging values (7.26 to 8.57, averaging at 7.99). The TDS and TH of shallow groundwater samples in the study area ranged between 423.77 and 1366.50 and 194.33 and 785.34 mg/L, averaging at 706.11 and 310.23 mg/L, respectively (see Figure 2b). The spatial distribution of TDS and TH concentration is shown in Figure 3a,b. As shown in Figure 3a,b, both TDS and TH show some high concentration spots in the study area. The highest TDS concentration was observed in Chengji and the northeast part of the area (Figure 3a), and the highest TH was found in Chengji and Chahua (Figure 3b). TDS and TH exceeded recommended limits in nine sample sites and accounted for 11.8 and 14.7% of the entire samples, respectively. Yet, the averages of both TDS and TH were within the prescribed limit, indicating that mixing can effectively control their concentrations, as reported in [15]. The TDS and TH enrichment in groundwater may be due to soluble salts and minerals dissolving into groundwater, and human intervention [8,46,49,55]. As shown in Table 2, the concentrations of major cations, namely K^+, Ca^{2+}, and Mg^{2+}, were in the range of 0.28 to 0.72, 44.89 to 222.70, and 17.28 to 73.31, with an average of 0.45, 78.33, and 27.84 mg/L, respectively, signifying that all concentrations were within the acceptable limit. The $HCO_3{}^-$ concentration ranged between 266.39 and 821.58 with a means of 458.59 mg/L and it exceeded the recommended value of drinking water quality of WHO standards. Sodium (Na) plays an important role in a certain amount, but the excessive use of Na concentration may lead to various health problems such as hypertension, vomiting, and kidney disorders [55,56]. In this study, Na concentration varied from 20.34 to 184.50 mg/L with an average of 70.65 mg/L and was acceptable for drinking purpose. The concentration of Cl^- and $SO_4{}^{2-}$ anions ranged between 1.77 and 160.23 mg/L and 0.25 and 160.22 mg/L, with an average of 28.52 and 31.99 mg/L, respectively, and both were within the permissible limits of the Chinese standards (250 mg/L for each).

Table 2. Statistical analysis of the physicochemical parameters of shallow groundwater samples in the study area.

Variable	Unit	Min	Max	Mean	SD	CV	Chinese Standard	WHO
pH	-	7.26	8.57	7.99	0.28	0.04	6.5–8.5	6.5–8.5
TDS	mg/L	423.77	1366.50	706.11	218.31	0.31	1000	1500
TH	mg/L	194.33	785.34	310.23	142.09	0.46	450	500
K^+	mg/L	0.28	0.72	0.45	0.12	0.26	-	12 **
Na^+	mg/L	20.34	184.50	70.65	44.04	0.62	200	200
Ca^{2+}	mg/L	44.89	222.70	78.33	43.48	0.56	400 ***	-
Mg^{2+}	mg/L	17.28	73.31	27.84	12.35	0.44	-	150 **
Cl^-	mg/L	1.77	160.23	28.52	39.23	1.38	250	250
SO_4^{2-}	mg/L	0.25	160.22	31.99	40.63	1.27	250	250
HCO_3^-	mg/L	266.39	821.57	458.89	116.13	0.25	-	600 **
NH_4-N	mg/L	0.04	0.24	0.06	0.05	0.91	0.5	0.5
NO_3-N	mg/L	0.50	92.44	8.25	23.95	2.90	20	50
NO_2-N	mg/L	0.004 *	0.22	0.01	0.04	3.04	1	3
F^-	mg/L	0.04 *	1.20	0.69	0.33	0.48	1	1.5
Fe^{3+}	mg/L	0.04 *	4.02	0.30	0.68	2.31	0.3	0.3
Mn	mg/L	0.01 *	0.89	0.23	0.17	0.73	0.1	0.1

*: limit value of detection; ** refer to [57]; *** refer to [49].

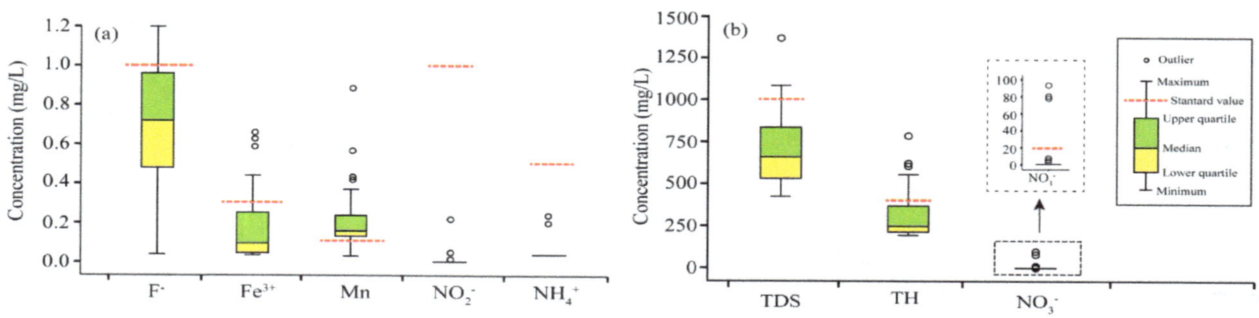

Figure 2. Box plots with respect to the reference Chinese national standards values for the selective chemical parameters (in mg/L) of the groundwater. (**a**) concentrations of relatively low-content chemical constituents; (**b**) concentrations of relatively high-content chemical constituents.

As several researchers have reported, an intensive amount of nitrogenous fertilizer utilization in agricultural practices is one of the most common sources of nitrogen in groundwater [6,8,56]. The results for NH_4-N and NO_2-N were found to vary from 0.04 to 0.24 and 0.004 to 0.22 mg/L, with a mean of 0.06 and 0.01 mg/L, respectively (see Figure 2a). The range concentration of nitrate-nitrogen reported in this study was varied from 0.5 to 92.44 mg/L, with an average value of 8.25 mg/L (Figure 2b). In this study, high NO_3-N pollution of groundwater may be related to extensive agricultural activities. The safeguard limits set for NH_4-N, NO_3-N, and NO_2-N are 0.5, 20, and 1.0 mg/L, respectively, out of which only NO_3-N was shown to deviate out and, in turn, can be considered as a major contaminant. The spatial distribution of NO_3-N concentration was shown in Figure 3c. As can be seen from Figure 3c, NO_3-N concentration is not uniformly distributed in the study region, and also very higher NO_3-N concentration was observed in the Datian, Chahua, and Central parts of the area. Three samples (accounting for 8.82%) were beyond acceptable limits according to the national Chinese drinking water standards.

Fluoride in drinking water within the permissible limit (1.0 mg/L) is essential to protect dental caries (tooth decay). However, taking too much fluoride concentration frequently can cause health problems such as dental and skeletal fluorosis [58–61]. The obtained fluoride water sample ranged between 0.04 and 1.2 mg/L, with a mean of 0.69 mg/L (Figure 2a), and the spatial distribution was shown in Figure 3d. Sites with high concentrations of fluoride were mainly found in the southwest, southeast, northwest, and northeastern parts of the study area. The high concentration of fluoride in groundwater mainly related to lithological reasons and related to fluoride-bearing minerals [62].

Seven samples (20.6% of all samples) exceeded the permissible limit according to the national Chinese drinking water standard in the present study.

Measured values of Fe^{3+} and Mn were in the range of 0.04–4.02 and 0.01–0.89 mg/L, respectively, whose concentration distributions were shown as in Figure 3e,f. According to the national standard, this stipulates that both Fe^{3+} and Mn are not meeting the requirement for drinking water (Figure 2a). Eight samples (about 23.5% of the total samples) and thirty samples (about 88.2% of the total samples) for Fe^{3+} and Mn concentration, respectively, exceeded permissible limits. Fe^{3+} with high concentrations was mainly distributed in the southeast part of the study area, whereas Mn with high concentrations mainly existed in Chengji, Chahua, and southeast regions. The enrichment of Fe^{3+} and Mn trace elements in groundwater is considered through natural interactions of water with rocks and soils [34]. Generally, according to the Chinese standard values, pH, TDS, TH, NO_3-N, F^-, Fe^{3+}, and Mn in this study area exceed the standard concentration for drinking purposes [39], and their spatial distributions are depicted in Figure 3.

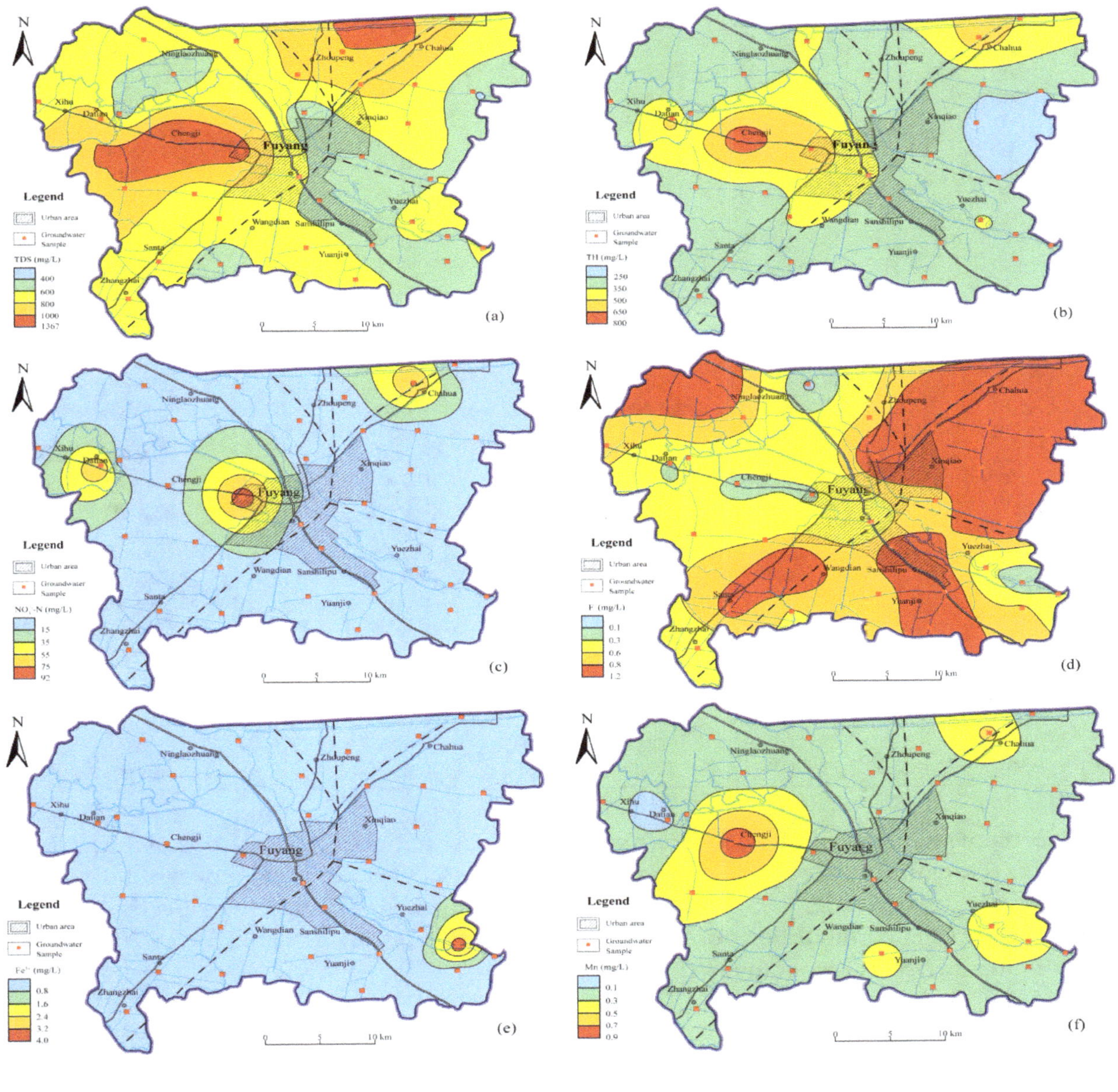

Figure 3. Spatial distribution of contaminants in groundwater: (**a**) TDS, (**b**) TH, (**c**) NO_3-N, (**d**) F^-, (**e**) Fe^{3+}, and (**f**) Mn.

5.2. Groundwater Quality Assessment

Fuzzy comprehensive assessment method was applied to evaluate the overall groundwater quality in this study. The obtained results of NH_4-N, NO_3-N, NO_2-N, Fe^{3+}, Mn, Cl^-, F^-, TDS, and TH were selected as parameters to investigate groundwater quality assessment in this study. This selection is based on regular monitoring by local environmental authorities for their critical importance to drinking water quality and their potential impacts on human health [11]. The calculated results are presented in Table 3. As shown in Table 3, the assessment results of the study area exhibit of groundwater samples were classified into grade II, III, IV, and V.

Table 3. Assessment results of shallow groundwater quality based on fuzzy comprehensive approach.

Sample	Membership Degree					Grade	Water Quality
	$j = I$	$j = II$	$j = III$	$j = IV$	$j = V$		
S01	0.169	0.461	0.365	0.005	0.000	II	good
S02	0.160	0.630	0.170	0.040	0.000	II	good
S03	0.075	0.031	0.118	0.162	0.614	V	very poor
S04	0.060	0.119	0.106	0.399	0.316	IV	poor
S05	0.057	0.039	0.131	0.070	0.703	V	very poor
S06	0.186	0.542	0.251	0.021	0.000	II	good
S07	0.180	0.261	0.468	0.092	0.000	III	fair
S08	0.094	0.437	0.371	0.098	0.000	II	good
S09	0.137	0.407	0.454	0.002	0.000	III	fair
S10	0.122	0.300	0.498	0.080	0.000	III	fair
S11	0.224	0.377	0.382	0.017	0.000	III	fair
S12	0.038	0.308	0.559	0.095	0.000	III	fair
S13	0.075	0.497	0.401	0.026	0.000	II	good
S14	0.161	0.199	0.633	0.008	0.000	III	fair
S15	0.055	0.223	0.622	0.100	0.000	III	fair
S16	0.027	0.079	0.202	0.123	0.570	V	very poor
S17	0.181	0.201	0.454	0.165	0.000	III	fair
S18	0.168	0.210	0.617	0.005	0.000	III	fair
S19	0.230	0.229	0.448	0.092	0.000	III	fair
S20	0.203	0.296	0.501	0.000	0.000	III	fair
S21	0.228	0.250	0.434	0.088	0.000	III	fair
S22	0.144	0.208	0.574	0.074	0.000	III	fair
S23	0.026	0.153	0.097	0.057	0.667	V	very poor
S24	0.130	0.361	0.405	0.104	0.000	III	fair
S25	0.251	0.463	0.271	0.015	0.000	II	good
S26	0.205	0.403	0.393	0.000	0.000	II	good
S27	0.129	0.521	0.279	0.071	0.000	II	good
S28	0.071	0.368	0.452	0.108	0.000	III	fair
S29	0.174	0.335	0.469	0.022	0.000	III	fair
S30	0.150	0.565	0.262	0.023	0.000	II	good
S31	0.192	0.597	0.202	0.009	0.000	II	good
S32	0.180	0.405	0.412	0.002	0.000	III	fair
S33	0.217	0.288	0.456	0.039	0.000	III	fair
S34	0.235	0.331	0.432	0.002	0.000	III	fair

In the present study region, as compared with the Chinese national classification of the groundwater quality standards values, the observed parameters of NO_3-N, TH, and Fe^{3+} content fall within very poor water quality (V) in 8.82, 5.88, and 2.94% of the groundwater samples, respectively, while Mn, Fe^{3+}, F^-, TDS, and TH come under poor water quality (IV) in 88.24, 20.59, 20.59, 11.76, and 8.82% of the groundwater samples, respectively (Figure 4).

Out of the 34 sampling sites, ten groundwater sampling sites (accounting for 29.4%) belong to grade II (good quality), which is suitable for various purposes. The water quality samples at nineteen sampling sites (about 55.9% of all samples) are classified as grade III (fair quality), which can be used as drinking water with precaution. Only five samples are categorized as poor and very poor quality, accounting for 2.9% and 11.8% of the total groundwater samples, respectively. The poor and very poor quality water is not suitable for drinking, but after proper treatment, it might be used for recreation and irrigation purposes.

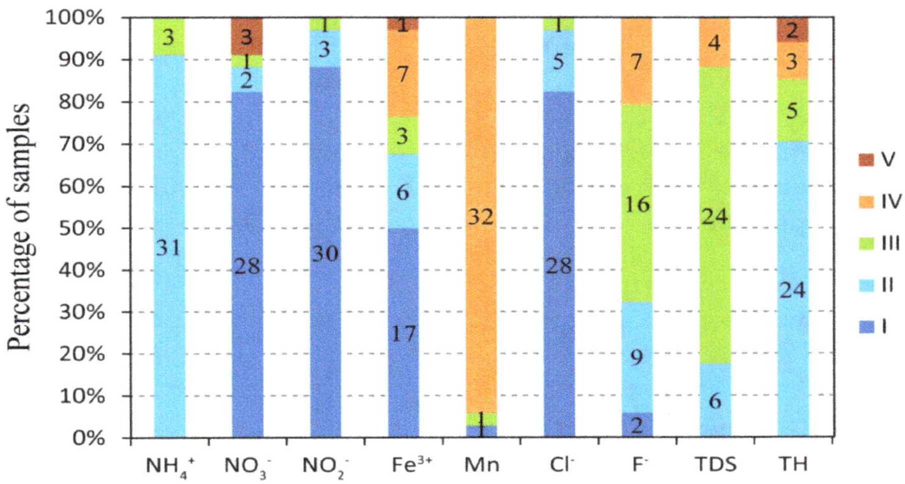

Figure 4. Percentage of the shallow groundwater samples for drinking purposes in five water quality categories.

The assessment results of groundwater quality were used for spatial interpolation. The spatial distribution of groundwater quality assessment was obtained as shown in Figure 5. As shown in Table 3 and Figure 5, the evaluated groundwater quality of more than half of the total samples, is dominated by fair quality water and may require some treatment measures to improve the quality of drinking water. The main pollutants in deterioration of groundwater quality at sampling site S04 (poor quality) are TH, TDS, and Mn. The very poor quality (S03, S05, S16, and S23,) are severely polluted with NO_3-N, TH, TDS, and Fe^{3+} (Figure 4). In recent years, with the infiltration of fertilizers, discharge of industrial and domestic sewage was the main reason for the increase in NO_3-N concentration in groundwater. TH, TDS, Fe^{3+}, and Mn could be both the geogenic process that causes groundwater pollution through water–rock interaction and relatively strong evaporation, and the area with poor and very poor quality water was mainly observed in the Chahua, Datian, Chengji, central and southeast areas.

From the spatial distribution of groundwater quality assessment, we can easily identify the contaminated site and also provide an insight into potentially contaminated areas, in which serious efforts should be taken to mitigate or reduce the deterioration of groundwater quality and to satisfy the municipal groundwater demand. In addition, regular monitoring should be carried out to prevent pollution and consumption by residents. That could be useful for government officials and local residents.

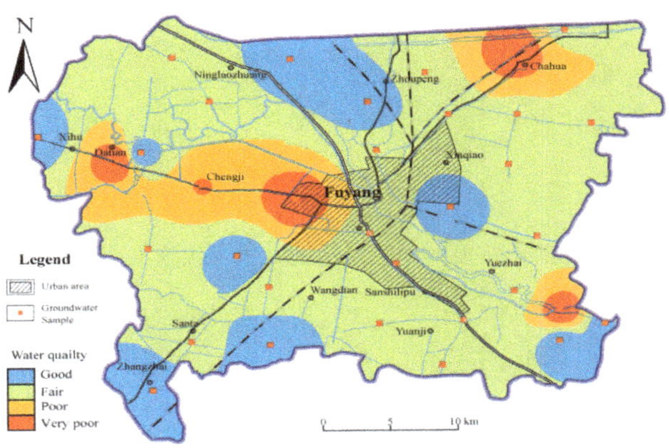

Figure 5. Spatial distribution of shallow groundwater quality based on fuzzy comprehensive assessment result.

5.3. Health Risk Assessment

Seriously polluted groundwater by nitrogen is unsafe for drinking and poses a high risk to human health. In this study, the concentration of nitrate, nitrite, and ammonium was investigated. As mentioned above, the obtained concentrations of ammonium and nitrite in the present study were within acceptable limits [39,63]. According to the results, only the concentration of nitrate was higher than the acceptable standard values, and therefore, the noncarcinogenic risk assessment through drinking water intake is calculated only for it. The result of the health risk assessment was calculated based on the approach mentioned above. The calculated results of health risk incurred by noncarcinogenic nitrate pollution through drinking water intake are shown in Table 4, for adults and children, respectively.

Table 4. The exposure dosage through drinking water intake (*E*) and the hazard quotient of nitrate-nitrogen (mg/L) for the drinking water in Fuyang City.

Sample	Nitrate-Nitrogen	RfD	*E*		*HQ*	
			Adult	Child	Adult	Child
S01	0.5	1.6	0.013	0.022	0.008	0.014
S02	0.5	1.6	0.013	0.022	0.008	0.014
S03	78.66	1.6	2.077	3.463	1.298	2.164
S04	0.56	1.6	0.015	0.025	0.009	0.015
S05	92.44	1.6	2.441	4.070	1.526	2.544
S06	0.5	1.6	0.013	0.022	0.008	0.014
S07	0.5	1.6	0.013	0.022	0.008	0.014
S08	0.5	1.6	0.013	0.022	0.008	0.014
S09	0.5	1.6	0.013	0.022	0.008	0.014
S10	3.17	1.6	0.084	0.140	0.052	0.087
S11	0.5	1.6	0.013	0.022	0.008	0.014
S12	0.5	1.6	0.013	0.022	0.008	0.014
S13	6.94	1.6	0.183	0.306	0.115	0.191
S14	0.5	1.6	0.013	0.022	0.008	0.014
S15	4.73	1.6	0.125	0.208	0.078	0.130
S16	80.18	1.6	2.117	3.530	1.323	2.206
S17	0.5	1.6	0.013	0.022	0.008	0.014
S18	0.5	1.6	0.013	0.022	0.008	0.014
S19	0.5	1.6	0.013	0.022	0.008	0.014
S20	0.5	1.6	0.013	0.022	0.008	0.014
S21	0.5	1.6	0.013	0.022	0.008	0.014
S22	0.5	1.6	0.013	0.022	0.008	0.014
S23	0.5	1.6	0.013	0.022	0.008	0.014
S24	0.52	1.6	0.014	0.023	0.009	0.014
S25	0.5	1.6	0.013	0.022	0.008	0.014
S26	0.5	1.6	0.013	0.022	0.008	0.014
S27	0.5	1.6	0.013	0.022	0.008	0.014
S28	0.5	1.6	0.013	0.022	0.008	0.014
S29	0.5	1.6	0.013	0.022	0.008	0.014
S30	0.5	1.6	0.013	0.022	0.008	0.014
S31	0.69	1.6	0.018	0.030	0.011	0.019
S32	0.65	1.6	0.017	0.029	0.011	0.018
S33	0.5	1.6	0.013	0.022	0.008	0.014
S34	0.5	1.6	0.013	0.022	0.008	0.014

As shown in Table 4, the range of HQ values of NO_3-N for adults is 0.008–1.526, with means of 0.136. For children, it varies from 0.014 to 2.544, with an average value of 0.227. The hazard quotient of nitrate through drinking water intake was higher than the acceptable standards [15,52]. This indicates that NO_3-N is the most deleterious effect for high human health risk in the present study. The highest risks of nitrate are 1.526 and 2.544, with the means of 0.136 and 0.227 for adults and children, respectively, which demonstrated that children are more susceptible to groundwater nitrogen contamination as mentioned in Table 3, when the groundwater quality of wells S03, S05, and S16, and categorized as being grade V, which is classified as very poor groundwater quality, was also deemed to be unsuitable for drinking purposes. Also, these three sites will pose noncarcinogenic health risks to adults and children. This indicates that the noncarcinogenic risks of NO_3-N in very poor groundwater quality are beyond the permissible limit. Regions of poor and very poor groundwater quality can be targeted for more detailed investigation and tight monitoring programs. On the contrary, the sample sites of S04 and S23 fall under poor and very poor groundwater quality, but the noncarcinogenic health risks of NO_3-N were within an acceptable limit. For NO_3-N, the HQ values of about 88.2% of all samples were less than 1 for both adults and children when groundwater was classified as good and fair groundwater quality in the present study. In general, compared to the overall groundwater quality assessment results, it can be found that some water samples which are classified as unsuitable water for drinking purposes are not associated with a high risk to adults and children, suggesting that even groundwater with unacceptable quality cannot pose acceptable potential health risks on humans. Therefore, health risk assessment should always be accompanied by overall groundwater quality assessment.

Therefore, government officials and public health workers should pay more attention to groundwater polluted by nitrogen pollution, especially the noncarcinogenic risk induced by nitrate in the present study. Furthermore, the contaminated groundwater should be urgently treated before direct consumption by local residents. Similarly, several studies were conducted with regard to health risk assessment all over the world, which indicates that the noncarcinogenic risk of nitrate via drinking water should be of high concern and cannot be ignored.

In the present study, the content of NO_3-N and other chemical components in groundwater was analyzed. Then, the risk assessment results were obtained. Health risk assessment has some sources of uncertainty resulting from parameter and input data uncertainty [15], such as unevenly distributed sampling, sample analyses, and strong spatial variability of groundwater quality. Despite uncertainty in health risk assessment due to the uncertainty of parameters and input data, the research results are still of significance and practical value for scientific decision-making. For instance, Chen et al. used the USEPA human health risk assessment method to evaluate health risks of nitrate contamination in an agricultural area of Ningxia, northwest China, only by drinking water pathways to humans [6]. They found that eight and four collected samples would have unacceptable noncarcinogenic risks for infants and adults, respectively. Similarly, Su et al. (2018) conducted a detailed study on the health risk of nitrogen pollution in the Shenfu mining area. They found that 30.3% and 40.8% of all collected samples have noncarcinogenic risk exceeding the permissible limit recommended by the USEPA for adults and children, respectively. The study results are valid and meaningful, despite these uncertain factors in the research process and results.

6. Conclusions

In this study, 34 groundwater samples from Fuyang City were collected and analyzed for physicochemical parameters based on national technical regulations. The overall groundwater quality assessment was conducted using fuzzy comprehensive assessment method, and the noncarcinogenic health risk due to groundwater pollution was assessed for adults and children through drinking water intake.

The pH values of shallow groundwater in this study area are generally neutral to slightly alkaline in nature. The TDS and TH fall under the brackish and very-hard water category, accounting for 11.8% and 14.7% of total samples, respectively, and very-higher concentration was found in Chengji and the northeast part of the area. According to the national groundwater quality standards, the analyzed parameter of water samples of pH, TDS, TH, NO_3-N, F^-, Fe^{3+}, and Mn were beyond the acceptable limits for drinking purposes. The results revealed that the concentrations of K^+, Na^+, Ca^{2+}, Mg^{2+}, Cl^-, $SO_4{}^{2-}$, $HCO_3{}^-$, NH_4-N, and NO_2-N of groundwater samples were within acceptable limits. In the study, the concentration of NO_3-N ranged from 0.5 to 92.4 mg/L; 8.8% of groundwater samples were above the acceptable limit, which mainly results from extensive agricultural activities in the region. Hence, higher NO_3-N concentrations were mainly distributed in the Datian, Chahua, and central part of the study area.

The map of evaluated groundwater quality results was categorized into four regions, in which 29.4%, 55.9%, 2.9%, and 11.8% of the area belong to good, fair, poor, and very poor quality water, respectively. The good and fair quality water can be used for drinking purposes, but poor and very poor quality water is unsuitable for drinking and requires deep treatment before utilization. The poor and very poor quality groundwater were mainly collected from the Chahua, Datian, Chengji, central and southeast areas, which are not recommended to be the water-source region for the study area. NO_3-N, TH, TDS, Fe^{3+}, and Mn are the major pollutants in the study area determined through fuzzy comprehensive evaluation, which mainly originates from natural and anthropogenic activities.

Nitrogen pollution of groundwater can have a potential negative impact on the health of the local residents through drinking water intake, particularly nitrate-nitrogen in the present study area. The noncarcinogenic risk of nitrate-nitrogen ranges from 0.008 to 1.526 and from 0.014 to 2.544 with the means of 0.136 and 0.227 for adults and children, respectively. This indicates that children are more susceptible to nitrate-nitrogen pollution. In addition, nitrate-nitrogen contributes the most to human risk in this study, which requires more attention from authorities and decision-makers. Therefore, polluted groundwater should be urgently treated before direct consumption by local residents.

Author Contributions: All authors contributed to the results of this work. N.K.W. wrote the original draft; L.M. designed the research and revised the manuscript; J.L. surveyed the hydrogeological conditions and sampled groundwater; T.H. drew maps; Q.L. analyzed the data; J.Q. improved the manuscript with constructive discussions. All authors have read and agreed to the published version of the manuscript.

Acknowledgments: Authors are also grateful to anonymous reviewers for their extremely valuable comments and suggestions to improve the quality of the paper.

References

1. Zabala, M.E.; Martínez, S.; Manzano, M.; Vives, L. Groundwater chemical baseline values to assess the Recovery Plan in the Matanza-Riachuelo River basin, Argentina. *Sci. Total Environ.* **2016**, *541*, 1516–1530. [CrossRef]
2. Li, P.; He, X.; Li, Y.; Xiang, G. Occurrence and health implication of fluoride in groundwater of Loess Aquifer in the Chinese Loess Plateau: A case study of Tongchuan, Northwest China. *Expo. Health* **2018**, *11*, 95–107. [CrossRef]
3. Zhang, Q.; Xu, P.; Qian, H. Groundwater quality assessment using improved water quality index (WQI) and human health risk (HHR) evaluation in a semi-arid region of Northwest China. *Expo. Health* **2020**, 1–14. [CrossRef]
4. Connor, R. *The United Nations World Water Development Report 2015: Water for a Sustainable World*; UNESCO Publishing: Paris, France, 2015; Volume 1.
5. Yin, S.; Xiao, Y.; Han, P.; Hao, Q.; Gu, X.; Men, B.; Huang, L. Investigation of groundwater contamination and health implications in a typical semiarid basin of North China. *Water* **2020**, *12*, 1137. [CrossRef]
6. Chen, J.; Wu, H.; Qian, H. Groundwater nitrate contamination and associated health risk for the rural communities in an agricultural area of Ningxia, Northwest China. *Expo. Health* **2016**, *8*, 349–359. [CrossRef]
7. Subba Rao, N.; Marghade, D.; Dinakar, A.; Chandana, I.; Sunitha, B.; Ravindra, B.; Balaji, T. Geochemical characteristics and controlling factors of chemical composition of groundwater in a part of Guntur district, Andhra Pradesh, India. *Environ. Earth Sci.* **2017**, *76*, 747. [CrossRef]
8. Adimalla, N. Controlling factors and mechanism of groundwater quality variation in semiarid region of South India: An approach of water quality index (WQI) and health risk assessment (HRA). *Environ. Geochem. Health* **2019**, *42*, 1–28. [CrossRef]

9. Amalraj, A.; Pius, A. Health risk from fluoride exposure of a population in selected areas of Tamil Nadu South India. *Food Sci. Hum. Wellness* **2013**, *2*, 75–86. [CrossRef]
10. He, S.; Wu, J. Hydrogeochemical characteristics, groundwater quality, and health risks from hexavalent chromium and nitrate in groundwater of Huanhe formation in Wuqi County, Northwest China. *Expo. Health* **2018**, *11*, 125–137. [CrossRef]
11. Mohamed, A.K.; Liu, D.; Mohamed, M.A.A.; Song, K. Groundwater quality assessment of the quaternary unconsolidated sedimentary basin near the Pi river using fuzzy evaluation technique. *Appl. Water Sci.* **2018**, *8*. [CrossRef]
12. Wu, J.; Zhou, H.; He, S.; Zhang, Y. Comprehensive understanding of groundwater quality for domestic and agricultural purposes in terms of health risks in a coal mine area of the Ordos basin, north of the Chinese Loess Plateau. *Environ. Earth Sci.* **2019**, *78*, 446. [CrossRef]
13. Feng, L.; Sang, G.; Hong, W. Statistical prediction of changes in water resources trends based on set pair analysis. *Water Resour. Manag.* **2014**, *28*, 1703–1711. [CrossRef]
14. Tian, R.; Wu, J. Groundwater quality appraisal by improved set pair analysis with game theory weightage and health risk estimation of contaminants for Xuecha drinking water source in a loess area in Northwest China. *Hum. Ecol. Risk Assess. Int. J.* **2019**, *25*, 132–157. [CrossRef]
15. Li, P.; Li, X.; Meng, X.; Li, M.; Zhang, Y. Appraising groundwater quality and health risks from contamination in a Semiarid region of Northwest China. *Expo. Health* **2016**, *8*, 361–379. [CrossRef]
16. Li, P.; He, S.; He, X.; Tian, R. Seasonal hydrochemical characterization and groundwater quality delineation based on matter element extension analysis in a paper wastewater irrigation area, northwest China. *Expo. Health* **2018**, *10*, 241–258. [CrossRef]
17. Rabeiy, R.E. Assessment and modeling of groundwater quality using WQI and GIS in Upper Egypt area. *Environ. Sci. Pollut. Res.* **2018**, *25*, 30808–30817. [CrossRef] [PubMed]
18. Zhang, B.; Song, X.; Zhang, Y.; Han, D.; Tang, C.; Yu, Y.; Ma, Y. Hydrochemical characteristics and water quality assessment of surface water and groundwater in Songnen plain, Northeast China. *Water Res.* **2012**, *46*, 2737–2748. [CrossRef]
19. Mujumdar, P.; Sasikumar, K. A fuzzy risk approach for seasonal water quality management of a river system. *Water Resour. Res.* **2002**, *38*, 5-1–5-9. [CrossRef]
20. Rapant, S.; Cveckova, V.; Fajcikova, K.; Sedlakova, D.; Stehlikova, B. Impact of calcium and magnesium in groundwater and drinking water on the health of inhabitants of the Slovak Republic. *Int. J. Environ. Res. Public Health* **2017**, *14*, 278. [CrossRef]
21. Gao, Y.; Yu, G.; Luo, C.; Zhou, P. Groundwater nitrogen pollution and assessment of its health risks: A case study of a typical village in rural-urban continuum, China. *PLoS ONE* **2012**, *7*, e33982. [CrossRef]
22. Adimalla, N. Spatial distribution, exposure, and potential health risk assessment from nitrate in drinking water from semi-arid region of South India. *Hum. Ecol. Risk Assess. Int. J.* **2020**, *26*, 310–334. [CrossRef]
23. Chica-Olmo, M.; Peluso, F.; Luque-Espinar, J.A.; Rodriguez-Galiano, V.; Pardo-Iguzquiza, E.; Chica-Rivas, L. A methodology for assessing public health risk associated with groundwater nitrate contamination: A case study in an agricultural setting (southern Spain). *Environ. Geochem. Health* **2017**, *39*, 1117–1132. [CrossRef] [PubMed]
24. Elisante, E.; Muzuka, A.N.N. Occurrence of nitrate in Tanzanian groundwater aquifers: A review. *Appl. Water Sci.* **2015**, *7*, 71–87. [CrossRef]
25. Hudak, P. Regional trends in nitrate content of Texas groundwater. *J. Hydrol.* **2000**, *228*, 37–47. [CrossRef]
26. Rahmati, O.; Samani, A.N.; Mahmoodi, N.; Mahdavi, M. Assessment of the contribution of N-fertilizers to nitrate pollution of groundwater in western Iran (Case study: Ghorveh–Dehgelan aquifer). *Water Qual. Expo. Health* **2015**, *7*, 143–151. [CrossRef]
27. Li, P.; Wu, J.; Qian, H.; Lyu, X.; Liu, H. Origin and assessment of groundwater pollution and associated health risk: A case study in an industrial park, northwest China. *Environ. Geochem. Health* **2014**, *36*, 693–712. [CrossRef]
28. Wu, J.; Sun, Z. Evaluation of shallow groundwater contamination and associated human health risk in an Alluvial Plain impacted by agricultural and industrial activities, Mid-west China. *Expo. Health* **2015**, *8*, 311–329. [CrossRef]
29. Carey, R.O.; Migliaccio, K.W.; Brown, M.T. Nutrient discharges to Biscayne Bay, Florida: Trends, loads, and a pollutant index. *Sci. Total Environ.* **2011**, *409*, 530–539. [CrossRef]

30. Ji, X.; Xie, R.; Hao, Y.; Lu, J. Quantitative identification of nitrate pollution sources and uncertainty analysis based on dual isotope approach in an agricultural watershed. *Environ. Pollut.* **2017**, *229*, 586–594. [CrossRef]
31. Wu, J.; Lu, J.; Wen, X.; Zhang, Z.; Lin, Y. Severe nitrate pollution and health risks of coastal aquifer simultaneously influenced by saltwater intrusion and intensive anthropogenic activities. *Arch. Environ. Contam. Toxicol.* **2019**, *77*, 79–87. [CrossRef]
32. Bao, Z.; Hu, Q.; Qi, W.; Tang, Y.; Wang, W.; Wan, P.; Chao, J.; Yang, X.J. Nitrate reduction in water by aluminum alloys particles. *J. Environ. Manag.* **2017**, *196*, 666–673. [CrossRef] [PubMed]
33. Golekar, R.B.; Patil, S.N.; Baride, M. Human health risk due to trace element contamination in groundwater from the Anjani and Jhiri river catchment area in northern Maharashtra, India. *Earth Sci. Res. J.* **2013**, *17*, 17–23.
34. Bempah, C.K.; Ewusi, A. Heavy metals contamination and human health risk assessment around Obuasi gold mine in Ghana. *Environ. Monit. Assess.* **2016**, *188*, 261. [CrossRef] [PubMed]
35. Li, P.; Feng, W.; Xue, C.; Tian, R.; Wang, S. Spatiotemporal variability of contaminants in lake water and their risks to human health: A case study of the Shahu Lake tourist area, Northwest China. *Expo. Health* **2017**, *9*, 213–225. [CrossRef]
36. Batayneh, A. Toxic (aluminum, beryllium, boron, chromium and zinc) in groundwater: Health risk assessment. *Int. J. Environ. Sci. Technol.* **2012**, *9*, 153–162. [CrossRef]
37. Ryu, H.; Han, J.K.; Jung, J.W.; Bae, B.; Nam, K. Human health risk assessment of explosives and heavy metals at a military gunnery range. *Environ. Geochem. Health* **2007**, *29*, 259–269. [CrossRef]
38. Su, X.; Wang, H.; Zhang, Y. Health risk assessment of nitrate contamination in groundwater: A case study of an agricultural area in Northeast China. *Water Resour. Manag.* **2013**, *27*, 3025–3034. [CrossRef]
39. General Administration of Quality Supervision, Inspection and Quarantine of the People's Republic of China; Standardization Administration of China. *Standards for Groundwater Quality (GB/T 14848-2017)*; Standards Press of China: Beijing, China, 2017. (In Chinese)
40. Rezaei, H.; Jafari, A.; Kamarehie, B.; Fakhri, Y.; Ghaderpoury, A.; Karami, M.A.; Ghaderpoori, M.; Shams, M.; Bidarpoor, F.; Salimi, M. Health-risk assessment related to the fluoride, nitrate, and nitrite in the drinking water in the Sanandaj, Kurdistan County, Iran. *Hum. Ecol. Risk Assess. Int. J.* **2018**, *25*, 1242–1250. [CrossRef]
41. Qasemi, M.; Afsharnia, M.; Farhang, M.; Bakhshizadeh, A.; Allahdadi, M.; Zarei, A. Health risk assessment of nitrate exposure in groundwater of rural areas of Gonabad and Bajestan, Iran. *Environ. Earth Sci.* **2018**, *77*, 551. [CrossRef]
42. Ferchichi, H.; Ben Hamouda, M.F.; Farhat, B.; Ben Mammou, A. Assessment of groundwater salinity using GIS and multivariate statistics in a coastal Mediterranean aquifer. *Int. J. Environ. Sci. Technol.* **2018**, *15*, 2473–2492. [CrossRef]
43. El-Hames, A.S.; Al-Ahmadi, M.; Al-Amri, N. A GIS approach for the assessment of groundwater quality in Wadi Rabigh aquifer, Saudi Arabia. *Environ. Earth Sci.* **2010**, *63*, 1319–1331. [CrossRef]
44. Islam, A.R.M.T.; Shen, S.; Haque, M.A.; Bodrud-Doza, M.; Maw, K.W.; Habib, M.A. Assessing groundwater quality and its sustainability in Joypurhat district of Bangladesh using GIS and multivariate statistical approaches. *Environ. Dev. Sustain.* **2017**, *20*, 1935–1959. [CrossRef]
45. Ma, L.; Qian, J.; Yan, X.; Luo, Q.; Chen, L.; Zhao, W. Using GIS methods for determining the spatial distribution of groundwater hydrochemical facies: A case study for a shallow aquifer in Fuyang, China. *Environ. Earth Sci.* **2019**, *78*, 335. [CrossRef]
46. Mohamed, A.K.; Dan, L.; Kai, S.; Mohamed, M.A.A.; Aldaw, E.; Elubid, B.A. Hydrochemical analysis and fuzzy logic method for evaluation of groundwater quality in the North Chengdu Plain, China. *Int. J. Environ. Res. Public Health* **2019**, *16*, 302. [CrossRef] [PubMed]
47. Dahiya, S.; Singh, B.; Gaur, S.; Garg, V.K.; Kushwaha, H.S. Analysis of groundwater quality using fuzzy synthetic evaluation. *J. Hazard. Mater.* **2007**, *147*, 938–946. [CrossRef] [PubMed]
48. Zhang, X.-H. A study on the water environmental quality assessment of Fenjiang River in Yaan city of Sichuan Province in China. *IERI Procedia* **2014**, *9*, 102–109. [CrossRef]
49. Su, H.; Kang, W.; Xu, Y.; Wang, J. Assessing groundwater quality and health risks of nitrogen pollution in the Shenfu mining area of Shaanxi Province, Northwest China. *Expo. Health* **2017**, *10*, 77–97. [CrossRef]
50. Su, H.; Kang, W.; Xu, Y.; Wang, J. Evaluation of groundwater quality and health risks from contamination in the north edge of the Loess Plateau, Yulin City, Northwest China. *Environ. Earth Sci.* **2017**, *76*, 467. [CrossRef]

51. MEP. *Technical Guidelines for Risk Assessment of Contaminated Sites*; HJ 25.3-2014; China Environmental Science Press: Beijing, China, 2014. (In Chinese)
52. USEPA. *Risk Assessment Guidance for Superfund: Human Health Evaluation Manual (Part A)*; Office of Emergency and Remedial Response: Washington, DC, USA, 1989; Volume 1.
53. Ji, Y.; Wu, J.; Wang, Y.; Elumalai, V.; Subramani, T. Seasonal variation of drinking water quality and human health risk assessment in Hancheng City of Guanzhong Plain, China. *Expo. Health* **2020**, *12*, 469–485. [CrossRef]
54. Li, P. Groundwater quality in Western China: Challenges and paths forward for groundwater quality research in Western China. *Expo. Health* **2016**, *8*, 305–310. [CrossRef]
55. Amalraj, A.; Pius, A. Assessment of groundwater quality for drinking and agricultural purposes of a few selected areas in Tamil Nadu, South India: A GIS-based study. *Sustain. Water Resour. Manag.* **2017**, *4*, 1–21. [CrossRef]
56. Adimalla, N.; Qian, H. Groundwater quality evaluation using water quality index (WQI) for drinking purposes and human health risk (HHR) assessment in an agricultural region of Nanganur, south India. *Ecotoxicol. Environ. Saf.* **2019**, *176*, 153–161. [CrossRef]
57. Raju, A.; Singh, A. Assessment of groundwater quality and mapping human health risk in Central Ganga Alluvial Plain, Northern India. *Environ. Process.* **2017**, *4*, 375–397. [CrossRef]
58. Haji, M.; Wang, D.; Li, L.; Qin, D.; Guo, Y. Geochemical evolution of fluoride and implication for F—Enrichment in groundwater: Example from the Bilate River Basin of Southern Main Ethiopian Rift. *Water* **2018**, *10*, 1799. [CrossRef]
59. Adimalla, N.; Li, P. Occurrence, health risks, and geochemical mechanisms of fluoride and nitrate in groundwater of the rock-dominant semi-arid region, Telangana State, India. *Hum. Ecol. Risk Assess. Int. J.* **2018**, *25*, 81–103. [CrossRef]
60. Mukherjee, I.; Singh, U.K. Groundwater fluoride contamination, probable release, and containment mechanisms: A review on Indian context. *Environ. Geochem. Health* **2018**, *40*, 2259–2301. [CrossRef] [PubMed]
61. Narsimha, A.; Rajitha, S. Spatial distribution and seasonal variation in fluoride enrichment in groundwater and its associated human health risk assessment in Telangana State, South India. *Hum. Ecol. Risk Assess. Int. J.* **2018**, *24*, 2119–2132. [CrossRef]
62. Xiao, J.; Jin, Z.; Zhang, F. Geochemical controls on fluoride concentrations in natural waters from the middle Loess Plateau, China. *J. Geochem. Explor.* **2015**, *159*, 252–261. [CrossRef]
63. WHO. *Guidelines for Drinking-Water Quality: Fourth Edition Incorporating the First Addendum*; World Health Organization: Geneva, Switzerland, 2017.

5

The use of a Polymer Inclusion Membrane for Arsenate Determination in Groundwater

Ruben Vera, Enriqueta Anticó and Clàudia Fontàs *

Department of Chemistry, University of Girona, C/Maria Aurèlia Capmany 69, 17003 Girona, Spain; ruben.vera@udg.edu (R.V.); enriqueta.antico@udg.edu (E.A.)

* Correspondence: claudia.fontas@udg.edu

Abstract: A polymer inclusion membrane (PIM) containing the ionic liquid methyltrioctylammonium chloride (Aliquat 336) as the carrier has been used satisfactorily for the preconcentration of arsenate present in groundwater samples, allowing its determination by a simple colorimetric method. The optimization of different chemical and physical parameters affecting the membrane performance allowed its applicability to be broadened. The transport of As(V) was not affected by the polymer used to make the PIM (cellulose triacetate (CTA) or poly(vinyl chloride) (PVC)) nor the thickness of the membrane. Moreover, the use of a 2 M NaCl solution as a stripping phase was found to allow the effective transport of arsenate despite the presence of other major anions in groundwater. Using the PIM for the analysis of different groundwaters spiked at 100 µg L^{-1} resulted in recoveries from 79% to 124% after only 5 h of contact time. Finally, the validated PIM-based method was successfully applied to the analysis of waters containing naturally occurring arsenate.

Keywords: arsenate; polymer inclusion membranes; Aliquat 336; groundwater

---

1. Introduction

Arsenic is a well-known pollutant that is present in high levels in soil and water in different countries around the world [1]. The World Health Organization (WHO) has set an upper limit of 10 µg L^{-1} in a guideline for concentrations in drinking water [2]. Due to the high toxicity of arsenic even at low concentrations, it is of paramount importance to perform routine analyses to monitor this pollutant in waters. Of the different separation techniques, functionalized membranes have attracted considerable attention as a valuable technology for many analytical purposes in recent years. This is the case of polymer inclusion membranes (PIMs), which are non-porous functionalized membranes that consist of a polymer, a plasticizer, and an extractant. These membranes are transparent, flexible and stable, and have been used in many applications such as sensing, both ion-selective electrodes (ISE) and optodes, sample pre-treatment (separation and pre-concentration), and electro-driven extraction. PIMs have also been used as passive samplers deploying the membrane device for a 7-day period without reporting any drawback due to membrane fouling [3–5]. With the proper selection of the carrier, these membranes can effectively transport different species, such as inorganic pollutants [5], organic compounds [6] and metallic species [7].

In a previous study [8], a PIM made of cellulose triacetate (CTA) as the polymer and Aliquat 336 as the carrier was used for the transport of arsenate from aqueous natural samples to an 0.1 M NaCl stripping solution. Arsenate was transported through the membrane by the formation of the ion-pair $[(R_3R'N^+)_2HAsO_4^{2-}]$, which was released in the stripping compartment by exchanging arsenate with the chloride present in this phase [8]. Under neutral pH conditions, inorganic As(III) transport was negligible since it is mainly present as a neutral species and, thus, the developed PIM allowed the quantitative separation of both inorganic As species. Moreover, it was found that even though other anions present in natural waters were also transported (e.g., chloride, phosphate, nitrate, sulphate and carbonate), As(V) transport efficiency was not hampered. The developed PIM-based separation system was later implemented in a special device, incorporating a PIM made of 69% (*w*/*w*) poly(vinyl chloride) (PVC) as the base polymer and 31% (*w*/*w*) Aliquat 336 as the carrier, which allowed the preconcentration of arsenate, thus providing easy arsenate detection by means of the formation of a blue complex [9]. This method provided a working range from 20 μg L^{-1} to 120 μg L^{-1} As(V) in ultrapure water, and a limit of detection (LOD) of 4.5 μg L^{-1} after 24 h of contact time using 5 mL 0.1 M NaCl as the stripping phase. It was successfully applied in the analysis of different waters from the Pyrenees region with low conductivity values. These two previous works allows us to establish the conditions for the effective transport of inorganic As(V) in a transport cell (without preconcentration) [8], and to apply the membrane system in a PIM-based device to allow As(V) preconcentration and detection with good results after 24 h contact time for water samples bearing low conductivity [9].

In the present study, in order to extend the applicability of this separation system, we have evaluated and optimized the chemical and physical parameters that can affect the PIM-based device, such as membrane composition, stripping phase characteristics, and membrane thickness, in order to accomplish the preconcentration of arsenic species in a more convenient timescale and to broaden the applicability of the method to more complex groundwater (GW) samples.

2. Methods

2.1. Reagents and Solutions

Stock solution (100 mg L^{-1}) of As(V) was prepared from solid Na_2HAsO_4. 7 H_2O purchased from Merck (Darmstadt, Germany). Working solutions of arsenate in ultrapure water and GW were prepared by dilution of the corresponding stock solution. Sodium chloride, obtained from Fluka (Bern, Switzerland), was used to prepare the stripping solution. Calibration standards of arsenic were prepared using the Spectrascan standard solution for atomic spectroscopy (Teknolab, Drubak, Norway).

The extractant Aliquat 336 and the polymer PVC were purchased from Sigma-Aldrich (Steinheim, Germany) and CTA from Acros Organics (Geel, Belgium). The organic solvents tetrahydrofuran (THF) and chloroform ($CHCl_3$) (Panreac, Castellar del Vallès, Spain) were used to prepare the polymeric films.

Simulated groundwater (SGW) was prepared by dissolving 0.17 g of $NaHCO_3$, 0.22 g of $CaCl_2{\cdot}6H_2O$ (Panreac, Castellar del Vallès, Spain) and 0.07 g of Na_2SO_4 (Merck, Darmstadt, Germany) in 1 L of ultrapure water.

All reagents and solvents were of analytical reagent grade. Ultrapure water from a MilliQ Plus water purification system (Millipore Ibérica S.A., Madrid, Spain) was used to prepare all solutions.

2.2. Colorimetric Detection of As(V)

The determination of As(V) preconcentrated in the stripping phase was performed using the molybdenum blue method, which is based on the formation of an arsenomolybdate complex. The reagent solutions were prepared in accordance with the latest, improved version of the method [10]. Ammonium molybdate was prepared by dissolving 5.2 g of $(NH_4)_6Mo_7O_{24}{\cdot}4H_2O$ (Scharlau, Barcelona, Spain) and 8.8 mg of potassium antimonyl tartrate, $K(SbO)C_4H_4O_6{\cdot}0.5H_2O$ (Merck, Darmstadt, Germany) in 30 mL of 9 M sulphuric acid and diluted with deionized water to 50 mL in a volumetric flask. A solution 10% (w/v) of ascorbic acid (Panreac, Castellar del Vallès, Spain), which was used as a reductant, was prepared daily. The reagents were added to As(V) samples or standard solutions in accordance with the recommended procedure. To account for the matrix effect, standard solutions were prepared both in ultrapure water and 2 M NaCl.

2.3. Polymer Inclusion Membrane (PIM) Preparation

PIMs were prepared by dissolving either CTA (200 mg) or PVC (400 mg) and the appropriate volume of a 0.5 M Aliquat 336 solution in chloroform or in THF, respectively. The solution was poured into a 9.0 cm diameter flat-bottom glass Petri dish which was set horizontally and covered loosely. The solvent was allowed to evaporate over 24 h at room temperature and the resulting film was then carefully peeled off the bottom of the Petri dish and circular 2.5 cm^2 pieces were cut from its central section and used in the experiments. PIMs of different thicknesses were prepared by reducing proportionally the total mass of polymer and Aliquat 336.

All PIM compositions are given in mass percentages for each component.

2.4. Preconcentration Experiments and Calibration Curve

The schematics and the whole set-up of the PIM-based device used in the preconcentration experiments are described elsewhere [11]. The device incorporates the PIM with an area of 2.5 cm^2 and contains the stripping solution. This device is immersed 1 cm in a vertical position in 100 or 50 mL of a water sample containing arsenic and placed on a magnetic stirrer. After a predetermined contact time, the device was removed from the solution and a selected volume of the stripping solution (usually 2 mL) was taken for the colorimetric analysis.

Arsenic transport efficiency (TE) was determined by using Equation (1):

$$\mathrm{TE}(\%) = \frac{[As]_{strip\ (t)}}{[As]_{feed\ (0)}} \times \frac{1}{Vr} \times 100 \tag{1}$$

where $[As]_{strip(t)}$ denotes the arsenic concentration in the stripping compartment at the end of the contact time, whereas $[As]_{feed(0)}$ is the initial arsenic concentration in the water sample. The volume ratio between the feed solution and stripping solutions is denoted by *Vr*.

The calibration curve for the preconcentration method was prepared by using the final selected conditions in accordance with the TE results (50 mL of aqueous feed solution, 2.5 mL of 2 M NaCl as the stripping phase and 5 h contact time).

All experiments were conducted at 22 $\pm$ 1 °C, and were done per duplicate as minimum. Standard deviation (SD) is shown for each case with the corresponding number of replicates (n).

2.5. Apparatus

A Cary ultraviolet-visible (UV-Vis) (Agilent Technologies, Tokyo, Japan) instrument was used to measure the absorbance of As(V) complex at λ = 845 nm.

Arsenic concentration in the source solution of two GW samples with naturally occurring arsenic was measured using an inductively coupled plasma optical emission spectroscopy system (Agilent 5100 Vertical Dual View ICP-OES, Agilent Technologies, Tokyo, Japan).

PIM thickness was measured using a Digimatic Micrometer 0–25 mm (Mitutoyo, Takatsu-ku Japan).

The pH and conductivity values were determined with a Crison Model GLP 22 pH meter and Ecoscan, Entech Instruments, portable conductimeter, respectively. A magnetic multistirrer 15 (Fischer Scientific, Hampton, NH, USA) was also used.

2.6. Water Samples

Seven GW samples were collected from different locations in north-east Catalonia (Spain). Table 1 indicates the location of the different sampling spots as well as the main chemical characteristics of the different waters. GW samples 1–5 do not contain As and were used to study the effect of the water matrix by adding arsenate at different concentrations. GW samples 6 and 7 contained naturally occurring arsenic. All waters were used without any treatment (filtration or pH adjustment) except for sample GW7, which was brought to neutral pH by adding HCl.

Table 1. Characteristics and georeferences of the water samples used in this study.

Samples	Georeferences of Sampling Point (Coordinates)	pH	Conductivity (μS cm^{-1})	Ions (mg L^{-1})							Arsenic Concentration (μg L^{-1})
				$[NO_3^-]$	$[Cl^-]$	$[SO_4^{2-}]$	$[HCO_3^-]$	$[Na^+]$	$[Mg^+]$	$[Ca^+]$	
GW1 (Pujarnol)	42°6′16.907″ N Lat., 2°42′34.64″ E Long.	7.21	684	1.2	15.4	63.9	269	19.0	30.1	96.1	n.f.
GW2 (Mongai)	41°47′59.047″ N Lat., 0°57′38.832″ E Long.	7.76	470	9.5	15	38.3	n.m.	10.8	14.9	68.9	n.f.
GW3 (St. Hilari)	41°53′16.46″ N Lat., 2°31′11.867″ E Long.	7.98	275	21.9	11.4	8.7	172	16.6	9.0	50.0	n.f.
GW4 (Cerdanya)	42°21′16.059″ N Lat., 1°42′17.742″ E Long.	7.5	423	0.3	4.3	2.5	349	11.4	13.2	70.7	n.f.
GW5 (Setcases)	42°22′22.208″ N Lat., 2°18′3.026″ E Long.	7.56	110	3.3	0.9	8.0	88	3.7	2.7	23.6	n.f.
GW6 (Cerdanya)	42°22′16.393″ N Lat., 1°40′41.159″ E Long.	7.5	236	2.2	2.7	12.4	140	n.m.	n.m.	n.m.	67.1
GW7 (Cerdanya)	42°22′12.595″ N Lat., 1°40′54.456″ E Long.	9.69	185	0.4	2.2	11.8	n.m.	n.m.	n.m.	n.m.	70.4
SGW	-	7.5	459	n.a.	71	47.0	123	70.2	n.a.	40.3	n.a.

n.m.: not measured; n.a.: not added; n.f.: not found.

3. Results and Discussion

3.1. Parameters Affecting the Preconcentration System

3.1.1. Stripping Composition

As reported in our previous study [9], As(V) transport through the PIM containing Aliquat 336 is based on an anionic exchange in which the chloride present in the striping phase is the driving force behind the up-hill transport of the arsenate anion. Despite other anions such as sulphate or nitrate also being transported through the PIM, the effectiveness of the system was not affected as long as the total concentration of anions in GW did not exceed the chloride concentration in the stripping solution. However, the applicability of the method was severely hampered in the case of water samples with high conductivity. Thus, to broaden the application of this PIM-based preconcentration system, we tested the use of a 2 M NaCl solution instead of 0.1 M NaCl, maintaining all the other experimental parameters (e.g., PIM: 69% PVC–31% Aliquat 336 (*w*/*w*), time: 24 h, feed volume: 100 mL and stripping volume: 5 mL). The results are presented in Figure 1 as As(V) transport efficiency for both ultrapure water and GW1 and the two stripping solution compositions tested. It is worth mentioning that the conductivity of GW1 (648 µS) is between 2–6 times higher than the conductivity of natural waters studied in our previous work (in the 120–194 µS range) [9]. As can be observed, As(V) transport efficiency is dramatically affected if 0.1 M NaCl is used as the stripping phase, since only 34% of arsenate is transported. TE for GW1 increases up to 70% when the concentration of NaCl in the stripping phase is increased to 2 M. Similar results were observed in the work of Garcia-Rodríguez et al. [12], where the same PIM device was used for the monitoring of sulfamethoxazole (SMX) in natural waters, where a 2 M NaCl solution allowed a more efficient mass transfer across the membrane than an 0.5 M NaCl solution. Hence, a stripping phase consisting of a 2 M NaCl solution was fixed for subsequent experiments.

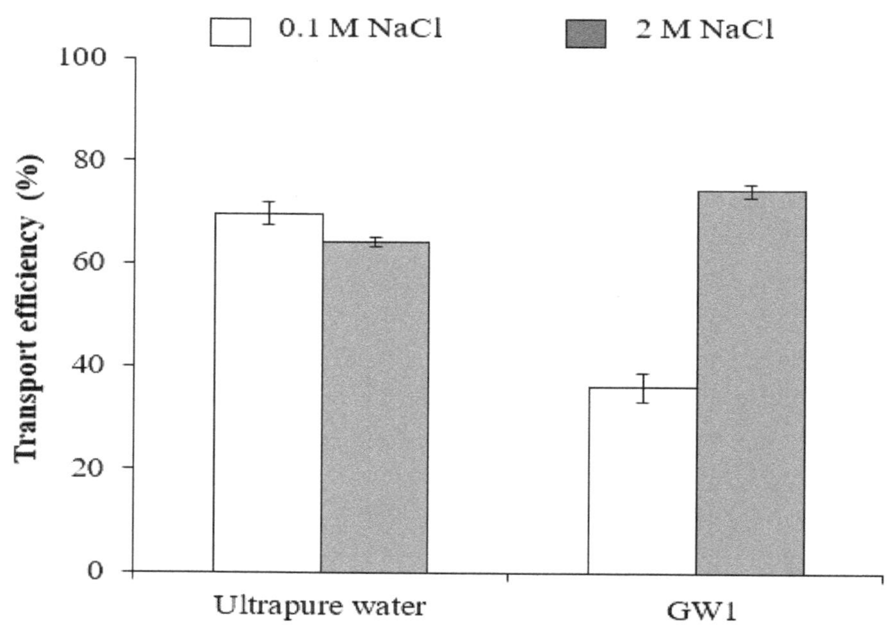

Figure 1. The effect of NaCl concentration used as a stripping solution on As(V) transport in both ultrapure and groundwater GW1 (spiked at 100 µg L^{-1} As(V)). Polymer inclusion membrane (PIM): 69% poly(vinyl chloride) (PVC)–31% Aliquat 336 (*w*/*w*); time: 24 h; feed volume: 100 mL; stripping volume: 5 mL (n = 2).

3.1.2. Contact Time and Sample and Stripping Volume

As(V) transport was assessed by varying the volume of both feed and stripping solutions (volume ratio fixed at 20) and using two different water samples, ultrapure and GW1. As shown in Figure 2, better transport efficiencies are obtained for both water samples tested, when 50 mL volumes for the feed and 2.5 mL for the stripping solution (50/2.5) were used in the PIM-based device system after 24 h contact time.

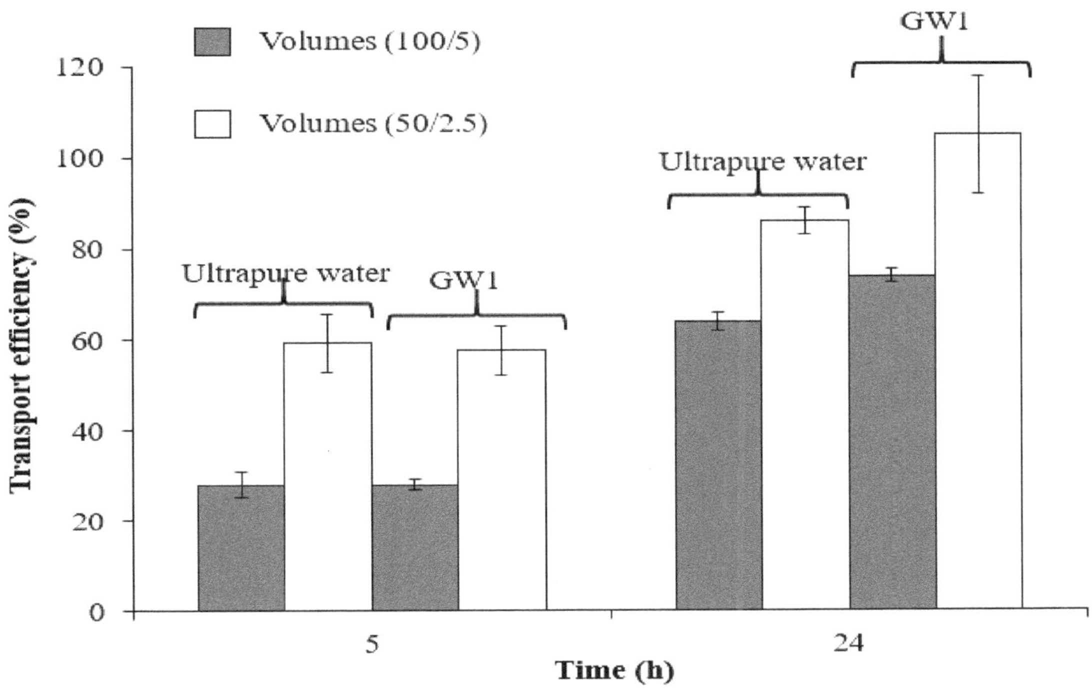

Figure 2. As(V) transport efficiency using a PIM-device with different volumes as feed and stripping solutions in both ultrapure and GW1 (spiked at 100 μg L^{-1} As(V)), after 24 h (a) and 5 h (b) contact time (n = 2). PIM composition was 69% PVC–31% Aliquat 336 (*w*/*w*) and 2 M NaCl was used as the stripping phase.

Moreover, As(V) transport efficiencies at 5 h and 24 h are compared in the same figure. It can be observed that TE are higher when the contact time is 24 h but satisfactory results, around 60% of TE, are also obtained at a time as short as 5 h. In our previous work [9], at a contact time of 5 h and using 0.1 M NaCl as the stripping phase and the 100/5 volumes, the TE obtained was around 20%. Besides, at 5 h contact time, the difference in terms of As(V) transport between the two different water samples is negligible. Therefore, 50/2.5 volumes for feed and stripping solutions and a contact time of 5 h were selected for further experiments.

3.1.3. Membrane Characteristics

As reported by other authors [8,12], transport efficiency can be affected by both the amount of extractant in the PIM and the nature of the polymer. Therefore, various membrane compositions with different thickness were studied in terms of arsenate preconcentration in the stripping phase, after a 5 h contact time. As can be seen in Table 2, there was only slight variation for the different PIM compositions even though different polymers were employed. These results are in concordance with other publications where similar results were obtained with PIMs prepared with the two polymers, CTA and PVC [13,14].

Table 2. Effect of membrane composition and thickness on As(V) preconcentration in the stripping phase (n = 3). Feed composition: 100 μg L^{-1} As(V) in GW1 (50 mL).

Polymer	PIM Composition (*w/w*)	Thickness (μm)	[As] Stripping (μg L^{-1}) (± SD)
Poly(vinyl chloride) (PVC)	69% PVC–31% Aliquat 336	60	1160 (± 58)
		30	940 (± 94)
	50% PVC–50% Aliquat 336	96	900 (± 45)
		39	840 (± 53)
Cellulose triacetate (CTA)	70% CTA–30% Aliquat 336	38	880 (± 18)
	52% CTA–48% Aliquat 336	45	1060 (± 85)
		25	940 (± 19)

Additionally, different authors have reported the great influence of PIM thickness when metal transport is rate-limited by the diffusion of the metal across the membrane [15–17]. However, the reduction of the membrane thickness, using the PIM-based device under the selected conditions, did not enhance the As(V) preconcentration, which can be explained by diffusion through the membrane not being the only rate-limiting factor as diffusion in the acceptor solution is also involved [11].

The fact that only slight differences are obtained in terms of arsenate preconcentration using different PIMs composition highlights the great sturdiness of the system under these experimental conditions. PIM with a composition of 52% CTA–48% Aliquat 336 (*w*/*w*) and a thickness of 45 μm was used in further experiments since the amount of reagents necessary to prepare PIMs made of CTA is smaller than PIMs based on PVC.

It should be noted that the preconcentration system provides an arsenate enrichment of around 10 times the initial concentration in the feed solution, which is a clear improvement in facilitating the detection of arsenate in polluted GW samples at low levels.

3.2. *Analytical Application of the PIM-device*

3.2.1. Effect of Water Matrix

The matrix composition of the calibration standards must be considered for the application of the PIM-based preconcentration system in the determination of As(V) as this is a critical point which can affect the TE and, consequently, the sensitivity of the method. For this reason, different water samples, GW1 to GW5 and SGW, were tested (see Table 1 for composition) under the selected experimental conditions. As(V) transport efficiencies are compared in Table 3, where values ranging from 53% up to 72% for GW1–4 and a value of 81% for GW5 are presented. The highest TE of GW5 is clearly related to the lowest conductivity value, which enables higher arsenate transport across the PIM. Our results support the hypothesis that the accuracy of the method might be compromised by the matrix composition used for the preparation of the calibration standards and the conductivity of the target water sample. As SGW presents an intermediate TE between GW1 and GW5, this is finally selected for the calibration and validation of the PIM-based device method.

Table 3. Comparison of the different GW samples on arsenate transport efficiency with the proposed PIM-based method (n = 2).

Water Sample	Conductivity (μS cm^{-1})	Amount of As(V) Added (μg L^{-1})	TE (%) (± SD)
GW1	684	100	53 (± 8)
GW2	470	100	65 (± 6)
GW3	275	100	72 (± 6)
GW4	423	100	66 (± 23)
GW5	110	100	81 (± 11)
SGW	459	100	63 (± 8)

3.2.2. Analytical Parameters

Under selected conditions, the proposed PIM-based method was applied to standards containing As(V) in the range of 10 μg L^{-1} to 150 μg L^{-1} in SGW. Figure 3 shows the absorbance measured for each standard plotted versus known concentrations of As(V) in the initial feed phase, and a straight line was fitted to measured points by the least-square method. Parameters of the resulting calibration curve are also included in Figure 3. The fact that the regression coefficient was higher than 0.99 indicates a good linearity throughout the studied working range. It is worth mentioning that the LOD of 7 μg L^{-1} is in concordance with the maximum permitted in drinking waters set at 10 μg L^{-1} by the WHO [2], and acceptable relative standard deviation (RSD) values (below 20%) at two different levels (i.e., 10% at 30 μg L^{-1} and 13% at 100 μg L^{-1}) are also obtained.

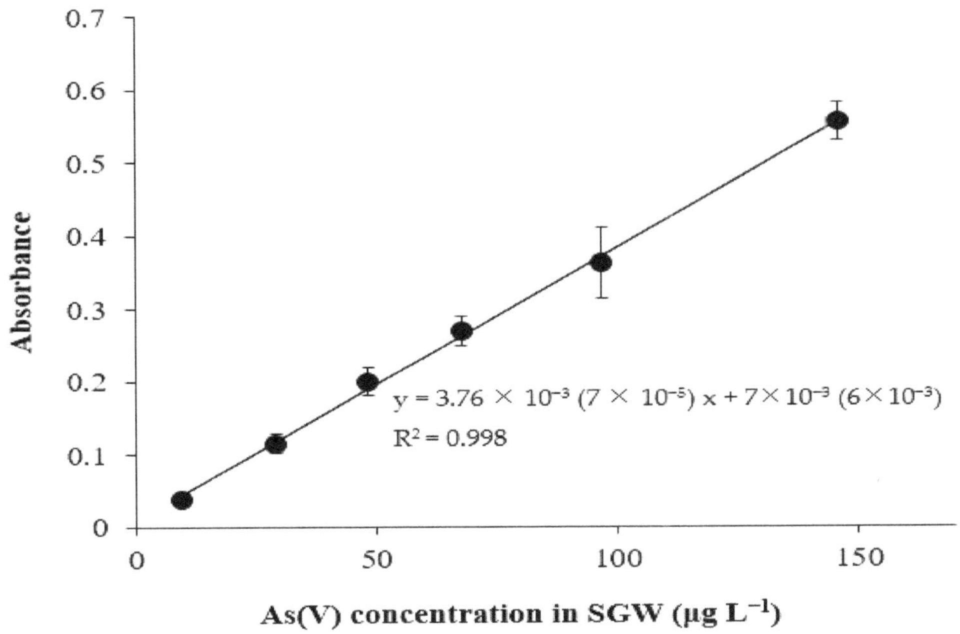

Figure 3. Calibration curve obtained with the PIM-based method. PIM, 52% CTA–48% Aliquat 336 (w/w) (n = 3). Feed solution, 50 mL of different As(V) concentrations in simulated groundwater (SGW); stripping solution 2.5 mL 2 M NaCl.

Recovery values of the proposed PIM-based method were calculated taking into account the calculated As(V) content based on the concentration found in the stripping and using the equation shown in Figure 3, in relation to the spiked level of different GW with 100 μg L^{-1} As(V). The results are collected in Table 4, where it can be seen that recovery values range from 79% up to 124% with the highest recovery corresponding to GW5.

Table 4. Effect of water sample on As(V) recovery (n = 2).

Water Sample	As(V) Recovery (%) (± SD)
GW1	79 (± 13)
GW2	98 (± 10)
GW3	109 (± 9)
GW4	100 (± 35)
GW5	124 (± 17)

All recovery values and standard deviations obtained using our proposed PIM-based device method can be considered satisfactory taking into account the μg L^{-1} concentration level, as reported in the guidelines for standard method performance requirements [18].

3.3. Application to Contaminated Groundwater (GW) Samples

The proposed method was used to analyse two naturally occurring arsenate GW samples from Catalonia (north-east Spain). A comparison between the direct analysis of the water sample by inductively coupled plasma optical emission spectroscopy (ICP-OES) and the proposed PIM-based device method is presented in Table 5. The good agreement shows that the method is suitable for the determination of As(V) in GW samples.

Table 5. As(V) concentration in GW samples determined by the ICP-OES reference method and the proposed PIM-based device method (n = 2).

Water Sample	As Concentration Measured (μg L^{-1}) (± SD)	
	ICP-OES	PIM-Based Device
GW6	67 (± 2)	82 (± 2)
GW7	70 (± 3)	67.4 (± 0.5)

4. Conclusions

An effective and simple methodology has been proposed employing a device incorporating a PIM made of 52% CTA–48% Aliquat 336 (*w*/*w*) and using the volumes of 50 mL and 2.5 mL for the feed and stripping solutions, respectively. The selection of a 2 M NaCl solution as the stripping phase and 5 h contact time resulted in TE around 53–81%. The type of polymer and the membrane thickness do not seem to influence the transport results under the studied experimental conditions. The use of an SGW as a matrix for the preparation of calibration standards enabled an improvement of the analytical parameters for the determination of As(V) in GWs with different chemical compositions.

The results obtained for the PIM-based method of two GW samples naturally containing As(V) is in concordance with the ICP-OES analysis. Hence, it is demonstrated that the proposed method can be used as an attractive alternative for the determination of arsenate within a range of different aqueous matrices with different conductivities.

Author Contributions: R.V. conducted the experimental work as a Ph.D. student under the supervision of C.F. and E.A., and was the responsible of writing the manuscript. C.F. and E.A. contributed to propose the research, the discussion of the results, data interpretation and the revision of the manuscript.

Acknowledgments: R. Vera acknowledges a grant from the Spanish Ministerio de Economía y Competitividad, ref. BES-2014-068314.

References

1. Villaescusa, I.; Bollinger, J.-C. Arsenic in drinking water: Sources, occurrence and health effects (a review). *Rev. Environ. Sci. Bio./Technol.* **2008**, *7*, 307–323. [CrossRef]
2. World Health Organization (WHO). *Guidelines for Drinking-Water Quality*, 4th ed.; WHO: Geneva, Switzerland, 2011.
3. Almeida, M.I.G.S.; Cattrall, R.W.; Kolev, S.D. Polymer inclusion membranes (PIMs) in chemical analysis—A review. *Anal. Chim. Acta* **2017**, *987*, 1–14. [CrossRef] [PubMed]
4. Almeida, M.I.G.S.; Chan, C.; Pettigrove, V.J.; Cattrall, R.W.; Kolev, S.D. Development of a passive sampler for Zinc (II) in urban pond waters using a polymer inclusion membrane. *Environ. pollut.* **2014**, *193*, 233–239. [CrossRef] [PubMed]

5. Almeida, M.I.G.S.; Silva, A.M.L.; Coleman, R.A.; Pettigrove, V.J.; Cattrall, R.W.; Kolev, S.D. Development of a passive sampler based on a polymer inclusion membrane for total ammonia monitoring in freshwaters. *Anal. Bioanal. Chem.* **2016**, *408*, 3213–3222. [CrossRef] [PubMed]
6. Garcia-Rodríguez, A.; Matamoros, V.; Kolev, S.D.; Fontàs, C. Development of a polymer inclusion membrane (PIM) for the preconcentration of antibiotics in environmental water samples. *J. Membr. Sci.* **2015**, *492*, 32–39. [CrossRef]
7. Pont, N.; Salvadó, V.; Fontàs, C. Selective transport and removal of cd from chloride solutions by polymer inclusion membranes. *J. Membr. Sci.* **2008**, *318*, 340–345. [CrossRef]
8. Güell, R.; Anticó, E.; Kolev, S.D.; Benavente, J.; Salvadó, V.; Fontàs, C. Development and characterization of polymer inclusion membranes for the separation and speciation of inorganic as species. *J. Membr. Sci.* **2011**, *383*, 88–95. [CrossRef]
9. Fontàs, C.; Vera, R.; Batalla, A.; Kolev, S.D.; Anticó, E. A novel low-cost detection method for screening of arsenic in groundwater. *Environ. Sci. Pollut. Res.* **2014**, *21*, 11682–11688. [CrossRef] [PubMed]
10. Tsang, S.; Phu, F.; Baum, M.M.; Poskrebyshev, G.A. Determination of phosphate/arsenate by a modified molybdenum blue method and reduction of arsenate by $S_2O_4^{2-}$. *Talanta* **2007**, *71*, 1560–1568. [CrossRef] [PubMed]
11. Vera, R.; Fontàs, C.; Galceran, J.; Serra, O.; Anticó, E. Polymer inclusion membrane to access Zn speciation: Comparison with root uptake. *Sci. Total Environ.* **2018**, *622–623*, 316–324. [CrossRef] [PubMed]
12. Garcia-Rodríguez, A.; Fontàs, C.; Matamoros, V.; Almeida, M.I.G.S.; Cattrall, R.W.; Kolev, S.D. Development of a polymer inclusion membrane-based passive sampler for monitoring of sulfamethoxazole in natural waters. Minimizing the effect of the flow pattern of the aquatic system. *Microchem. J.* **2016**, *124*, 175–180. [CrossRef]
13. Vázquez, M.I.; Romero, V.; Fontàs, C.; Anticó, E.; Benavente, J. Polymer inclusion membranes (PIMs) with the ionic liquid (IL) Aliquat 336 as extractant: Effect of base polymer and IL concentration on their physical–chemical and elastic characteristics. *J. Membr. Sci.* **2014**, *455*, 312–319. [CrossRef]
14. Kebiche-Senhadji, O.; Tingry, S.; Seta, P.; Benamor, M. Selective extraction of Cr (VI) over metallic species by polymer inclusion membrane (PIM) using anion (Aliquat 336) as carrier. *Desalination* **2010**, *258*, 59–65. [CrossRef]
15. Konczyk, J.; Kozlowski, C.; Walkowiak, W. Removal of chromium (III) from acidic aqueous solution by polymer inclusion membranes with D2EHPA and Aliquat 336. *Desalination* **2010**, *263*, 211–216. [CrossRef]
16. Mohapatra, P.K.; Lakshmi, D.S.; Bhattacharyya, A.; Manchanda, V.K. Evaluation of polymer inclusion membranes containing crown ethers for selective cesium separation from nuclear waste solution. *J. Hazardous Mater.* **2009**, *169*, 472–479. [CrossRef] [PubMed]
17. Kusumocahyo, S.P.; Kanamori, T.; Sumaru, K.; Aomatsu, S.; Matsuyama, H.; Teramoto, M.; Shinbo, T. Development of polymer inclusion membranes based on cellulose triacetate: Carrier-mediated transport of cerium (III). *J. Membr. Sci.* **2004**, *244*, 251–257. [CrossRef]
18. AOAC Official Methods of analysis. *Appendix F: Guidelines for Standard Method Performance Requirements*; AOAC International: Rockville, Maryland, 2012.

6

Coupled Thermally-Enhanced Bioremediation and Renewable Energy Storage System: Conceptual Framework and Modeling Investigation

Ali Moradi [1,*], Kathleen M. Smits [2] and Jonathan O. Sharp [3]

[1] Department of Environmental Resources Engineering, Humboldt State University, Arcata, CA 95521, USA
[2] Department of Civil Engineering, the University of Texas Arlington, Arlington, TX 76019, USA; ksmits@mines.edu
[3] Department of Civil and Environmental Engineering, Colorado School of Mines, Golden, CO 80401, USA; jsharp@mines.edu
* Correspondence: ali.moradi@Humboldt.edu

Abstract: This paper presents a novel method to couple an environmental bioremediation system with a subsurface renewable energy storage system. This method involves treating unsaturated contaminated soil using in-situ thermally enhanced bioremediation; the thermal system is powered by renewable energy. After remediation goals are achieved, the thermal system can then be used to store renewable energy in the form of heat in the subsurface for later use. This method can be used for enhanced treatment of environmental pollutants for which temperature is considered a limiting factor. For instance, this system can be used at a wide variety of petroleum-related sites that are likely contaminated with hydrocarbons such as oil refineries and facilities with above- and underground storage tanks. In this paper, a case-study example was analyzed using a previously developed numerical model of heat transfer in unsaturated soil. Results demonstrate that coupling energy storage and thermally-enhanced bioremediation systems offer an efficient and sustainable way to achieve desired temperature–moisture distribution in soil that will ultimately enhance the microbial activity.

Keywords: thermally enhanced bioremediation; renewable energy storage; sustainability; heat and mass transfer in unsaturated soil

---

1. Introduction

Soil Borehole Thermal Energy Storage (SBTES) systems are a promising renewable energy storage option. An opportunity to enhance the efficiency of SBTES systems, thus, making them more effective is to link their infrastructure costs with thermal bioremediation. As both SBTES and thermal remediation require the installation of boreholes that can either deliver heat to the subsurface (i.e., thermal remediation) or store heat for later use (i.e., SBTES), linking the two technologies offers a unique opportunity to assist in environmental clean-up and enhance the efficiency of renewable energy storage systems. In this introduction, we will first describe thermal bioremediation, followed by SBTES. How to link the two systems in a practical application is then discussed.

1.1. Bioremediation of Contaminated Soil and Groundwater

One promising technology to clean-up petroleum-contaminated soil and groundwater is bioremediation. Bioremediation uses microbes to degrade, transform, and ultimately remove target pollutants (e.g., petroleum hydrocarbons) from contaminated soil. Bioremediation can be performed in-situ, requiring the targeting of a remediation strategy to the specific subsurface environment, which inherently has site-specific challenges including differences in soil type, moisture, heterogeneities, and resident microorganisms. Despite these difficult to control variables, in-situ bioremediation has many economic and environmental benefits [1].

Environmental conditions affecting bioremediation efficiency include (1) climate (diurnal temperature, precipitation); (2) carbon, nutrient, and oxygen supply; (3) soil conditions (texture, type, moisture, and layering), and (4) contaminant composition and concentration. For example, Mori et al. [2] performed a series of experiments to investigate the effect of soil moisture on the bioremediation efficiency of oil-contaminated soils. They showed that unsaturated conditions prevented bypass flow and allowed dispersion of injected nutrients, resulting in higher bioremediation efficiency than for saturated soil conditions. However, microbes in unsaturated soil systems are more sensitive to the availability of nutrients and changes in temperature than those in comparatively stable saturated soil systems [3,4]. Furthermore, the hydrocarbon concentration trends appear to be season-dependent. For instance, in wet weather, since the water content of soil pore space is higher (i.e., effect of soil moisture), limitation in oxygen diffusion can result in reductions in the activity of aerobic petroleum-degrading microorganism [5]. These findings suggest that bioremediation technology is well suited to the vadose zone but could be enhanced by addressing variables inherent to this region.

Optimum levels of critical environmental factors for in-situ bioremediation can vary with the environmental conditions and contaminated site-characterizations. However, the Environmental Protection Agency (EPA) [6] provides general recommendations for in-situ bioremediation of contaminated unsaturated subsurface soils as listed in Table 1.

Table 1. Optimum levels of some important environmental factors for in-situ bioremediation of contaminated unsaturated subsurface soils (table amended from [6]).

Environmental Factor	Optimum Range *
Soil moisture	25–85% of soil porosity
Oxygen	Aerobes > 0.2 mg/L and Anaerobes thrive in the absence of oxygen
Redox potential	Aerobes > 50 mV and Anaerobes < 50 mV
pH	5.5–8.5
Nutrients	Sufficient for microbial growth

* As stated in the Environmental Protection Agency (EPA) report [6], these ranges are obtained from References [7–10].

Based on the temperature ranges that are optimal for respiration and for growth in the environment, microorganisms that are used for bioremediation of hydrocarbon-contaminated soil can be grouped into two categories; thermophilic and mesophilic. Thermophilic bacteria thrive at relatively high temperatures (~45–75 °C) whereas mesophilic bacteria only survive at moderate temperatures (~15–45 °C). While thermophilic bacteria are also present in colder environments, their population is limited under these conditions [11]. A multitude of bacteria (e.g., *B. thermoleovorans*, *P. aeruginosa*, etc.) are capable of biodegrading different categories of petroleum hydrocarbons as detailed in Reference [12] and some listed in Table 2. Many of these species have been isolated from oil-rich environments or geothermally heated regions and have been well studied for petroleum degradation capabilities and characteristics in both laboratory and field settings [13].

Table 2. Potential microorganism for hydrocarbon biodegradation.

Microorganism Name	Ideal Temperature
Consortium mainly of *Pseudomonas* sp. [14]	40–42 °C
P. aeruginosa AP02-1 [12]	45 °C
B. thermoleovorans DSM 10561 [15]	45–60 °C
G. thermoleovorans T80 [13]	60 °C
Thermus & Bacillus sp. [16]	60–70 °C

It has been shown that temperature plays an important role in controlling the nature and extent of microbial metabolism in the presence of most contaminants, including low-solubility hydrocarbons [17], partly because kinetic processes are temperature driven. Moreover, certain types of bacteria (i.e., thermophilic bacteria) are optimized for elevated temperature conditions.

The temperature dependence of soil microbial activity is usually investigated by measuring soil respiration rate [18]. Several field and laboratory studies have been conducted on measuring the effect of temperature on microbial respiration. Previous studies confirmed the effect of temperature on microbial respiration [18–22]. Lin et al. [19] showed that a temperature rise could increase soil respiration. Their analysis for samples incubated at 35 °C suggested that bacterial structure is related to soil temperate while in samples incubated at 15 °C and 20 °C, it correlates with time. Dijkstra et al. [20] used metabolic tracers and modeling to evaluate the response of soil metabolism to an abrupt change in temperature from 4 to 20 °C. Their results showed that respiration increases almost 10-fold two hours after temperature increases. Abed et al. [21] studied the effect of different temperatures and salinities on respiration activities, oil mineralization and bacterial community composition in desert soils. They monitored CO_2 evolution at different temperatures and showed an increase in CO_2 evolution and oil mineralization rates with increasing temperature. As mentioned by Boopathy [22], the rate of contaminant conversion during bioremediation depends on the rate of contaminant uptake and metabolism as well as the rate of mass transfer to the cell.

In addition to affecting microorganisms, elevated temperatures also affect the contaminant properties. Elevated temperature results in decreasing contaminant viscosity, increasing its solubility and enhancing diffusivity, all of which are all favorable to increased biodegradation rates. Using a heated and humidified biopile system, Sanscartier et al. [23] showed that raising the temperature of soil by only 5 °C enhanced bioremediation of a soil contaminated by diesel fuel. Using a mathematical model to study dissolution, biodegradation, and diffusion limited desorption of Dense Non-Aqueous Phase Liquid (DNAPL)-contaminated groundwater, Kosegi et al. [24] demonstrated that thermally enhanced bioremediation sites contaminated with DNAPL could reduce effluent concentrations (i.e., the amount of contaminant mass not degraded in-situ) by 94% when temperature increased from 15 °C to 35 °C. They also showed the thermally enhanced bioremediation can reduce clean-up time by 70% compared to ambient conditions. Perfumo et al. [25] experimentally demonstrated increases in temperature in the presence of thermophilic and mesophilic bacteria significantly enhanced the degradation rates of hexadecane-contaminated soil. In unsaturated soil systems, it has been shown that there is a strong correlation between microbial activity and the amount of carbon dioxide released within the soil (i.e., an indicator of microbial respiration) and seasonal fluctuations in air/soil temperature [26,27]. Indeed, microbial activity can be quantified as a function of carbon dioxide flux from soils [28]. According to Hendry et al. [26], the highest soil carbon dioxide concentrations are reported during summer months, indicating the highest biological activity of the year while the minimum concentrations occur in winter months. In terms of depth below the soil surface, carbon dioxide concentrations exhibit a profile of relative decrease with depth in the summer and increase in the winter. This is due to changes to the vertical temperature profile in summer and winter months. Thus, compensating for seasonal and diurnal temperature changes within the unsaturated soil by artificially heating the subsurface has the potential to enhance bioremediation efficiency over time.

Thermally enhanced bioremediation has been previously used in a variety of environmental remediation scenarios. These systems deliver heat to the subsurface using electrical resistance and radio frequency techniques as well as hot fluid injection as reviewed by Hinchee and Smith [29]. Although capable of producing ample heat for remediation enhancement, they demand high amounts of energy, making them very expensive. Perfumo et al. [25] highlighted the importance of further investigation into cost-effective methods to provide thermal energy. One cost-effective method is using renewable energy (e.g., solar, wind) to generate energy requirements for thermally enhanced bioremediation systems as proposed by Nakamura et al. [30] and Rossman et al. [31]. A limitation of renewable energy is its intermittency [31]. An important issue limiting the implementation and use of renewable sources is energy storage as it is not possible to control the timing of the supply of solar or wind energy in spite of their abundance. Therefore, the traditional way of using renewable energy would not provide a continuous heat source to enhance bioremediation. In addition, this would require that systems be linked into electrical grids, thus limiting the deployment of thermally enhanced bioremediation systems to areas with ample power supplies (i.e., not remotely deployable).

1.2. Renewable Energy Storage

Renewable energy resources continue to gain attention as the gap between energy consumption and production grows. Although there are many benefits to renewable energy resources, one unresolved issue is energy storage for use when demand is high. For example, wind or solar energy is produced intermittently and oftentimes at off-peak times of the day. A considerable amount of research has been done on producing renewable energy, yet the storage of renewable energy is oftentimes overlooked. Some research has been undertaken in ways to store renewable energy in the form of electricity; storage of energy as heat can oftentimes be more cost-effective. SBTES is a promising energy storage option in which heat, generated from renewable energy sources, is stored through circulating heated fluid (e.g., water) in geothermal borehole arrays in the subsurface. One of the main concerns hindering the widespread use of SBTES systems is their efficiency. Currently, efficiency ranges between 27–30% over the lifespan of these systems. Recently, McCartney et al. [32] proposed installing SBTES systems in the shallow subsurface above the water table (i.e., unsaturated zone). Installation in the unsaturated zone provides an opportunity to enhance system efficiency by taking advantage of latent and convective heat transfer, resulting in greater heat injection and extraction rates [33,34].

1.3. Coupled Bioremediation and Renewable Energy Storage System

In this paper, we explore the application of a novel approach to treat contaminated soil using an in-situ, thermally enhanced bioremediation system. This approach addresses the intermittency of energy supply to compensate for diurnal and seasonal temperature fluctuations coupled with a long-term energy storage system for subsequent energy requirements after remediation is achieved. The paper specifically focuses on: (a) the improvement of current renewable thermal remediation systems by addressing the intermittency of energy supply (phase I), and (b) the secondary use of a thermal remediation system as a renewable energy storage system (phase II).

2. System Characteristics

The schematics of the proposed enhanced bioremediation/energy storage system can be seen in Figure 1. The system operates in two phases; remediation (phase I) followed by energy storage (phase II). First, heat generated from a renewable energy source (e.g., solar or wind) is stored in the form of hot water or fluid in insulated tanks. The heated fluid is then circulated within the subsurface through a series of closed-loop, u-shaped tubes known as borehole heat exchangers to raise the temperature of the soil. The temperature of the injected fluid can be adjusted based on the application phase. For instance, during the remediation phase with mesophilic bacteria, lower temperatures (i.e., 40–50 °C) in the introduced fluid will be used to achieve temperatures for maximal soil biological activity. During the period of circulation, water within the boreholes returns to the water storage tank to be reheated.

As illustrated in Figure 1, a typical coupled enhanced bioremediation–energy storage system would include several sub-systems. A hot water/fluid storage tank, connected to the renewable energy source and pumping system, will be used to adjust the temperature of heated water or fluid to reach the desired soil temperature range for microbial activity or the heat storage. Solar thermal panels or wind turbines are used to heat up the fluid.

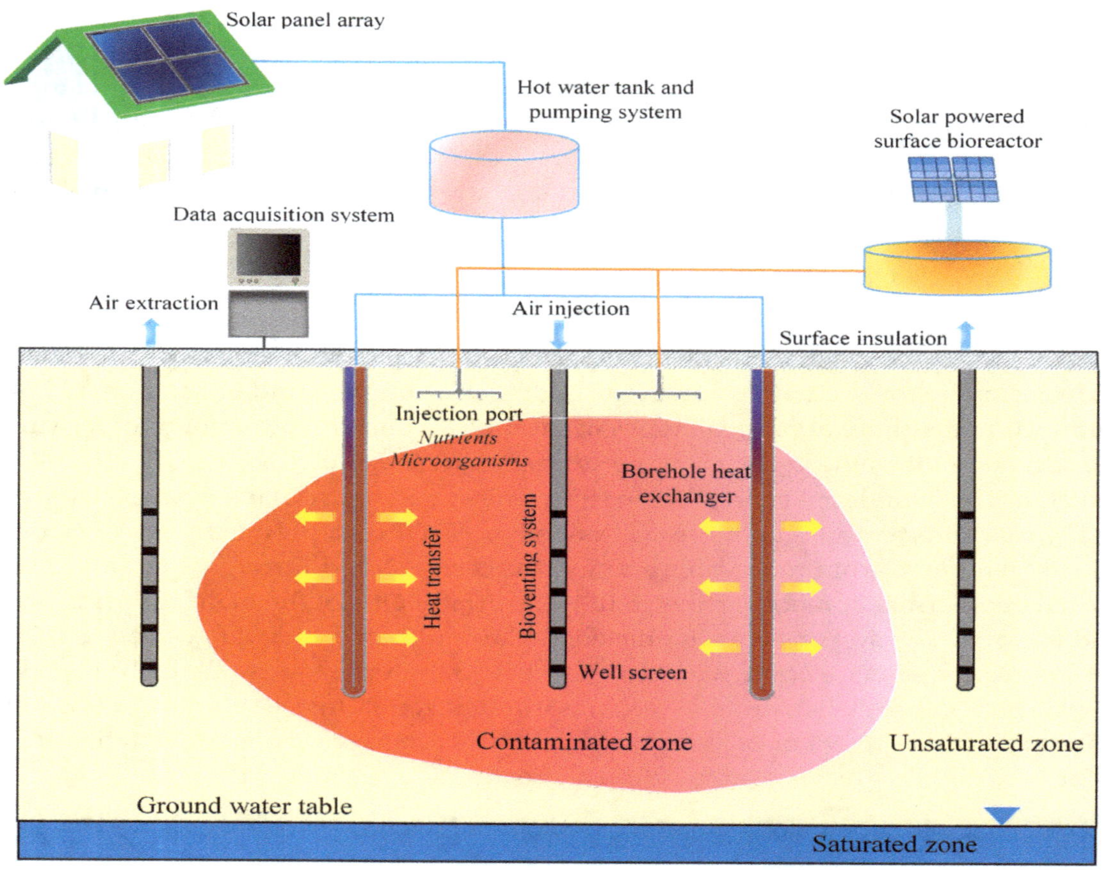

Figure 1. Schematic of the proposed coupled enhanced bioremediation–energy storage system.

A transition period might be necessary to repress microbial activity in subsurface soil. However, it depends on the state of the microbial activity when the remediation phase is completed. For thermophilic microorganisms, this is achieved by circulating cold fluid through heat exchangers to lower the soil temperature. The decrease in temperature can shift the population of microorganisms and decrease their overall numbers and activity to assure they do not pose any threat to the subsurface environment. In case of mesophilic microorganisms, circulating hot fluid during a transition period can select for microorganisms' population and growth since these types of microorganisms cannot survive under high temperatures. This provides an analogous pressure that "shocks" the community and represses activity. Organic carbon, nutrient, and oxygen injection should also be ceased during this phase. It is anticipated that the natural population will rebound once temperature, electron donors and acceptors rebound to the prior unperturbed steady state conditions. Subsequently, phase II (soil energy storage) will commence after this transition period. After achieving clean-up goals, the thermal remediation system can serve as a renewable energy storage system (i.e., storage period). During this phase (phase II), hotter fluid (i.e., 80–100 °C) can be circulated within the boreholes. Ultimately, the stored heat can be extracted in the winter by circulating cold water/fluid (i.e., 10 °C).

The solar-powered surface bioreactor is used to provide essential nutrients for microbial growth similar to American Type Culture Collection (ATTC) media described in Reference [35] (US patent 5753122). In most cases, indigenous bacteria are already present in the soil, and this reactor will only be used for biostimulation (adding constituents that enhance microbial growth such as organic carbon and nutrients). The reactor could also be used for bioaugmentation (adding microorganisms) if deemed appropriate. Surface insulation can reduce heat loss through the system by decreasing the soil-atmospheric interaction, especially in cold climates. The insulation layer can include layers of sand, waterproof membrane and insulation materials. The injection ports located below the soil surface are used to deliver a suspension of water, and nutrients to the soil. Percolating downward by gravity, the addition of a suspension enables both biostimulation and bioaugmentation to the system. The type of nutrients and microorganisms should be determined based on contaminant type, properties of the site and other technical considerations.

Moreover, the injection ports can also be used to adjust soil moisture during Phase II and compensate for drying out effect due to heat sources, thus, maximizing the heat transfer rate and system efficiency. As demonstrated in previous studies [33], for each soil type, there might be a critical degree of saturation (i.e., ratio of water content to pore volume) in which overall heat transfer (conduction and convection and latent heat transfer) is maximized. Similarly, as reported by Bear et al. [36], there is also a critical degree of saturation (which depends on the soil type) that causes no considerable drying at hot boundaries.

The bioventing system contains air injection and extraction wells to provide sufficient oxygen for aerobic microorganisms. As depicted in Figure 1, an extraction well is located in the middle of the contaminated zone. The negative pressure that is applied in the extraction well develops in the soil and can enhance remediation through two different mechanisms, volatilization and bioventing. First, the negative pressure gradient can accelerate volatilization of hydrocarbon sorbed to the soil particles. The extraction well collects the volatilized contaminant and provides opportunities for additional treatment before the gas being emitted into the atmosphere. Second, bioventing can help overcome oxygen deficits for aerobic bioremediation through delivering air to the subsurface [37]. More information on the site-specific design of bioventing wells can be found in Reference [38]. In addition, soil samples collected from bioventing wells can provide simpler tools to evaluate the performance and efficiency of the remediation system.

Borehole heat exchangers are u-shape tubes installed in the soil and used to circulate heated water/fluid. After installation, the area around u-tubes is backfilled with grout (e.g., mixture of silica sand and bentonite clay), assuring maximum heat transfer between the heat exchanger and the soil. Spacing and configuration of boreholes are determined based on soil domain properties, soil type, desired temperature, and moisture distribution in the system and the overall efficiency of the bioremediation processes based on the results of mathematical modeling.

The data acquisition system includes data loggers/computer to collect data from the sensor network for real-time analysis of environmental conditions. A series of thermocouples and soil moisture sensors can be strategically installed along with the borehole heat exchangers to simultaneously monitor soil temperature and moisture. Data can be reviewed and used to adjust system inputs (moisture, temperature, flow rate, etc.) to achieve better system efficiency.

Collected data from the sensor network, as well as results of the soil analysis for estimating biodegradation rates can be used to establish appropriate monitoring processes. Long-term monitoring will help to evaluate the performance of bioremediation through which termination time of phase (I) can be determined.

To evaluate remediation efficiency and the transition timing from phase I to phase II (I), the guidelines provided by EPA can be used. Based on these guidelines, soil samples should be collected and analyzed for Total Petroleum Hydrocarbon (TPH) and other contaminants of concerns at least bi-annually basis using standard spectrophotometry methods [39]. Moreover, soil gas samples collected from the bioventing system (extraction wells) should be regularly monitored for O_2, CO_2, and methane.

If volatile organic compounds (VOCs) are present in the contaminated soil, the soil gas can be further analyzed to evaluate the degradation level of such chemicals. The soil gas samples can indicate the rate of microbial activity in the soil. For instance, reduced oxygen levels and higher CO_2 concentrations compared to background, support bioremediation activity. Detailed information on long-term performance monitoring can be found in EPA guidelines [40].

This method builds upon pillars of in-situ treatment that minimizes disturbance and economic inputs while using physical, chemical and biological advantages of thermal treatment. It offers some valuable advantages as listed below. It should be mentioned that some of these advantages are also general to any thermally enhanced remediation method but herein with the advantage of the dual purpose of both remediation and energy storage.

- Elevated temperatures in the soil domain during the bioremediation period (phase I) enhance contaminant attenuation rates: As discussed earlier, temperature can play an important role in increasing the efficacy and rates of bioremediation. In this system, the injected heat can compensate for diurnal and seasonal variations in the soil temperature profile, allowing for more consistent and longer heating periods and thus a shorter remediation time.
- Uniform distribution of nutrients/oxygen through moisture redistribution increases biostimulation in unsaturated zone: As discussed in Section 4, moisture movement occurs in the presence of thermal gradients. The moisture circulation in both the liquid and vapor forms/phases can help redistribute oxygen and nutrients that are delivered from injection wells, allowing for the increase in contact between microorganisms and contaminant throughout the domain. This is important as in bioremediation injection wells, the injected nutrients/biomass is consumed very quickly and in the vicinity of wells. Thus, nutrients/biomass is rarely distributed far from the injection wells.
- Minimal disruption of the site: Installing borehole heat exchangers does not require any excavation and can be done with minimal disturbance of the soil. This is also known to be one of the important advantages of traditional and thermally enhanced bioremediation as well [22].
- Applicable to both populated and rural areas: Enhanced bioremediation/energy storage systems can be implemented in domestic areas (e.g., under building foundations), and remote/rural locations.
- Renewable energy consumption: Except for the initial installation costs and routine maintenance, there is minimal energy cost associated with this system, resulting in a considerably cheaper remediation technique than traditional thermal remediation systems.
- Environmentally friendly: This method links a remediation initiative with a clean and renewable energy storage system. The clean-up has minimal impact to the environment while implementing a sustainable system that allows the long-term use of the renewable energy system. Historically, bioremediation and renewable energy alternatives are well accepted with the public.
- The proposed method can be implemented in colder environments above freezing point where natural attenuation rates are unacceptably slow. Temperature will enhance the movement of contaminants through the soil which could increase bioavailability.
- In this method, the elevated soil temperature is considerably lower and easier to control compared to, for instance, electrical resistance or radio frequency methods. Therefore, the potential adverse effect of high temperatures (e.g., mobilizing contaminants, sterilizing microorganism, etc.) is minimal.
- Long-term energy storage: When the remediation goals are achieved, the system can still be used to store renewable energy without any additional investment or modification.
- Higher energy storage efficiency during phase II: Continuous heating of soil domain during phase (I) without a cooling period in the wintertime will likely increase the efficiency during energy storage phase. Although the transition may involve a cooling phase for the central regions of the contaminated domain (only in case of using thermophilic bacteria), it is expected that the

surrounding soil will still have a slightly higher temperature than background temperature. Therefore, it results in a lower temperature gradient between core of the system and the surrounding soil, thus, decreasing the heat loss from the system.

- The system has limited footprint, and it is not expected to have an extensive environmental impact in upper soil layers.

3. Numerical Modeling

In this section, we provide a brief case study, introducing the numerical model used to determine system efficiency. In addition, we provide an example of the application of this model to a hypothetical site.

To develop the mathematical model, three physicochemical/biological areas in the soil should be taken into account: (a) coupled heat and mass transfer, (b) metabolic activity and rates, and (c) contaminant fate and transport. Figure 2 shows a general mathematical modeling framework in designing a coupled thermally enhanced bioremediation–renewable energy storage system and the interrelationships between each component of the framework. It should be mentioned that including or excluding certain processes/assumptions from modeling depends on the problem at hand. The mathematical model is an entirely coupled model in which important parameters are functions of other processes and parameters. For instance, biological activity and growth can alter hydraulic properties of the soil or the temperature can affect both biological processes and multiphase flow in the soil.

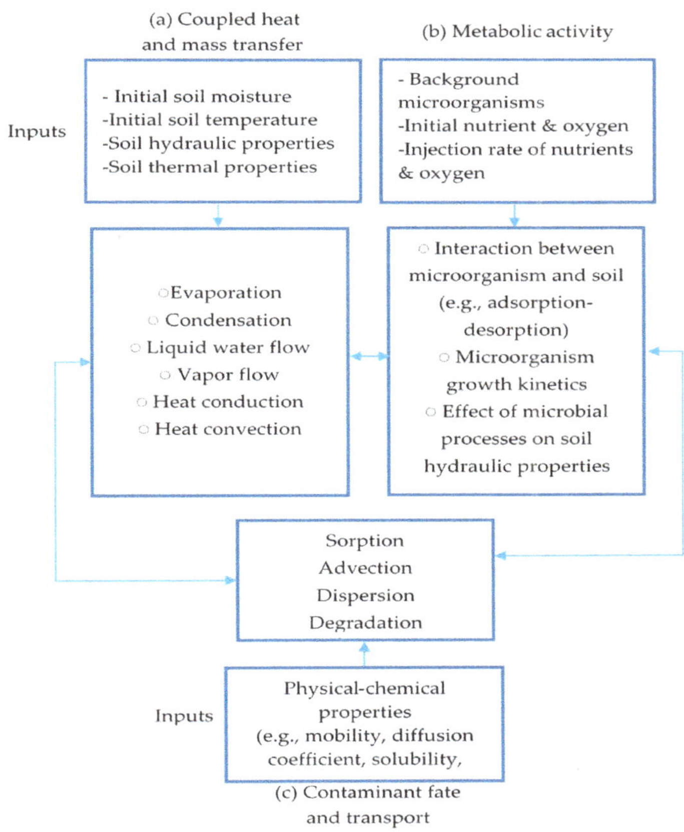

Figure 2. Mathematical modeling framework for a fully coupled thermally enhanced bioremediation–energy storage system.

As a reminder, the purpose of this paper is to present the concept of coupling enhanced bioremediation and renewable energy storage, it is not to discuss the details of the numerical modeling process. Our intent is not to model the entire system as shown in Figure 1 but rather focus on the enhancement of the system due to temperature effects (i.e., (a) model of Figure 2). In other words, not all of the affecting parameters and their interactions (e.g., effect of biological activity in soil hydraulic properties) are considered in this case study but will rather be presented in future works.

3.1. Simulation of Heat and Mass Transfer

Although much work has been done to numerically study ground heat exchangers [41,42], there is little to no work that investigates the effect of coupled heat and mass (water vapor and liquid water) transfer in presence of heat gradients. Instead, a common assumption in most modeling efforts is to consider soil as a conductive material with constant thermal properties. Although this assumption can be valid in some cases, it will not provide accurate results when modeling heat transfer in the vadose zone. Therefore, in the current study, a non-isothermal numerical model that simulates coupled heat, water vapor, and liquid water flux through the soil and considers non-equilibrium liquid/gas phase change should be used to simulate heat and moisture transfer in the domain. This model has been validated using the data collected from laboratory-scale tank tests that involved heating an unsaturated sand layer. Details of the numerical model development, experimental procedure and results can be found in References [33,34].

Figure 3 shows the domain and boundary conditions used for this case study. As depicted in the figure, no-flux boundary conditions were assumed for both liquid and vapor transfer for all the boundaries. However, constant temperature boundary conditions were applied on the surface of the heat sources whereas the top boundary was assumed to be insulated. For the bottom boundary as well as side boundaries, convective heat flux boundaries were applied. An initial temperature of 15 °C was assumed for the entire domain. Furthermore, the groundwater table was assumed to be 5 m below the bottom of the simulated domain as schematically shown in Figure 3. Natural soil type (Bonny silt) properties were used to perform the simulation. General properties of this soil are available in Appendix A (Table A1 and Figure A1).

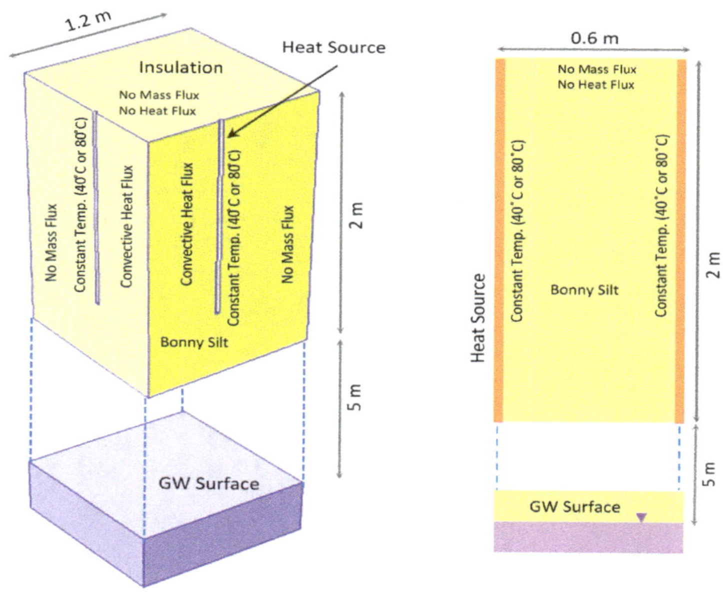

Figure 3. Three- and two-dimensional schematics of the simulated domain as well as boundary conditions used in the simulation. Due to symmetry, only a quarter of the domain is modeled. The line heat sources have a constant temperature (40 °C or 80 °C). The side boundaries are modeled as convective heat transfer boundaries. It is assumed that the surface of the domain is insulated (GW: Ground Water).

The system of differential equations was solved using the COMSOL Multiphysics software package (COMSOL Multiphysics® v. 5.2. www.comsol.com. COMSOL AB, Stockholm, Sweden). The numerical model was developed and validated in previous studies [33,34] and was slightly amended to use here. Two scenarios are modeled: (a) a constant temperature of 40 °C was applied in heat boundaries to achieve desired temperature range (i.e., 20–30 °C) assuming a mesophilic bacteria is used for bioremediation. This scenario helps to illustrate how the system operates during the first phase (remediation process). (b) Assuming the system operates for thermal energy storage purpose, the inlet temperature was increased to 80 °C. Only the first four days of each phase have been simulated.

3.2. Simulation of Bioremediation Process

A significant amount of research has been devoted to developing quantitative relationships between physicochemical and biological processes in polluted soils. A brief review of these relationships can be found in the paper presented by Murphy and Ginn [43]. As they pointed out, there is a linkage between the subsurface transport of bacteria and the biodegradation of dissolved contaminants. Most previous studies are for remediation in saturated soil systems. There are very few focusing on the modeling of the bioremediation process in unsaturated soils. A review of mathematical models to simulate bioremediation in a homogenous soil under unsaturated conditions is available in Reference [44].

As mentioned previously, since direct quantification of microorganism in soil and sediment samples requires invasive sampling and can be biased by system heterogeneity. However, carbon dioxide measurements provide a convenient and rapid analytical tool to estimate bulk microbial heterotrophic activity [45]. A van't Hoff-Arrhenius-type relationship is usually used to mathematically describe the enhanced microbial activity as a function of temperature [46,47]:

$$\alpha_m = \alpha_{mo} \exp\left[k(T - T_0)\right] \tag{1}$$

where α_m is the microbial carbon dioxide production rate (or microbial activity) at temperature, T, α_{mo} is the microbial carbon dioxide production rate at reference temperature T_0 and k is a constant.

4. Results and Discussion

Figure 4 shows the surface plots of the temperature distribution in the soil domain after 4 days of heating using four heat sources for both the remediation and energy storage phases. As seen in Figure 4a, a considerable portion of the domain reaches a temperature ~20–30 °C after 4 days, which is desirable for mesophilic microbial activity. As Figure 4b shows, temperature considerably increases in the center of the domain during the thermal energy storage phase (phase II) because of the injection of a higher temperature fluid (80 °C). The capability to model the heat transfer and storage allows for the design of the most efficient well configuration as well as environmental conditions (e.g., moisture content) to achieve overall system efficiency.

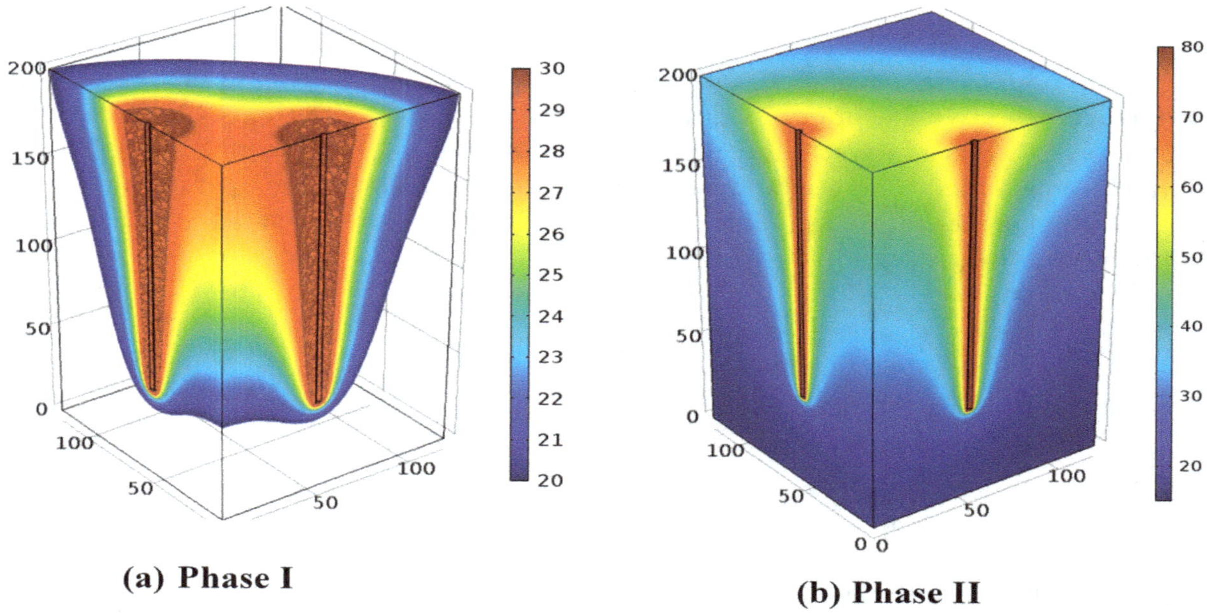

Figure 4. Temperature (°C) distribution in the domain after 4 days for (**a**) remediation phase and (**b**) storage phase. For the remediation phase, the part of the domain that has a temperature between 20–30 °C is shown. Only a quarter of the domain is presented.

Figure 5 shows the initial (t = 0) and final (t = 4 days) degrees of saturation in both phase I and II. The arrows in Figure 5b,c represent the gas-phase velocity field. A variable initial condition of the degree of saturation in the domain was associated with the gravity drainage of initially saturated soil as seen in Figure 5a. Figure 5b clearly shows how the soil moisture conditions are affected by a temperature rise in the domain (Figure 4). During phase I, the injected fluid temperature is 40 °C. Therefore, in phase I, a limited decrease in moisture is observed as demonstrated in Figure 5b. In addition, the gas-phase velocity field shows the occurrence of convective flow around the heat sources. As mentioned before, moisture redistribution due to convective mass transfer and evaporation/condensation processes in the system can improve the biological processes. In phase II, where the temperature of the injected fluid is higher (80 °C), extended drying occurred in the domain (Figure 5c). This is not a concern from a remediation standpoint (remediation has already finished when phase II begins) but can affect system efficiency. However, using the real-time monitoring data, moisture injection ports installed in the surface of the system can be used to compensate for the moisture decrease in the domain.

To demonstrate the effect of temperature on microbial activity for phase I of the example case, we calculated the ratio of carbon dioxide production compared to a reference value of carbon production at the initial conditions. Constant values for Equation (1) were selected from [44] (k = 0.10555 and $\alpha_{mo} = 1.5925 \times 10^{-17}$). Figure 6 shows the α_m/α_{mo} ratio at a single point located in the middle of the domain. As shown in Figure 6, the microbial activity increases more than twice its reference value due to a temperature rise of 10 °C in the domain. The temperature gradient reaches steady state condition in a short amount of time (i.e., few days) that could boost microbial activity until the transition period. During the transition period, when the temperature will increase above a critical value (i.e., 40 °C for mesophiles) that leads to a regime shift of the microorganism. The timing in ecological shift across the temperature domain during phase I and the transition period could be considered as a design constraint.

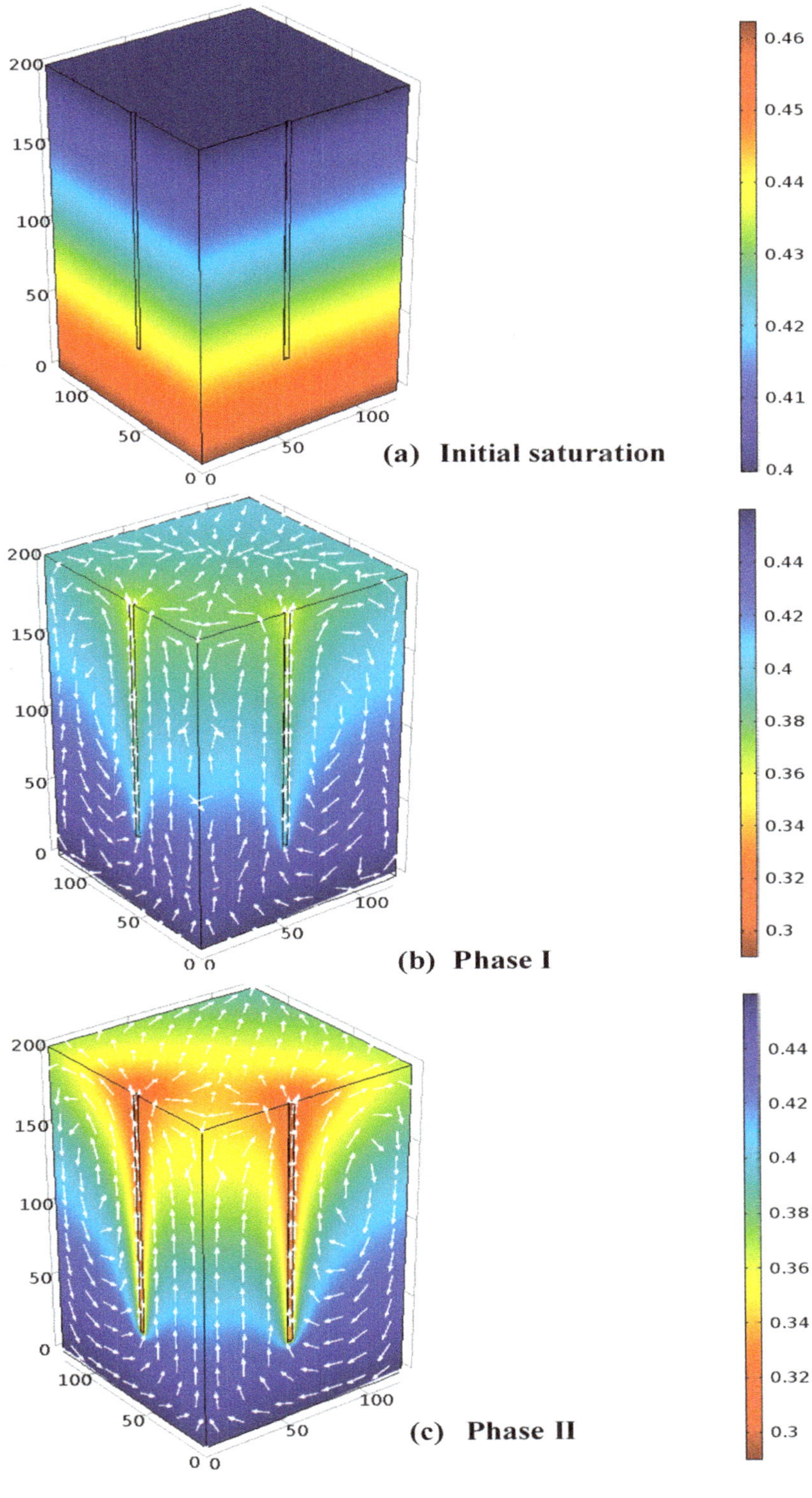

Figure 5. Saturation distribution: (**a**) before applying heat (t = 0 days); (**b**) after applying heat in phase I (t = 4 days) and (**c**) after applying heat in phase (II) (t = 4 days). Arrows in figure (**b**,**c**) show the gas phase velocity field.

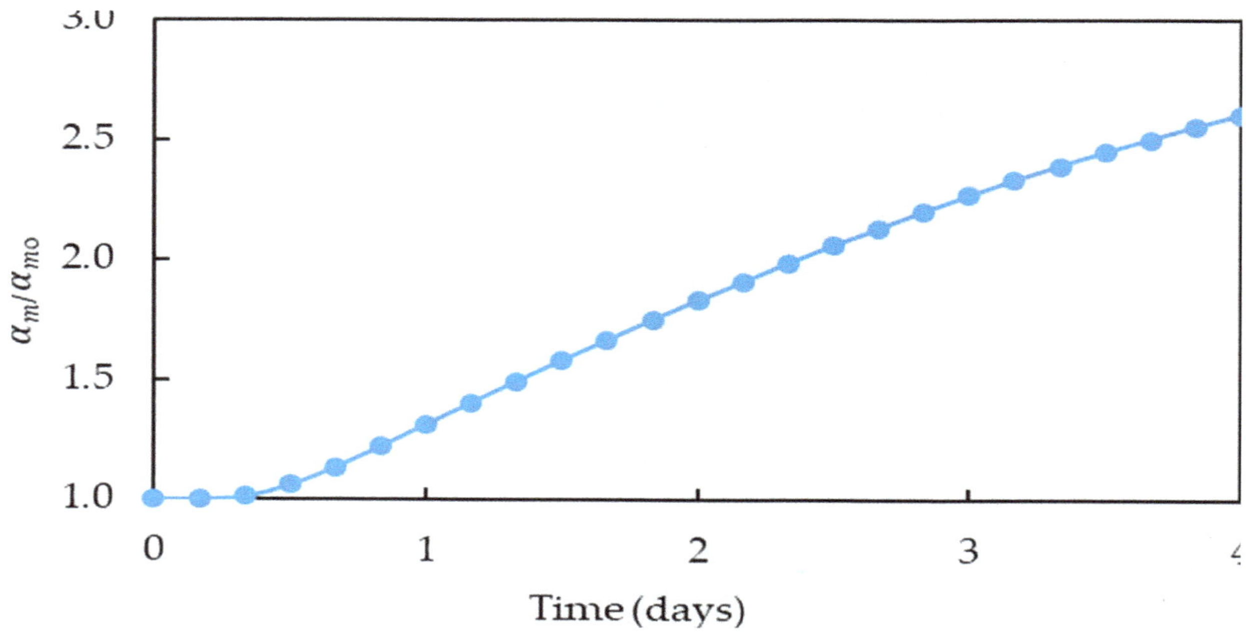

Figure 6. Ratio of carbon dioxide production after applying heat with respect to the reference value of carbon production at initial temperature (α_m / α_{mo}) at a single point located in the middle of the domain.

5. Conclusions

This paper presents the conceptual model of a coupled renewable energy storage and thermally-enhanced bioremediation system. The results of this study suggest that such a coupled system offers more efficient and sustainable way to achieve desired temperature–moisture distribution in soil that can be used to optimize desired microbial activity. The numerical simulation of a simple case study showed that by adjusting the inlet temperature, desired temperature distribution for both enhanced bioremediation and heat storage could be achieved. The proposed method of coupling these two concepts allows for a more cost-effective and sustainable alternative than implementing the systems individually. It is noted that for more accurate and efficient design, realistic domain size, number, and configuration of heat exchangers, boundary and initial conditions should be used in field applications.

6. Patents

A provisional patent application directed to a coupled thermally enhanced bioremediation and energy storage system and has been filed with the US Patent and Trademark Office and assigned Application No. 62/353475.

Author Contributions: Conceptualization, A.M.; Methodology, A.M.; Software, A.M.; Validation, A.M.; Formal analysis, A.M.; Investigation, A.M.; Resources, A.M. and K.M.S.; Data Curation, A.M.; Writing-Original Draft preparation, A.M.; Writing-Review and editing, A.M., K.M.S. and J.O.S.; Visualization, A.M.; Supervision, K.M.S. and J.O.S.; Project administration, K.M.S.; Funding acquisition, K.M.S.

Appendix A

This appendix contains information on hydraulic and thermal properties of soil that was used to perform numerical simulation.

Table A1. Selected hydraulic properties of soil used to perform numerical simulations.

d50 (mm)	Porosity	Residual Volumetric Water Content (m/m)	Saturated Hydraulic Conductivity, Ks, ($m \cdot s^{-1}$)	van Genuchten Parameters	
				Alpha (kPa^{-1})	n
0.039	0.430	0.030	1.3×10^{-6}	0.0863	1.58

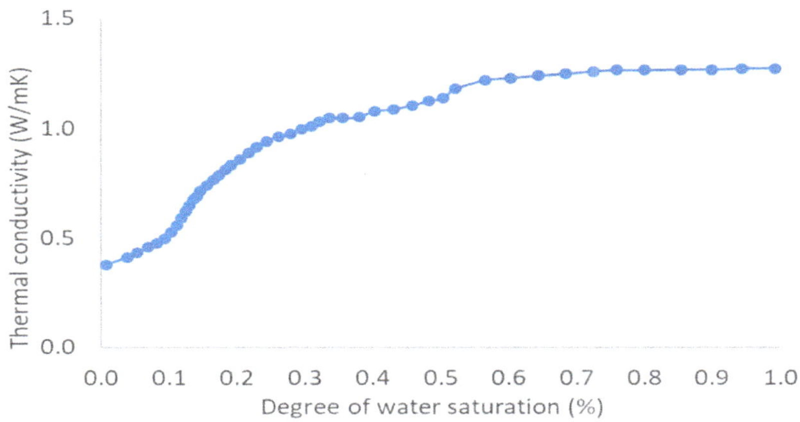

Figure A1. Thermal conductivity-Saturation curve for Bonny Silt.

References

1. Balba, M.T.; Al-Awadhi, N.; Al-Daher, R. Bioremediation of oil-contaminated soil: Microbiological methods for feasibility assessment and field evaluation. *J. Microbiol. Methods* **1998**, *32*, 155–164. [CrossRef]
2. Mori, Y.; Suetsugu, A.; Matsumoto, Y.; Fujihara, A.; Suyama, K. Enhancing bioremediation of oil-contaminated soils by controlling nutrient dispersion using dual characteristics of soil pore structure. *Ecol. Eng.* **2013**, *51*, 237–243. [CrossRef]
3. Fierer, N.; Allen, A.S.; Schimel, J.P.; Holden, P.A. Controls on microbial CO_2 production: A comparison of surface and subsurface soil horizons. *Glob. Chang. Biol.* **2003**, *9*, 1322–1332. [CrossRef]
4. Fierer, N.; Jackson, R.B. The diversity and biogeography f soil bacterial communities. *Proc. Natl. Acad. Sci. USA* **2006**, *103*, 626–631. [CrossRef] [PubMed]
5. Franzmann, P.D.; Robertson, W.J.; Zappia, L.R.; Davis, G.B. The role of microbial populations in the containment of aromatichydrocarbons in the subsurface. *Biodegradation* **2002**, *13*, 65–78. [CrossRef] [PubMed]
6. Sims, J.L.; Sims, R.C.; DuPont, R.R.; Matthews, E.; Russell, H.H. *In-Situ Bioremediation of Contaminated Unsaturated Subsurface Soils*; Ann Arbor Press: Ann Arbor, MI, USA, 1993.
7. Sims, R.C.; Sorensen, D.L.; Sims, J.; McLean, J.E.; Mahmood, R.; Dupont, R.R. *Review of In-Place Treatment Techniques for Contaminated Surface Soils—Volume 2: Background Information for In-Situ Treatment*; Utah State University: Logan, UT, USA, 1984.
8. Huddleston, R.L.; Bleckmann, C.A.; Wolfe, J.R. Land treatment—Biological degradation processes. In *Land Treatment: A Hazardous Waste Management Alternative*; Loehr, R.C., Malina, J.F., Jr., Eds.; Center for Research in Water Recourses, University of Texas: Austin, TX, USA, 1986; pp. 41–62.
9. Rochkind, M.L.; Blackburn, J.W.; Sayler, G.S. *Microbial Decomposition of Chlorinated Aromatic Compounds*; National Technical Information Service: Springfield, VA, USA, 1986; pp. 2–86.
10. Paul, E.A.; Clark, F.E. *Soil Microbiology and Biochemistry*; Academic Press, Inc.: San Diego, CA, USA, 1998.
11. Corwin, P. What are high-temperature bacteria doing in cold environments? *Trends Microbiol.* **2002**, *10*, 120–121.
12. Perfumo, A.; Banat, I.M.; Marchant, R. The use of thermophilic bacteria in accelerated hydrocarbon bioremediation. In *Environmental Problems in Coastal Regions VI: Including Oil Spill Studies*; Brebbia, C.A., Ed.; WIT Press: Southampton, UK, 2006; pp. 67–77.
13. Marchant, R.; Sharkey, F.H.; Banat, I.M.; Rahman, T.J.; Perfumo, A. The degradation of n-hexadecane in soil by thermophilic geobacilli. *FEMS Microbiol. Ecol.* **2006**, *56*, 44–54. [CrossRef] [PubMed]

14. Lugowsky, A.J.; Palamteer, G.A.; Boose, T.R.; Merriman, J.E. Biodegradation Process for Detoxifying Liquid Streams. U.S. Patent 5,656,169, 12 August 1997.
15. Markl, H.; Antranikian, G.; Becker, P.; Markossian, S. Aerobic Biodegradation of Aromatic Compounds Having Low Water Solubility Using Bacillus Thermoleovorans Strain DSM 10561. U.S. Patent 5,965,431, 12 October 1999.
16. Feitkenhauer, H.; Müller, R.; MAuml, H. Degradation of polycyclic aromatic hydrocarbons and long chain alkanes at 6070 C by Thermus and *Bacillus* spp. *Biodegradation* **2003**, *14*, 367–372. [CrossRef] [PubMed]
17. Margesin, R.; Schinner, F. Biodegradation and bioremediation of hydrocarbons in extreme environments. *Appl. Microbiol. Biotechnol.* **2001**, *56*, 650–663. [CrossRef] [PubMed]
18. Pietikäinen, J.; Pettersson, M.; Bååth, E. Comparison of temperature effects on soil respiration and bacterial and fungal growth rates. *FEMS Microbiol. Ecol.* **2005**, *52*, 49–58. [CrossRef] [PubMed]
19. Lin, Y.T.; Jia, Z.; Wang, D.; Chiu, C.Y. Effects of temperature on the composition and diversity of bacterial communities in bamboo soils at different elevations. *Biogeosciences* **2017**, *14*, 4879–4889. [CrossRef]
20. Dijkstra, P.; Thomas, S.C.; Heinrich, P.L.; Koch, G.W.; Schwartz, E.; Hungate, B.A. Effect of temperature on metabolic activity of intact microbial communities: Evidence for altered metabolic pathway activity but not for increased maintenance respiration and reduced carbon use efficiency. *Soil Biol. Biochem.* **2011**, *43*, 2023–2031. [CrossRef]
21. Abed, R.M.; Al-Kharusi, S.; Al-Hinai, M. Effect of biostimulation, temperature and salinity on respiration activities and bacterial community composition in an oil polluted desert soil. *Int. Biodeterior. Biodegrad.* **2015**, *98*, 43–52. [CrossRef]
22. Boopathy, R. Factors limiting bioremediation technologies. *Bioresour. Technol.* **2000**, *74*, 63–67. [CrossRef]
23. Sanscartier, D.; Zeeb, B.; Koch, I.; Reimer, K. Bioremediation of diesel-contaminated soil by heated and humidified biopile system in cold climates. *Cold Reg. Sci. Technol.* **2009**, *55*, 167–173. [CrossRef]
24. Kosegi, J.M.; Minsker, B.S.; Dougherty, D.E. Feasibility study of thermal in-situ bioremediation. *J. Environ. Eng.* **2000**, *126*, 601–610. [CrossRef]
25. Perfumo, A.; Banat, I.M.; Marchant, R.; Vezzulli, L. Thermally enhanced approaches for bioremediation of hydrocarbon-contaminated soils. *Chemosphere* **2007**, *66*, 179–184. [CrossRef] [PubMed]
26. Hendry, M.J.; Mendoza, C.A.; Kirkland, R.A.; Lawrence, J.R. Quantification of transient CO_2 production in a sandy unsaturated zone. *Water Resour. Res.* **1999**, *35*, 2189–2198. [CrossRef]
27. Hiraishi, A.; Yamanaka, Y.; Narihiro, T. Seasonal microbial community dynamics in a flowerpot-using personal composting system for disposal of household biowaste. *J. Gen. Appl. Microbiol.* **2000**, *46*, 133–146. [CrossRef] [PubMed]
28. Brouillard, B.M.; Mikkelson, K.M.; Bokman, C.M.; Berryman, E.M.; Sharp, J.O. Extent of localized tree mortality influences soil biogeochemical response in a beetle-infested coniferous forest. *Soil Biol. Biochem.* **2017**, *114*, 309–318. [CrossRef]
29. Hinchee, R.E.; Smith, L.A. *In-Situ Thermal Technologies for Site Remediation*; CRC Press: Boca Raton, FL, USA, 1992.
30. Nakamura, T.; Senior, C.L.; Burns, E.G.; Bell, M.D. Solar-powered soil vapor extraction for removal of dense non-aqueous phase organics from soil. *J. Environ. Sci. Health Part A* **2000**, *35*, 795–816. [CrossRef]
31. Rossman, A.J.; Hayden, N.J.; Rizzo, D.M. Low-temperature soil heating using renewable energy. *J. Environ. Eng.* **2006**, *132*, 537–544. [CrossRef]
32. McCartney, J.S.; Ge, S.; Reed, A.; Lu, N.; Smits, K. Soil-borehole thermal energy storage systems for district heating. In Proceedings of the European Geothermal Congress, Pisa, Italy, 3–7 June 2013; pp. 1–10.
33. Moradi, A.; Smits, K.M.; Cihan, A.; Massey, J.; McCartney, J.M. Impact of coupled heat transfer and water flow on soil borehole thermal energy storage (SBTES) systems: Experimental and modeling investigation. *Geothermics* **2015**, *57*, 56–72. [CrossRef]
34. Moradi, A.; Smits, K.M.; Lu, N.; McCartney, J.S. Heat transfer in unsaturated soil with application to borehole thermal energy storage. *Vadose Zone J.* **2016**, *15*. [CrossRef]
35. Taylor, R.T.; Jackson, K.J.; Duba, A.G.; Chen, C.I. In Situ Thermally Enhanced Biodegradation of Petroleum Fuel Hydrocarbons and Halogenated Organic Solvents. U.S. Patent 5,753,122, 19 May 1998.
36. Bear, J.; Bensabat, J.; Nir, A. Heat and mass transfer in unsaturated porous media at a hot boundary: I. One-dimensional analytical model. *Transp. Porous Med.* **1991**, *6*, 281–298. [CrossRef]

37. Hoeppel, R.E.; Hinchee, R.E.; Arthur, M.F. Bioventing soils contaminated with petroleum hydrocarbons. *J. Ind. Microbiol.* **1991**, *8*, 141–146. [CrossRef]
38. Hinchee, R.E. Bioventing of Petroleum Hydrocarbons. In *Handbook of Bioremediations*; CRC Press: Boca Raton, FL, USA, 1993.
39. United States Environmental Protection Agency (USEPA). *Test Method for Evaluating Total Recoverable Petroleum Hydrocarbon, Method 418.1 (Spectrophotometric, Infrared)*; Government Printing Office: Washington, DC, USA, 1978.
40. United States Environmental Protection Agency (USEPA). *How to Evaluate Alternative Cleanup Technologies for Underground Storage Tank Sites. A guide for Corrective Action Reviewers*; EPA 510-R-04-002; Office of Solid Waste and Emergency Response: Washington, DC, USA, 2004.
41. Rees, S.J.; He, M. A three-dimensional numerical model of borehole heat exchanger heat transfer and fluid flow. *Geothermics* **2013**, *46*, 1–13. [CrossRef]
42. Zeng, H.; Diao, N.; Fang, Z. Efficiency of vertical geothermal heat exchangers in the ground source heat pump system. *J. Ther. Sci.* **2003**, *12*, 77–81. [CrossRef]
43. Murphy, E.M.; Ginn, T.R. Modeling microbial processes in porous media. *Hydrogeol. J.* **2000**, *8*, 142–158. [CrossRef]
44. Borsi, I.; Fasano, A. A general model for bioremediation processes of contaminated soils. *Int. J. Adv. Eng. Sci. Appl. Math.* **2009**, *1*, 33–42. [CrossRef]
45. Chapatwala, K.D.; Babu, G.R.V.; Vijaya, O.K.; Armstead, E.; Palumbo, A.V.; Zhang, C.; Phelps, T.J. Effect of temperature and yeast extract on microbial respiration of sediments from a shallow coastal subsurface and vadose zone. *Appl. Biochem. Biotechnol.* **1996**, *57*, 827–835. [CrossRef] [PubMed]
46. Wood, B.D.; Keller, C.K.; Johnstone, D.L. In-situ measurement of microbial activity and controls on microbial CO2 production in the unsaturated zone. *Water Resour. Res.* **1993**, *29*, 647–659. [CrossRef]
47. Or, D.; Smets, B.F.; Wraith, J.M.; Dechesne, A.; Friedman, S.P. Physical constraints affecting bacterial habitats and activity in unsaturated porous media—A review. *Adv. Water Resour.* **2007**, *30*, 1505–1527. [CrossRef]

7

Support Tool for Identifying In-Situ Remediation Technology for Sites Contaminated by Hexavalent Chromium

Gabriele Beretta, Andrea Filippo Mastorgio *, Lisa Pedrali, Sabrina Saponaro and Elena Sezenna

Department of Environmental and Civil Engineering, Politecnico di Milano, Piazza Leonardo da Vinci 32, 20133 Milan, Italy; gabriele.beretta@polimi.it (G.B.); lisa.pedrali@polimi.it (L.P.); sabrina.saponaro@polimi.it (S.S.); elena.sezenna@polimi.it (E.S.)

* Correspondence: andreafilippo.mastorgio@polimi.it

Abstract: Sites contaminated by hexavalent chromium raise concerns relating to the toxicity of the pollutant, as well as for the increased solubility of its compounds, which helps it to seep into aquifers. Chemical and biological in situ treatment technologies, with good potential in terms of environmental sustainability, have recently been designed and implemented on a wide scale. A useful support tool is shown in the manuscript in the preliminary phase of assessing possible technologies applicable according to the site-specific characteristics of sites. The actual efficacy of the technologies identified should nevertheless be verified in laboratory trials and pilot tests.

Keywords: hexavalent chromium; decision support tool; remediation technologies

---

1. Introduction

Wide scale industrial use of hexavalent chromium and its compounds has caused serious environmental pollution, generally relating to accidental or unlawful leakage of waste from production processes or illegal dumping of slags [1]. The presence of Cr (VI) in soil and groundwater has also been linked to geogenic processes, namely, weathering of ultramafic and mafic rocks in various areas around the world [2–4].

The increasing availability of scientific studies has progressively drawn attention to in situ remediation technologies. These are innovative compared to the "Dig and Dump" (D&D) of unsaturated soil and to the "Pump and Treat" (P&T) of groundwater. They enable the risks of the movement of contaminated matrices to be limited and a reduction in the remediation times, above all for the groundwater. Technologies for in situ treatment of Cr (VI), including the injection of reducing substances and bioremediation processes, do seem to ensure better results in terms of efficiency, with generally lower costs [5]. Full-scale application of these technologies is continuously growing, especially in the United States, with results appearing to confirm what has been illustrated on a smaller scale [6].

There are no written "Decision Guides" available for hexavalent chromium to refer to in choosing potentially the most suitable remediation technology depending on the site-specific conditions. Some tips are found in documents, such as those drawn up by the US Environmental Protection Agency [7,8] or the Savannah River National Laboratory [9], in which scenarios for sites contaminated by inorganic pollutants are set out. The scenario of greatest interest, on which this manuscript concentrates, is that of soil and groundwater in oxidising conditions, where the chromium remains in hexavalent form if not properly treated. The purpose is to provide a support tool useful in the preliminary assessment of the remedial options to address further investigations on technologies with potential feasibility. In fact, due to the highly complex behaviour of inorganic pollutants in the environment and the numerous chemical species with which they can interact, a definite choice can only be made after site-specific tests.

2. Behaviour of Chromium in Soil and Groundwater

Chromium can have several oxidation states, but the most common forms in the soil are Cr (III) and Cr (VI) [10,11]. Cr (III) tends to form insoluble and low polluting compounds in water. Cr (VI) is generally present as hydrogen-chromate ion ($HCrO_4^-$) and chromate ion (CrO_4^{2-}) [2]; it has high mobility and high toxicity in a broad pH range [1,12,13], and is classified as a Class A carcinogen [14,15].

The state of oxidation and the chemical form of the chromium in the ground are jointly influenced by the pH and by the redox potential, as shown in the diagram of Pourbaix [16] (Figure 1a). The pH range of interest includes values between 5 (acidic) and 9 (alkaline), which can be considered the possible extremes for soil in natural conditions [17]; the potentials typically encountered in an aquifer are included in the range between −100 and +600 mV (vs. Standard Hydrogen Electrode—SHE) [18]. With reference to the area within the red box in Figure 1a, the prevalence of chemical species of Cr (VI) is located in the portion relating to the most basic pH and redox higher than +200 mV. However, the theoretical Pourbaix diagram of Cr had to be properly adjusted to site-specific conditions, taking into account groundwater and soil composition.

Through redox processes, chromium changes dynamically from one state of oxidation to another (Figure 1b). Reducing species, which serve as electron donors (e.g., organic substances, such as carbohydrates, proteins, and humic acids), facilitate the reduction process of Cr (VI) to Cr (III); humic acids also form complexes with Cr (III) [9].

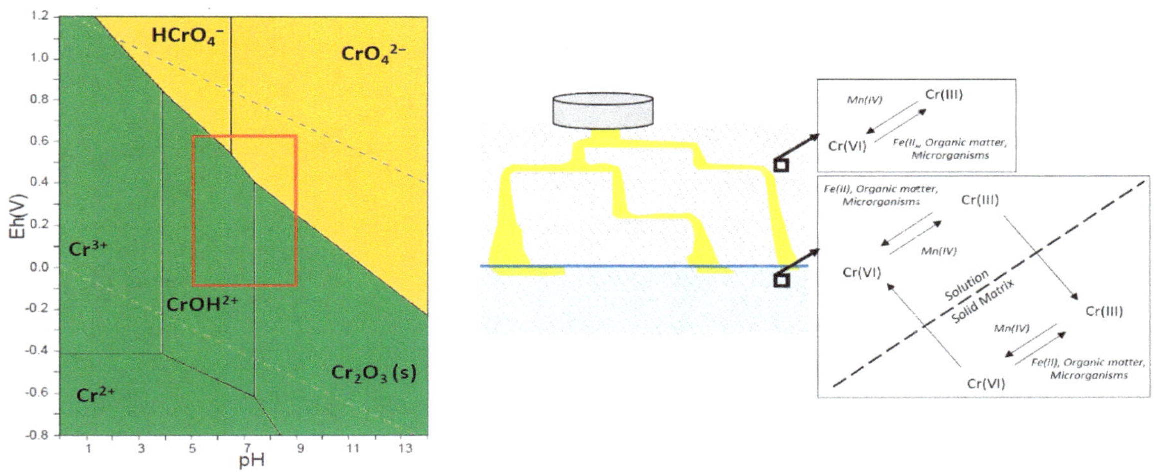

(a) (b)

Figure 1. (**a**) Diagram of Pourbaix (redox potential Eh vs. SHE) for the chromium (in yellow Cr (VI) species, in green Cr (III) species); the red rectangle encloses the area of natural environmental conditions; (**b**) mechanisms of action on the chemical species of the chromium (in yellow) in the subsoil (in grey the unsaturated zone, in pale blue the saturated zone).

Amongst the most widespread electron donors, Fe (II) assumes special importance [19]. In aerated soil, with high redox potential, the iron has a trivalent form. In asphyxial soil, with low redox potential, the Fe (II) ions in solution are plentiful, depending also on the chemical composition of the soil, and are prone to react with hexavalent chromium. At pH 5–6, the redox reaction is [20]:

$$3Fe^{2+} + HCrO_4^- + 3H_2O \rightarrow 3Fe(OH)_2^+ + CrOH^{2+} \quad (1)$$

At pH > 7, the reduction mechanism of the hexavalent chromium follows the reaction [21]:

$$3Fe^{2+} + CrO_4^{2-} + 4H_2O \rightarrow 3Fe^{3+} + Cr^{3+} + 8OH^- \quad (2)$$

The formation of Cr (III) and Fe (III) species result from reactions (1) and (2). Reacting with each other, or with further dissolved Fe (II), means they do not remain in solution, but are removed in the form of hydroxides.

Strong oxidising conditions, generated, for example, by the presence of Mn (IV) oxides, can boost the transformation of Cr (III) precipitates into chemical Cr (VI) species [22]. That said, the significant instances of contamination by Cr (VI) are essentially linked to soils/aquifers in oxidising conditions, with greater intrinsic permeability of 10^{-14} m^2 (coarser lithologies of fine silty sands); in fact, at a redox potential of around +500 mV (vs. SHE), the natural reduction of Cr (VI) to Cr (III) is widely disadvantaged [23]. Conditions of this type are typical of glacial/alluvial deposits with low organic substance and of fragmented rocks.

3. Technologies

In the last decade, numerous studies have been carried out, mainly based on laboratory scale and pilot tests, to assess the efficacy and sustainability of new technologies for the in-situ treatment of Cr (VI). Sustainability integrates many different, and sometimes competing, factors [24]; environmental, social, and economic factors must be considered and the final selected remediation plan will result in a balance of them [25].

In this chapter and in Table 1, the principal innovative technologies, which have reached full-scale application, are presented, subdivided according to the typology of mechanism used and potentially treatable zone. Some only apply in saturated or unsaturated zones; both unsaturated and saturated zones should also be further separated to take account of the fact that the full involvement of the contaminated matrix in the treatment is generally tied to the depth of the contamination from ground level (g.l.).

Table 1. Potential applicability (x: yes; -: no) of innovative technologies depending on the zone and maximum depth of soil to be treated.

Technology	Unsaturated 0–1 m	Unsaturated 1–10 m	Unsaturated > 10 m	Saturated < 10 m	Saturated 10–25 m	Saturated > 25 m
Chemical process with solutions or slurry	-	-	-	x	x	x
Chemical process with gaseous reagent	-	x	x	-	-	-
Indirect biological process	-	-	-	x	x	x
Biological process-Phytoremediation	x	-	-	-	-	-
Chemical-physical process-Electrokinetics	x	x	-	x	-	-
Chemical-physical process-Flushing	x	x	-	-	-	-

3.1. *Innovative Technologies for Cr (VI) Remediation*

3.1.1. Chemical Process

In general, applicable reagents for the chemical reduction of Cr (VI) in a saturated zone have an iron or sulphur based composition, and can act either directly or indirectly [26,27]. Amongst the iron-based species that act directly on the reduction of Cr (VI), the most commonly used is zerovalent iron in the form of nano-particles. Acid conditions facilitate Cr (VI) reduction with Fe (0) [28–30]. The calcium polysulphurs (CaS_4, CaS_5) are also in common use [31]; Chrysochoou et al. [32] have shown that, when polysulphurs are used, the reducing conditions remain in the soil for a long time; neutral or basic pH values have provided greater reducing capacities [33]. Sodium dithionite ($Na_2S_2O_4$) acts mainly indirectly, converting Fe (III) to Fe (II) [34–36]. This ion plays an active role in the reduction of the pollutant, according to reactions (1) and (2). Under acidic conditions, the process is favoured. Ascorbic acid, or vitamin C ($C_6H_8O_6$), like other organic acids, certainly represents a promising alternative, as it does not exhibit any toxic features. At pH ≤ 7, it reduces Cr (VI) efficiently, transforming itself into dehydroascorbic acid [37]. Bianco Prevot et al. [38] have, however, encountered high levels of Cr (VI) reduction in the environment with a pH up to 9.

The in situ reduction of Cr (VI) in an unsaturated zone by means of gaseous injections [39] is an approach which has been little developed so far and principally concerns the use of hydrogen sulphide (H_2S) diluted in air. The efficacy is limited to acid or more neutral environments; for pH > 7.5, a significant collapse in the efficiency of the process may occur [40]. The technology is especially suited to permeable soil, where the circulation of the gaseous reagent is enhanced. To promote the reduction of Cr (VI) to Cr (III) in unsaturated soil, there needs to be adequate moisture content in the soil or in the gas current injected [41].

3.1.2. Biological Process

With reference to biological processes in a saturated zone, the administration of carbonaceous substances aimed at supporting an indirect bacterial action can be assessed [38,42,43]. The mechanism is designed to create reducing conditions, with possible releases of Fe (II) from the solid phase of the soil. The quality of the treated soil is generally higher than that treated with chemicals [44]. The efficiency of Cr (VI) reduction through indirect biological processes tends to diminish as the concentration of the contaminant increases, because of the rise in toxicity [45]. This type of process is therefore not advised for environments with high concentrations of Cr (VI) and where there is a lack of iron. Chemical processes occasionally complement biological technologies, as, for example, in Němeček et al. [46], where there is the combined use of zerovalent iron and iron lactate. Many registered trademark reagents on the market incorporate the advantages of the two approaches.

For shallower unsaturated soil, phytoremediation treatment should be mentioned [47,48]. The process is certainly slow, but recent studies have shown how it can be accelerated, for example, by boosting the growth of the plants [49]. To define a phytoremediation treatment, it is crucial to evaluate whether the physical chemical features of the soil and the meteorological/climatic conditions of the site are compatible with the plant species to be used. In the case of phytoextraction, it is necessary also to take into account the periodic discharge of biomass, which contains chromium mainly in trivalent form [50]. The technology is not advised for environments with high concentrations of Cr (VI).

The injection of selected bacterial suspensions to reduce the Cr (VI) directly (chromium-reducing microrganisms) appears difficult to apply both in saturated and unsaturated zones. In fact, the in situ development and maintenance of these microrganisms is difficult [51–54]. As further proof, there is a lack of literature recording encouraging experiences.

3.1.3. Chemical-Physical Process

Electrokinetics is a remediation technique for both the saturated and the unsaturated soil zone, based on the application of a low constant electric field between two or more electrodes (positive/anode and negative/cathode) [55–57]. The field causes two important transport mechanisms, almost independent from soil intrinsic permeability: (a) Electromigration (transport of ionic species in bulk solution, according to the electric field direction); (b) electroosmosis (bulk pore fluid migration, including neutral or charged dissolved species, from the positive to negative electrode). The cathodic flow must be pumped out, whereas chromium can accumulate at the anode as a precipitate. Full-scale applications have been satisfactory, although significant limitations in the process were observed when

Cr (VI) concentration was low compared to non-target ion concentrations [58]. The equipment had to be optimised to reduce costs [59].

Soil flushing is used to treat unsaturated soil contaminated by leachable pollutants through suitable chemical agents [60,61]. In the case of Cr (VI), given its high solubility in water, the use of these latter may not be necessary [62]. Soil flushing for Cr (VI), in general, has satisfactory results in alkaline, permeable, and homogeneous soils whereas rocky formations or layers of less permeable soil help to create preferential flows that leave the untreated zone.

3.2. Full Scale Implementation

With reference to the above mentioned in situ remediation technologies for Cr (VI) requiring the injection of chemicals, they can be implemented in full scale using a range of approaches, depending on the technology, the zone to be treated, and the geological/hydrogeological features of the site. Reactive zones (RZ) and permeable reactive barriers (PRB) are discussed.

RZs involve the generation of a zone with suitable physical-chemical features in the portion of ground/aquifer to be treated by injecting appropriate reagents, without soil excavation. They are the most widely used, in view of their versatility and the possibility of reaching considerable depths [63]. The injections can take place upstream from the source of the contamination, next to it, and/or downstream. To intercept a plume in a saturated zone, lines of injection points can be used, perpendicular to the direction of the flow [64]. Within 10 m from g.l. and with lithology that is not excessively coarse (therefore, excluding gravel and pebbles), the reagents can be administered using "direct push" type systems, which require significant injection pressure to facilitate the distribution [65,66]. The zone of influence tends to diminish significantly as the viscosity of the fluid to be administered increases. The injection wells in a saturated zone can also reach very considerable depths, provided they use adequate pressure [67]. It is advisable to distribute the injection points along the vertical. It is also necessary to carry out pilot tests to evaluate the distribution of the chemicals in the subsoil [68].

In very heterogeneous soil, the creation of RZs can result in treatments of the contamination that are not homogeneous, with zones of finer lithology barely involved in the process [64]. The PRBs, which can only be used in saturated zones, consist of the substitution of the aquifer material with allocthonous material, through which the groundwater has to pass for the decontamination. This enables the achievement of a homogeneous treatment zone, regardless of the heterogeneity of the aquifer under examination. The PRBs are technically and economically sustainable if the depth of the installation does not exceed 25 m from g.l. [69–72]. Aquifers with high hydraulic conductivity are difficult to treat with this type of installation, because the reactive layer must have permeability of at least an order of magnitude greater than the aquifer to intercept effectively the contaminated plume. To increase the permeability of the barrier, it is necessary to increase its thickness so that the contaminant has an adequate hydraulic residence time in the RZs [73,74]. The use of reactive chemicals in the PRBs must consider possible problems of progressive fouling of the barrier. This is the case with the use of iron-based reagents, with the precipitation of the chemical species of Cr (III) and Fe (III) [75,76]. PRBs are well suited to the implementation of biological processes, in which case they are called "Biobarriers" [77].

4. Scenarios and Decision Support Tool

As already mentioned, in reducing environments (typically soil with low permeability, rich in organic substance), the redox conditions encourage the abundance of chemical species of Cr (III) rather than of Cr (VI); it is therefore rare to encounter significant contamination of Cr (VI) in these contexts [78]. The support tool proposed for the decisions, therefore, focuses on saturated or unsaturated permeable soil, in aerobic or, at most, anoxic conditions.

Table 2 shows the most influential factors on the choice of potentially applicable technologies: pH, concentration of Cr (VI), availability of iron in the soil, and homogeneity of the soil.

Table 2. Factors which influence the choice of technology.

Factor	Scenario	Value
Soil pH	Acid	5 ÷ 7
	Alkaline	7 ÷ 9
Cr (VI) concentration	Low	$< 10^2$ mg unsaturated soil; < 10 mg L^{-1} in aquifer
	High	$> 10^2$ mg kg^{-1} unsaturated soil; > 10 mg L^{-1} in aquifer
Fe concentration in soil	Low	< 1 g Fe kg^{-1}
	High	> 1 g Fe kg^{-1}
Soil homogeneity	Yes	Variation of hydraulic conductivity or intrinsic permeability within 2 orders of magnitude
	No	Variation of hydraulic conductivity or intrinsic permeability more than 2 orders of magnitude

It is useful to subdivide soils according to their pH. The use of some reactive chemicals, for example, is advised for acid or neutral environments, in view of the significant loss of efficiency for basic pHs, or vice-versa. Soil flushing for chromium is not suitable in acid soils because of the lower mobility of its chemical species.

High concentrations of hexavalent chromium can limit the feasibility of some technologies. Regarding biological treatments, the capacity of the microorganisms to survive at high concentrations of Cr (VI) (up to a few grams per liter in water) could be mediated, not just by enzymes and/or very specific transport proteins, but also by sub-cellular structures, which interact with the metals themselves [79]. Many microorganisms are able to grow and survive at high concentrations of Cr (VI), developing mechanisms of resistance and tolerance to the pollutant [45,48,80]. The use of selected inoculations, if able to remain in situ, would therefore not have limitations, even in contexts with a high level of contamination. Vice-versa, action in indirect biological treatments could be inhibited with dissolved contamination above 10 mg L^{-1} Cr (VI) or soil contamination above 10^2 mg kg^{-1} [81].

The presence of Fe (II) ions allows redox reactions, with the reduction of Cr (VI). Releases of iron in solution are possible at the moment in which changes in the redox conditions promote the development of reducing conditions. For the technologies that promote this change, the presence of iron in the solid matrix is a determining factor. In general, the matrix is considered to be at a high iron content if the concentration exceeds 0.1% in weight, or 1 mg kg^{-1} [82]; below this threshold, it becomes necessary to exclude technologies that use the iron as an essential element of the action mechanism. The releases of Fe (II) from the solid matrix must be sufficient to balance the quantity of Cr (VI) to be reduced; according to reactions (1) and (2), the indicative ratio in solution is Cr (VI): Fe (II) = 1:3 by weight [83].

Almost all natural soils are highly variable in their properties. The heterogeneity of the soil is linked to the presence of different lithologies [84]. The presence of heterogeneity limits the efficacy of the technologies, which envisage injection and dispersal of a reactive. As mentioned, the PRBs appear to overcome these problems. It is possible to quantify the homogeneity of the layer to be treated using parameters, such as the hydraulic conductivity or the intrinsic permeability of the layer to be treated [85].

Having taken the above into account, the support tool for the decisions is reflected in Table 3 where, for each innovative technology mentioned, the conditions of the factors of Table 2, which advise against/exclude its application, are specified; the zone of applicability of the different technologies reported in Table 1 is implied.

Table 3. Low applicability/inapplicability (shown by an "X") of the innovative technologies examined, according to the factors in Table 2; the zone of potential application shown in Table 1 is implied.

pH[1]	Cr (VI) Concentration[2]	Fe Concentration in Soil[3]	Soil Homogeneity	Fe (0), $C_6H_8O_6$, H_2S	Sodium Dithionite	Calcium Polysulphurs	Indirect Biological Process	Phytoremediation	Electrokinetics	Soil Flushing
A	L	L	Yes	-	X	X	X	-	X	X
A	L	L	No	X[4]	X	X	X	-	X	X
A	L	H	Yes	-	-	X	-	-	X	X
A	L	H	No	X[4]	X[4]	X	X[4]	-	X	X
A	H	L	Yes	-	X	X	X	X	-	X
A	H	L	No	X[4]	X	X	X	X	-	X
A	H	H	Yes	-	-	X	X	X	-	X
A	H	H	No	X[4]	X[4]	X	X	X	-	X
B	L	L	Yes	X	X	-	X	-	X	-
B	L	L	No	X	X	X[4]	X	-	X	X
B	L	H	Yes	X	X	-	-	-	X	
B	L	H	No	X	X	X[4]	X[4]	-	X	X
B	H	L	Yes	X	X	-	X	X	-	-
B	H	L	No	X	X	X[4]	X	X	-	X
B	H	H	Yes	X	X	-	X	X	-	-
B	H	H	No	X	X	X[4]	X	X	-	X

[1] A = Acid, B = Alkaline; [2] H = High; L = Low; [3] H = High; L = Low; [4] Not recommended/Excluded only in case of injections.

5. Discussion and Conclusions

The intention of this manuscript is to provide support to operators and decision makers that wish to undertake the remediation of a site more directed towards a concept of sustainability [86]. The innovative technologies considered aim at greater sustainability than traditional approaches persisting more for established practice than for real advantage.

Frequently, in sites where the contamination from Cr (VI) is mainly found in shallow unsaturated soil, the remediation is limited to the removal of the soil and the transfer to a recycling centre or landfill site. The method is very onerous, particularly when the costs of transport are taken into consideration. On the other hand, despite inapplicability under certain conditions (see Table 3), phytremediation, electrokinesis, and flushing can potentially be used. In relation to sustainability, time, and logistical limitations, phytoremediation is an interesting option from an economic and environmental standpoint, but cannot be used in sites with a high level of contamination, structures and/or land cover, and restricted remediation times.

Injections of reducing gases, electrokinesis, and flushing can potentially be used for unsaturated sub-surface soil; the latter two, however, are excluded for treating material more than 10 m from g.l. because the consequent technical-operational difficulties. Deep unsaturated ground (more than 10 m g.l.) therefore remains among the zones, with limited alternatives to treatment of the contamination by Cr (VI).

The contamination in a saturated zone can potentially be treated with all the innovative chemical and biological technologies mentioned and, in the case of depths within 10 m g.l., also with electrokinesis. In aquifers with low concentrations of Cr (VI), the indirect biological processes generally have lower costs, even if remediation times are usually longer than the one of purely chemical processes.

For all zones, among the technologies that are the most innovative and without significant site-specific limitations, electrokinesis is promising. Starting with this, there are also small-scale studies of remediation technologies underway, based on the application of low intensity electrical fields, for the reduction of chromium using electrochemical, biochemical, or bioelectrochemical processes.

From the perspective of full-scale implementation, the administration of chemical agents can be carried out using injections (in wells and/or with the direct push technique) and/or in PRBs. In groundwater within 25 m g.l., PRBs offer greater advantages in heterogeneous soils. However, there are significant implications in terms of the cost and time taken to excavate as well as the disposal of the material resulting from the installation of the work.

The final choice of the best remediation option in a site depends, in any case, on additional sustainability factors other than those considered in this manuscript, including the results of site-specific and laboratory tests. A lack of economic resources may lead in the direction of less onerous, but slower, technologies, just as the necessity of achieving quickly the remediation objectives for social purposes may lead to the exclusion of other technologies. Traditional treatment techniques should also not be excluded a priori from the assessment.

Author Contributions: Data curation: A.F.M., G.B. and L.P.; Supervision: S.S.; Writing—review & editing: E.S. and A.F.M.

Acknowledgments: The authors thank Marco Albertini for his technical support in developing the tool.

References

1. Guertin, J.; Jacobs, J.A.; Avakian, C.P. *Chromium (VI) Handbook*; CRC Press: Boca Raton, FL, USA, 2005.
2. Panagiotakis, I.; Dermatas, D.; Vatseris, C.; Chrysochoou, M.; Papassiopi, N.; Xenidis, A.; Vaxevanidou, K. Forensic investigation of a Chromium (VI) groundwater plume in Thiva, Greece. *J. Hazard. Mater.* **2015**, *281*, 27–34. [CrossRef] [PubMed]
3. Lilli, M.A.; Moraetis, D.; Nikolaidis, N.P.; Karatzas, G.P.; Kalogerakis, N. Characterization and mobility

of geogenic chromium in soils and riverbed sediments of asopos basin. *J. Hazard. Mater.* **2015**, *281*, 12–19. [CrossRef] [PubMed]
4. Chrysochoou, M.; Theologou, E.; Bompoti, N.; Dermatas, D.; Panagiotakis, I. Occurrence, origin and transformation processes of geogenic chromium in soils and sediments. *Curr. Pollut. Rep.* **2016**, *2*, 224–235. [CrossRef]
5. United States Environmental Protection Agency. *In Situ Treatment of Soil and Groundwater Contaminated with Chromium–Technical Resource Guide*; United States Environmental Protection Agency: Washington, DC, USA, 2000.
6. United States Environmental Protection Agency. *Introduction to In Situ Bioremediation of Groundwater–Technical Resource Guide*; United States Environmental Protection Agency, Office of Solid Waste and Emergency Response: Cincinnati, OH, USA, 2013.
7. Bekele, T.C.; Bucklin, K.; Burger, S.; Martinez, L.; Parnass, B.; Renzi, E.; Rodriguez, M.; Wong, P. *Investigation and Remediation of Plating Facilities–Technical Resource Guide*; United States Environmental Protection Agency, Department of Toxic Substances Control: Sacramento, CA, USA, 2011.
8. California Environmental Protection Agency. *Proven Technologies and Remedies Guidance–Remediation of Metals in Soil*; California Environmental Protection Agency, Department of Toxic Substances Control: Sacramento, CA, USA, 2008.
9. Savannah River National Laboratory. *The Scenarios Approach to Attenuation-Based Remedies for Inorganic and Radionuclide Contaminants*; Savannah River National Laboratory: Aiken, SC, USA, 2011.
10. Unceta, N.; Séby, F.; Malherbe, J.; Donard, O.F.X. Chromium speciation in solid matrices and regulation: A review. *Anal. Bioanal. Chem.* **2010**, *397*, 1097–1111. [CrossRef] [PubMed]
11. Dhal, B.; Thatoi, H.N.; Das, N.N.; Pandey, B.D. Chemical and microbial remediation of hexavalent chromium from contaminated soil and mining/metallurgical solid waste: A review. *J. Hazard. Mater.* **2013**, *250*, 272–291. [CrossRef] [PubMed]
12. United States Environmental Protection Agency. *Toxicological Review of Hexavalent Chromium*; United States Environmental Protection Agency: Washington, DC, USA, 1998.
13. US Department of Labor Occupational Safety and Health Administration. *Hexavalent Chromium*; US Department of Labor Occupational Safety and Health Administration: Washington, DC, USA, 2009.
14. Canadian Council of Ministers of the Environment. *Canadian Soil Quality Guidelines for the Protection of Environmental and Human Health—Chromium*; Canadian Council of Ministers of the Environment: Winnipeg, MB, Canada, 1999.
15. United States Department of Health and Human Services, Public Health Service, Agency for Toxic Substances and Disease Registry. Toxicological Profile for Chromium. 2012. Available online: https://www.atsdr.cdc.gov/toxprofiles/tp7.pdf (accessed on 15 July 2018).
16. Pradhan, D.; Sukla, L.B.; Sawyer, M.; Rahman, P.K.S.M. Recent bioreduction of hexavalent chromium in wastewater treatment: A review. *J. Ind. Eng. Chem.* **2017**, *55*, 1–20. [CrossRef]
17. US Department of Agriculture. Soil Science Division Staff, 2017. In *Soil Survey Manual—Handbook No. 18*; US Department of Agriculture: Washington, DC, USA, 2017.
18. Ohio Environmental Protection Agency. *Reduction-Oxidation (Redox) Control in Ohio's Ground Water Quality*; Ohio Environmental Protection Agency: Columbus, OH, USA, 2014.
19. Palmer, C.D.; Wittbrodt, P.R. Processes affecting the remediation of chromium contaminated sites. *Environ. Health Perspect.* **1991**, *92*, 25–40. [CrossRef] [PubMed]
20. Buerge, I.J.; Hug, S.J. Kinetics and pH dependence of Chromium (VI) reduction by Iron (II). *Environ. Sci. Technol.* **1997**, *3*, 1–7. [CrossRef]
21. He, Y. Chromate Reduction and Immobilization under High pH and High Ionic Strength Conditions. Ph.D. Thesis, The Ohio State University, Columbus, OH, USA, 2003.
22. Palmer, C.D.; Puls, R.W. *Natural Attenuation of Hexavalent Chromium in Groundwater and Soils*; USEPA, Office of Emergency and Remedial Response: Washington, DC, USA, 1994.
23. Sedlazeck, K.P.; Hollen, D.; Muller, P.; Mischitz, R.; Giere, R. Mineralogical and geochemical characterization of a chromium contamination in an aquifer—A combined analytical and modeling approach. *Appl. Geochem.* **2017**, *87*, 44–56. [CrossRef]
24. Network for Industrially Contaminated Land in Europe. How to Implement Sustainable Remediation in a

Contaminated Land Management Project? *2012*. Available online: http://www.nicole.org/uploadedfiles/wg-sustainableremediation-finalreport.pdf (accessed on 15 July 2018).

25. Department of Toxic Substances Control, California Environmental Protection Agency. *Interim Advisory for Green Remediation*; Department of Toxic Substances Control: Sacramento, CA, USA, 2009.
26. Beretta, G.P.; Pellegrini, R. *Linee Guida per la Verifica del Trattamento Chimico In Situ dei Terreni e Delle Acque Sotterranee*; Technical Report: Provincia di Milano, Italy, 2006. (In Italian)
27. Ludwig, R.D.; Su, C.; Lee, T.R.; Wilkin, R.T.; Acree, S.D.; Ross, R.R.; Keeley, A. In situ chemical reduction of Cr (VI) in groundwater using a combination of ferrous sulfate and sodium dithionite: A field investigation. *Environ. Sci. Technol.* **2007**, *41*, 5299–5305. [CrossRef] [PubMed]
28. Němeček, J.; Lhotský, O.; Cajthaml, T. Nanoscale zero-valent iron application for in situ reduction of hexavalent chromium and its effects on indigenous microorganism populations. *Sci. Total. Environ.* **2014**, *485*, 739–747. [CrossRef] [PubMed]
29. Li, X.; Lei, L.; Yang, C.; Zhang, W. Reduction of Cr (VI) by nanoscale zero valent Iron (nZVI): The reaction kinetics. In Proceedings of the 4th International Conference on Bioinformatics and Biomedical Engineering (IWBBIO 2016), Granada, Spain, 20–22 April 2016.
30. Yoon, I.H.; Kim, K.W. Effect of pH and Dissolved Oxygen on the Cr (VI) Removal by Zero-Valent Iron. Available online: https://www.researchgate.net/publication/242201190_Effect_of_pH_and_Dissolved_Oxygen_on_the_CrVI_Removal_by_Zero-Valent_Iron (accessed on 15 July 2018).
31. Storch, P.; Messer, A.; Palmer, D.; Pyrih, R. Pilot test for in situ geochemical fixation of chromium (VI) using calcium polysulfide. In *Proceedings of the Third International Conference on Remediation of Chlorinated and Recalcitrant Compounds: pts 1a–2b*; Battelle Press: San Diego, CA, USA, 2002.
32. Chrysochoou, M.; Ferreira, D.R.; Johnston, C.P. Calcium polysulfide treatment of Cr (VI)-contaminated soil. *J. Hazard. Mater.* **2010**, *179*, 650–657. [CrossRef] [PubMed]
33. Chrysochoou, M.; Ting, A. A kinetic study of Cr (VI) reduction by calcium polysulfide. *Sci. Total. Environ.* **2011**, *409*, 4072–4077. [CrossRef] [PubMed]
34. Xie, Y.; Cwiertny, D.M. Use of dithionite to extend the reactive lifetime of nanoscale zero-valent iron treatment systems. *Environ. Sci. Technol.* **2010**, *44*, 8649–8655. [CrossRef] [PubMed]
35. Khan, F.A.; Plus, R.W. In situ abiotic detoxification and immobilization of hexavalent chromium. *Ground Water Monit. Remediat.* **2003**, *23*, 77–84. [CrossRef]
36. Beukes, J.P; Pienaar, J.; J; Lachmann, G; Giesekke, E.W. The reduction of hexavalent chromium by sulphite in wastewater. *Water 1999, 25, 363*. Available online: http://www.wrc.org.za (accessed on 15 July 2018).
37. Xu, X.; Li, H.; Li, X.; Gu, J. Reduction of hexavalent chromium by ascorbic acid in aqueous solutions. *Chemosphere* **2004**, *57*, 609–613. [CrossRef] [PubMed]
38. Bianco Prevot, A.; Ginepro, M.; Peracaciolo, E.; Zelano, V.; De Luca, D.A. Chemical vs bio-mediated reduction of hexavalent chromium. An in-vitro study for soil and deep waters remediation. *Geoderma* **2018**, *312*, 17–23. [CrossRef]
39. Thornton, E.C.; Gilmore, T.J.; Olsen, K.B.; Giblin, J.T.; Phelan, J.M. Treatment of a chromate-contaminated soil site by in situ gaseous reduction. *Ground Water Monit. Remediat.* **2007**, *27*, 56–64. [CrossRef]
40. Kim, C.; Zhou, Q.; Deng, B.; Thorton, E.C.; Xu, H. Chromium (VI) reduction by hydrogen sulfide in aqueous media: stoichiometry and kinetics. *Environ. Sci. Technol.* **2001**, *35*, 2219–2225. [CrossRef] [PubMed]
41. Hua, B.; Deng, B. Influences of water vapor on Cr (VI) reduction by gaseous hydrogen sulfide. *Environ. Sci. Technol.* **2003**, *37*, 4771–4777. [CrossRef] [PubMed]
42. Ibbini, J.; Santharam, S.; Davis, L.C.; Erickson, L.E. Laboratory and Field Scale Bioremediation of Tetrachloroethene (PCE) Contaminated Groundwater. *Jordan J. Mech. Ind. Eng.* **2010**, *4*, 35–44.
43. Somenahally, A.C.; Mosher, J.J.; Yuan, T.; Podar, M.; Phelps, T.J.; Brown, S.D.; Yang, Z.K.; Hazen, T.C.; Arkin, A.P.; Palumbo, A.V.; et al. Hexavalent Chromium Reduction under Fermentative Conditions with Lactate Stimulated Native Microbial Communities. 2013. Available online: www.plosone.org (accessed on 15 July 2018).
44. Liao, Y.; Min, X.; Yang, Z.; Chai, L.; Zhang, S.; Wang, Y. Physicochemical and biological quality of soil in hexavalent chromium-contaminated soils as affected by chemical and microbial remediation. *Environ. Sci. Pollut. Res.* **2014**, *21*, 379–388. [CrossRef] [PubMed]
45. Viti, C.; Mini, A.; Ranalli, G.; Lustrato, G.; Giovannetti, L. Response of microbial communities to different doses of chromate in soil microcosms. *Appl. Soil Ecol.* **2006**, *34*, 125–139. [CrossRef]

46. Liu, J.; Duan, C.; Zhang, X.; Zhu, Y.; Lu, X. Potential of leersia hexandra swartz for phytoextraction of Cr from soil. *J. Hazard. Mater.* **2011**, *188*, 85–91. [CrossRef] [PubMed]
47. Nayak, A.K.; Panda, S.S.; Basu, A.; Dhal, N.K. Enhancement of toxic Cr (VI), Fe, and other heavy metals phytoremediation by the synergistic combination of native *Bacillus cereus* strain and *Vetiveria zizanioides* L. *Int. J. Phytoremediat.* **2018**, *20*, 682–691. [CrossRef] [PubMed]
48. Ashraf, M.A.; Hussain, I.; Rasheed, R.; Iqbal, M.; Riaz, M.; Arif, M. Advances in microbe-assisted reclamation of heavy metal contaminated soils over the last decade: A review. *J. Environ. Manag.* **2017**, *198*, 132–143. [CrossRef] [PubMed]
49. Ranieri, E.; Petros, G. Effects of plants for reduction and removal of hexavalent chromium from a contaminated soil. *Water Air Soil Pollut.* **2014**, 225. [CrossRef]
50. Chai, L.; Huang, S.; Yang, Z.; Peng, B.; Huang, Y.; Chen, Y. Cr (VI) remediation by indigenous bacteria in soils contaminated by chromium-containing slag. *J. Hazard. Mater.* **2009**, *167*, 516–522. [CrossRef] [PubMed]
51. Höhener, P.; Ponsin, V. In situ vadose zone bioremediation. *Curr. Opin. Biotechnol.* **2014**, *27*, 1–7. [CrossRef] [PubMed]
52. Thatoi, H.; Das, S.; Mishra, J.; Rath, B.P.; Das, N. Bacterial chromate reductase, a potential enzyme for bioremediation of hexavalent chromium: A review. *J. Environ. Manag.* **2014**, *146*, 383–399. [CrossRef] [PubMed]
53. Qu, M.; Chen, J.; Huang, Q.; Chen, J.; Xu, Y.; Luo, J.; Wang, K.; Guo, W.; Zheng, Y. Bioremediation of hexavalent chromium contaminated soil by a bioleaching system with weak magnetic fields. *Int. Biodeterior. Biodegrad.* **2018**, *128*, 41–47. [CrossRef]
54. Něměček, J.; Pokorný, P.; Lhotský, O.; Knytl, V.; Najmanová, P.; Steinová, J.; Černík, M.; Filipová, A.; Filip, J.; Cajthaml, T. Combined nano-biotechnology for in-situ remediation of mixed contamination of groundwater by hexavalent chromium and chlorinated solvents. *Sci. Total Environ.* **2016**, *563*, 822–834. [CrossRef] [PubMed]
55. Zhang, P.; Jin, C.; Zhao, Z.; Tian, G. 2D crossed electric field for electrokinetic remediation of chromium contaminated soil. *J. Hazard. Mater.* **2010**, *177*, 1126–1133. [CrossRef] [PubMed]
56. Wei, X.; Guo, S.; Wu, B.; Li, F.; Li, G. Effects of reducing agent and approaching anodes on chromium removal in electrokinetic soil remediation. *Front. Environ. Sci. Eng.* **2016**, *10*, 253–261. [CrossRef]
57. Yan, Y.; Xue, F.; Muhammad, F.; Yu, L.; Xu, F.; Jiao, B.; Li, D. Application of iron-loaded activated carbon electrodes for electrokinetic remediation of chromium-contaminated soil in a three-dimensional electrode system. *Sci. Rep.* **2018**, *8*, 1–11. [CrossRef] [PubMed]
58. Contaminated Land: Applications in Real Environments. Electrokinetic Ferric Ion Remediation and Stabilisation (FIRS) of Hexavalent Chromium Contaminated Soils: An Ex Situ Field Scale Demonstration. Available online: http://www.claire.co.uk (accessed on 15 July 2018).
59. Vocciante, M.; Caretta, A.; Bua, L.; Bagatin, R.; Ferro, S. Enhancements in ElectroKinetic Remediation Technology: Environmental assessment in comparison with other configurations and consolidated solutions. *Chem. Eng. J.* **2016**, *289*, 123–134. [CrossRef]
60. Tang, S.; Yin, K.; Lo, I.M. Column study of Cr (VI) removal by cationic hydrogel for in-situ remediation of contaminated groundwater and soil. *J. Contam. Hydrol.* **2011**, *125*, 39–46. [CrossRef] [PubMed]
61. Yan, D.Y.; Lo, I.M. Pyrophosphate coupling with chelant-enhanced soil flushing of field contaminated soils for heavy metal extraction. *J. Hazard. Mater.* **2012**, *199*, 51–57. [CrossRef] [PubMed]
62. Yaqiao, S.; Lei, D. The influence mechanism of ash-flushing water from power plant on groundwater environment. In Proceedings of the 4th International Conference on Bioinformatics and Biomedical Engineering (IWBBIO 2016), Granada, Spain, 2016.
63. Fruchter, J. In situ treatment of chromium-contaminated groundwater. *Environ. Sci. Technol.* **2002**, *36*, 464A–472A. [CrossRef] [PubMed]
64. Jeyasingh, J.; Somasundaram, V.; Philip, L.; Bhallamudi, S.M. Pilot scale studies on the remediation of chromium contaminated aquifer using bio-barrier and reactive zone technologies. *Chem. Eng. J.* **2011**, *167*, 206–214. [CrossRef]
65. Ohio Environmental Protection Agency. Use of direct push technologies for soil and ground water sampling. In *Technical Resource Guide for Ground Water Investigations*; Ohio Environmental Protection Agency: Columbus, OH, USA, 2005.

66. ASTM International. *Standard Guide for Direct-Push Groundwater Sampling for Environmental Site Characterization*; ASTM International: West Conshohocken, PA, USA, 2012.
67. United States Environmental Protection Agency. *Injection Wells: An Introduction to Their Use, Operation, and Regulation*; United States Environmental Protection Agency, Office of Drinking Water: Washington, DC, USA, 1989.
68. ASTM International. *Standard Guide for Development of Groundwater Monitoring Wells in Granular Aquifers*; ASTM International: West Conshohocken, PA, USA, 2013.
69. United States Environmental Protection Agency. *In Situ Permeable Reactive Barrier for Contaminated Groundwater at the U.S. Coast Guard Support Center Elisabeth City, North Carolina*; United States Environmental Protection Agency: Washington, DC, USA, 1988.
70. Sethi, R.; Day, S.; Di Molfetta, A. Clamshell vs. Backhoe excavation of permeable reactive barriers. *Am. J. Environ. Sci.* **2011**, *7*, 463–467. [CrossRef]
71. Obiri-Nyarko, F.; Grajales-Mesa, S.J.; Malina, G. An overview of permeable reactive barriers for in situ sustainable groundwater remediation. *Chemosphere* **2014**, *111*, 243–259. [CrossRef] [PubMed]
72. Faisal, A.A.; Rawaa, J.M. Performance of zero-valent iron barrier through the migration of lead-contaminated groundwater. *Assoc. Arab Univ. J. Eng. Sci.* **2018**, *25*, 132–144.
73. Scherer, M.M.; Richter, S.; Valentine, R.L.; Alvarez, P.J. Chemistry and microbiology of permeable reactive barriers for In Situ groundwater clean up. *Crit. Rev. Microbiol.* **2000**, *30*, 363–411. [CrossRef]
74. United States Environmental Protection Agency. A Citizen's guide to Permeable Reactive Barriers. 2012. Available online: https://clu-in.org/download/Citizens/a_citizens_guide_to_permeable_reactive_barriers.pdf (accessed on 15 July 2018).
75. Fuller, S.J.; Stewart, D.I; Ian, T.B. Chromate reduction in highly alkaline groundwater by zerovalent iron: implications for its use in a permeable reactive barrier. *Ind. Eng. Chem. Res.* **2013**, *52*, 4704–4714. [CrossRef]
76. Liu, Y.; Mou, H.; Chen, L.; Mirza, Z.A.; Liu, L. Cr (VI)-contaminated groundwater remediation with simulated permeable reactive barrier (PRB) filled with natural pyrite as reactive material: Environmental factors and effectiveness. *J. Hazard. Mater.* **2015**, *298*, 83–90. [CrossRef] [PubMed]
77. Careghini, A.; Saponaro, S.; Sezenna, E. Biobarriers for groundwater treatment: A review. *Water Sci. Technol.* **2013**, *67*, 453–468. [CrossRef] [PubMed]
78. Christoph, W.; Eggenberger, U.; Kurz, D.; Zink, S.; Mäder, U. A chromate-contaminated site in southern Switzerland—Part 1: Site characterization and the use of Cr isotopes to delineate fate and transport. *Appl. Geochem.* **2012**, *27*, 644–654.
79. Baldi, F.; Barbieri, P. Microbiologia Ambientale ed Elementi di Ecologia Microbica. 2008. Available online: https://iris.unive.it/handle/10278/22405?mode=full.156#.W62VsbgRVPY (accessed on 15 July 2018).
80. Kamaludden, S.P.B.; Arunkumar, K.R.; Avudainayagam, S.; Ramasamy, K. Bioremediation of chromium contaminated environments. *Indian J. Exp. Biol.* **2003**, *41*, 972–985.
81. Hassan, Z.; Ali, S.; Farid, M.; Rizwan, M.; Shahid, J. Effect of Chromium (Cr) on the Microbial and Enzymatic Activities in the Soil: A Review. 2017. Available online: https://www.researchgate.net/publication/317167651_Effect_of_Chromium_Cr_on_the_Microbial_and_Enzymatic_Activities_in_the_Soil_A_Review (accessed on 15 July 2018).
82. Dresel, P.E.; Wellman, D.M.; Cantrell, K.J.; Truex, M.J. Review: Technical and policy challenges in deep vadose zone remediation of metals and radionuclides. *Environ. Sci. Technol.* **2011**, *45*, 4207–4216. [CrossRef] [PubMed]
83. Mukhopadhyay, B.; Sundquist, J.; Schmitz, R.J. Removal of Cr (VI) from Cr-contaminated groundwater through electrochemical addition of Fe (II). *J. Environ. Manag.* **2007**, *82*, 66–76. [CrossRef] [PubMed]
84. Elkateb, T.; Chalaturnyk, R.; Robertson, P.K. An overview of soil heterogeneity: Quantification and implications on geotechnical field problems. *Can. Geotech. J.* **2003**, *40*, 1–15. [CrossRef]
85. Uzielli, M.; Simonini, P.; Cola, S. Statistical identification of homogeneous soil units for Venice lagoon soils. In Proceedings of the 3rd International Conference on Site Characterization, Taipei, Taiwan, 1–4 April 2008.
86. European Achievements in Soil Remediation and Brownfield Redevelopment. A Report of the European Information and Observation Network's National Reference Centers for Soil. 2017. Available online: http://publications.jrc.ec.europa.eu/repository/bitstream/JRC102681/kj0217891enn.pdf (accessed on 15 July 2018).

8

Bacterial Productivity in a Ferrocyanide-Contaminated Aquifer at a Nuclear Waste Site

Andrew Plymale [1,*], Jacqueline Wells [1], Emily Graham [2], Odeta Qafoku [3], Shelby Brooks [1] and Brady Lee [1]

1 Energy and Environment Directorate, Pacific Northwest National Lab (PNNL), Richard, WA 99354, USA; jacqueline.wells@pnnl.gov (J.W.); shelby.brooks@pnnl.gov (S.B.); Brady.Lee@pnnl.gov (B.L.)
2 Earth and Biological Sciences Directorate, Pacific Northwest National Lab (PNNL), Richard, WA 99354, USA; emily.graham@pnnl.gov
3 Physical and Computational Sciences Directorate, Pacific Northwest National Lab (PNNL); Richard, WA 99354, USA; odeta.qafoku@pnnl.gov
* Correspondence: plymale@pnnl.gov

Abstract: This study examined potential microbial impacts of cyanide contamination in an aquifer affected by ferrocyanide disposal from nuclear waste processing at the US Department of Energy's Hanford Site in south-eastern Washington State (USA). We examined bacterial productivity and microbial cell density in groundwater (GW) from wells with varying levels of recent and historical total cyanide concentrations. We used tritiated leucine (^{3}H-Leu) uptake as a proxy for heterotrophic, aerobic bacterial productivity in the GW, and we measured cell density via nucleic acid staining followed by epifluorescence microscopy. Bacterial productivity varied widely, both among wells that had high historical and recent total cyanide (CN^-) concentrations and among wells that had low total CN^- values. Standing microbial biomass varied less, and was generally greater than that observed in a similar study of uranium-contaminated hyporheic-zone groundwater at the Hanford Site. Our results showed no correlation between ^{3}H-Leu uptake and recent or historical cyanide concentrations in the wells, consistent with what is known about cyanide toxicity with respect to iron speciation. However, additional sampling of the CN^- affected groundwater, both in space and time, would be needed to confirm that the CN^- contamination is not affecting the GW biota.

Keywords: cyanide; ferrocyanide; bacteria; toxicity; groundwater; contamination; Hanford; nuclear waste; tritiated leucine; aquifer

---

1. Introduction

At the Hanford Site in semi-arid south-eastern Washington State, the US Department of Energy (DOE) and its contractors processed enriched uranium (^{235}U) into weapons-grade plutonium (^{239}Pu) from the 1940s through the 1980s [1]. The chemical extraction of Pu from U, as well as further chemical and physical processing of the plutonium fuel, was carried out in the 200 Area "Central Plateau" of Hanford, which lies south and west of the Columbia River, in a Pleistocene paleochannel (Figure S1). The groundwater (GW) in the 200 Area is an unconfined aquifer residing in Pliocene alluvial, lacustrine, and paleosol sediments, which lie below Pleistocene flood-deposited sands and gravels (Figure S2). The Central Plateau is underlain by Miocene basalt flows (Figure S2). The 200 Area aquifer is contaminated with a variety of organic and inorganic constituents, including radionuclides (Figure S1a). These contaminants have reached the aquifer after infiltration through the vadose zone from leaking waste tanks, unlined waste trenches, and other sources. Annual precipitation at the Hanford Site averages 177 mm [2]. Currently, flux to groundwater is controlled by natural recharge, which is expected to vary from near zero in undisturbed areas with native shrub vegetation, to nearly 100 mm/year for non-vegetated gravel-covered surfaces [3]. In the 200 Area unconfined aquifer, the predominant flow direction during the active liquid disposal period (1948–1974) was to the north-west. However, in more recent times, the flow direction appears to have reversed, and is now to the south-east, owing to lowering of the regional water table following cessation of active discharges [4].

In the south-eastern portion of the 200 Area, the BP-5 Operable Unit (OU) is contaminated with high levels of nitrate (NO_3^-), technetium-99 (^{99}Tc), tritium, uranium, iodine-129, and cyanide (CN^-) [5] (Figure S3). The BP-5 cyanide contamination has reached GW from a 1950s-era disposal trench [6]. Total cyanide concentrations in BP-5 GW have been detected as high ~1600 μg/L (parts per billion, ppb) [5], or ~8 times the US drinking water maximum contaminant level (200 μg/L total cyanide [7]). In addition, Washington State places a drinking water limit of 4.2 μg/L for free cyanide, since the free form (HCN + CN^-) is most toxic [8]. Groundwater from the OU currently is being extracted for physical, chemical, and biological processing at a nearby pump and treatment (P&T) facility, and the GW CN^- contamination is of particular concern to the operators of the 200 Area P&T facility. As a potential microbial toxin, CN^- has the potential to interfere with biological treatment processes at the plant, which primarily are anaerobic fluidized bed reactors for removing NO_3^-, ^{99}Tc-pertechntate ($^{99}TcO_4^-$), and other oxyanions. In addition, the effluent from the P&T plant is re-injected into the 200 Area vadose zone, so effluent CN^- concentrations are a concern [5]. Finally, bacterial CN^- toxicity may have the potential to inhibit microbially facilitated in situ contaminant attenuation and degradation, under either bioremediation or monitored natural attenuation scenarios.

Although results from pure culture, enrichment culture, and bioreactor studies indicate the potential for cyanide toxicity to bacteria [9–13], CN^- toxicity to native GW populations of bacteria has not been examined. In general, there are few studies of the toxicity of GW contaminants to in situ populations of bacteria under relevant conditions (one exception being Konopka et al. [14], for the case of uranium). In the case of cyanide, toxicity is determined by the proportion of free and complexed cyanide. For the latter, toxicity is determined by the stability of the metal–CN^- complex, with the strongest metal complexes showing the least toxicity with respect to CN^- [15]. When originally disposed to the 200 Area vadose zone, the cyanide was in the form of a ferrous iron [Fe(II)] complex, ferrocyanide [($Fe(CN)_6^{4-}$)], which was used to precipitate cesium-137 (^{137}Cs) from aqueous waste, with the ^{137}Cs-laden sludge being transferred to waste tanks, and the aqueous supernatant disposed of in unlined trenches [6]. Ferrocyanide is a highly stable complex, being considered "strong acid dissociable" (SAD), meaning that the complex will not appreciably dissociate above pH 2 [8]. In addition, both ferric [Fe(III)] cyanide and mixed iron-valence, ferric–ferro–cyanide complexes also are SAD. However, little is known with certainty about the current speciation of CN^- in the BP-5 operable unit, some 60 years after disposition.

This study sought to assess the GW microbiology of the 200-BP-5 operable unit in relation to cyanide (CN^-) contamination and other geochemical factors. Groundwater samples were collected from eight wells in the operable unit, and bacterial productivity in the GW samples was assessed via the tritiated leucine (^{3}H-Leu) incorporation assay [14,16–20]. In bacterial protein, leucine is present at a relatively constant level (~7%), but heterotrophic bacteria are auxotrophic for this amino acid, and so must acquire it from their environment. Thus, leucine uptake provides a reliable proxy of bacterial growth, and tritium labelling provides a high level of sensitivity [20]. Developed for marine and freshwater environments and widely used in studying these systems, the ^{3}H-Leu method recently has been adapted to GW environments [14,18,19]. We also assessed standing microbial biomass in the GW samples by total microscopic cell counts (nucleic-acid staining in conjunction with epifluorescence microscopy). Our biotic results were examined in relation to geochemical and hydrogeological factors, both from historical data and from results gathered as part of the current project.

2. Materials and Methods

2.1. Site Description and Sampling Procedure

Groundwater samples were collected in June and July of 2017 from 8 wells in the BP-5 Operable Unit of the Hanford Site 200 Area (Table 1, Figure S4). Wells were purged at a rate of 7.57 L/min. Samples were collected from within the area of the highest current total cyanide concentrations and from outside the current CN^- plume, from older monitoring wells, for comparison (Table 2, Figure 1). The latter included wells that had historical maximum total CN^- concentration that were either relatively high or low (Table 2, Figure 1), to account for possible long-term biotic effects. Within the plume, GW was collected from a recently drilled monitoring well (well #2, E33-360, see Table S1) and from a GW extraction well for the 200 West Area pump-and-treatment facility (well #1, E33-268, see Table S1). Groundwater was collected in 1-L aliquots in 1-L polypropylene bottles and was stored at 4 °C until use. To confirm that the tritiated leucine incorporation method would give appropriate sensitivity and response, the method (see below) was first tested on a Columbia River water sample taken downstream from the Hanford Site and on an unfiltered and untreated domestic well water sample from an aquifer near the site.

Table 1. Selected information on the wells sampled for this study (for more detailed information, see Table S1). Well #1 (E33-268) is an extraction well. Wells #2–#8 are monitoring wells.

Well #	Depth to Ground-Water when Sampled (ft bgs) [1]	Date of Well Construction (year) [1]	Elevation (m) [1]	Depth (ft) [1]	Screened Interval(s) (ft bgs) [1]
#1 (E33-268)	-	2012	198.0	263.5	241.9–252.4
#2 (E33-360)	-	2014	199.7	272.8	251.8–271.7
#3 (E28-27)	284.86	1987	208.5	301.5	269.8–289.8,
-	-	-	-	-	291.2–301.5
#4 (E32-9)	247.52	1991	197.1	254.6	-
#5 (E32-3)	281.01	1987	207.3	304.0	266.2–286.2,
-	-	-	-	-	291–301
#6 (E32-7)	262.54	1991	201.7	273.8	245.6–266.3
#7 (E32-8)	249.80	1990	197.8	256.7	234.7–255.0
#8 (E33-34)	237.71	1990	194.1	240.0	219.0–239.3

Notes: [1] Based on data from the Hanford Environmental Information System (HEIS) database, publicly accessible through the Environmental Dashboard Application (https://ehs.hanford.gov/EDA) and PHOENIX (http://phoenix.pnnl.gov). [2] Not applicable: casing 0–254 ft.

Table 2. Tritiated leucine uptake and microbial cell density observed in the wells examined, along with recent Eh values, recent total cyanide values, and historical maximum cyanide. Values for most recent pH ranged from 7.6 to 8.1 (see Table S1).

Well #	% ^{3}H-Leu Uptake [1]	Cells/mL [2]	Most Recent Eh (mV) [3]	Most Recent Total CN^- (μg/L) [3,4]	Historical Maximum Total CN^- (μg/L) [3]
#1 (E33-268)	8.42	$6.6 \times 10^{+5}$	+16	297	492
#2 (E33-360)	0.0110	$5.8 \times 10^{+5}$	+304	190	190
#3 (E28-27)	2.92	$6.6 \times 10^{+5}$	+8	4	12
#4 (E32-9)	2.33	$1.1 \times 10^{+6}$	+382	22	91
#5 (E32-3)	3.60	$2.3 \times 10^{+5}$	+202	4	9
#6 (E32-7)	1.95	$2.5 \times 10^{+5}$	+233	4	20
#7 (E32-8)	1.10	$3.1 \times 10^{+5}$	+169	40	40
#8 (E33-34)	0.0256	$1.4 \times 10^{+5}$	+369	100	558

Notes: [1] Based on an 8 h time point. [2] As determined by microscopic counting. [3] Based on data from the Hanford Environmental Information System (HEIS) database, publicly accessible through the Environmental Dashboard Application and PHOENIX. [4] Data closest in time to the date of groundwater sampling for the present study.

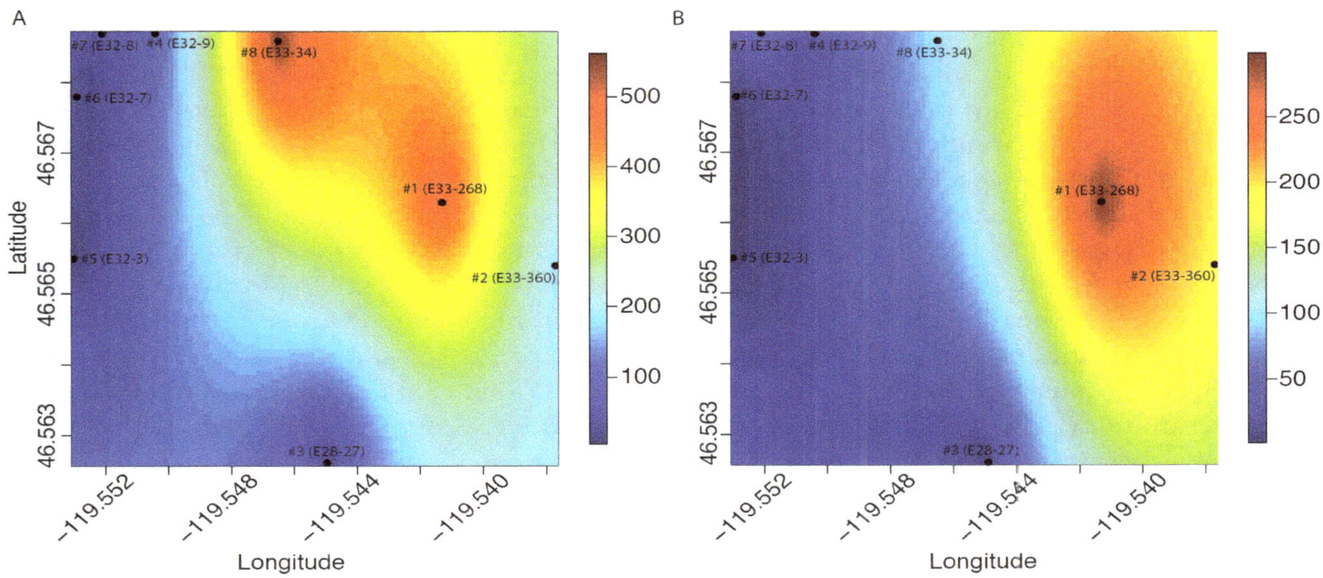

Figure 1. Historical maximum (**A**) and recent (**B**) [CN^-] values in spatial relation to the wells examined for this study, as interpolated by kriging.

2.2. Tritiated Leucine Incorporation Assay

Bacterial heterotrophic activity [20] was determined under aerobic conditions from ^{3}H-Leu incorporation using a method in which particulate cell material is collected by microcentrifugation [16,17], as implemented by Konopka et al. [14]. For activity assays, 1.45 mL of GW was added to a 1.7 mL Eppendorf plastic microcentrifuge tube, which was then amended with 14.5 µL of a stock of tritiated leucine to start the assay. The stock was amended with unlabeled L-leucine such that each sample would have a final total leucine (^{1}H-leucine + ^{3}H-leucine) concentration of 500 nM, the leucine concentration that Konopka et al. [14] found to produce maximum incorporation rates in bacteria from the Hanford Site 300 Area GW. The stock contained sufficient tritiated leucine to provide 2.5 µCi of ^{3}H-leucine (4,5-^{3}H(N), 56.2 Ci mmol-1, Perkin-Elmer) in each sample. Samples were incubated at ~20 °C without shaking, and activity was terminated by the addition of 0.075 mL of 100% trichloroacetic acid (TCA). For time-zero (T0) killed controls, 0.075 mL of TCA was added before adding the tritiated leucine stock. All treatments and time points were carried out in duplicate. Samples were stored at 4 °C in the dark until being extracted for tritium-labelled protein.

To extract the ^{3}H-labelled protein, the particulate material was collected by centrifugation (10,000 rcf for 10 min. at 4 °C) and then washed once (via centrifugation, as above) with an equal volume of cold (4 °C) 5% TCA. The labelled protein was precipitated with an equal volume of ice-cold 100% ethanol, and then collected by centrifugation as above, and finally dissolved in 0.2 mL of 1 N NaOH at 90 °C for 1 h. After cooling to room temperature, 1.25 mL of liquid scintillation cocktail (Hionic-Fluor, Perkin-Elmer) was added directly to the microcentrifuge tube, and the sample was evenly distributed in the cocktail by vortex mixing. The samples were then left at room temperature in the dark overnight to lower chemical luminescence. After removing the tab and hinge of the microcentrifuge tube with household toenail clippers, each vial was then placed in a separate 20 mL glass liquid scintillation vial, and the samples were then placed in a chilled (4 °C) liquid scintillation analyzer (Perkin-Elmer Tri-Carb 3180 TR/SL) and allowed to chill in the dark for at least 1 h. Each sample was then counted for 10 min. (with a 1 min. pre-count delay), using the spectral index of the sample (SIS) as the quench indicator, since previous work [14] had shown SIS to give more consistent results than other quench indicators for the microcentrifuge tube-contained samples.

2.3. *Microbial Cell Numbers*

Direct microscopic counts of microorganisms in GW were made on 10–30 mL subsamples that were collected on a 0.1 μm polycarbonate membrane (25 mm, Whatman Nucleopore Track-Etch). Groundwater was vacuum-filtered through the 0.1 μm membrane (with 0.45 μm backing filter) using a glass tower and portable vacuum pump, until ~1 mL of concentrated GW remained. The concentrated 1 mL GW sample was then stained with 60 μL of 4', 6-diamidino-2-phenylindole (DAPI; 50 μg/mL). Stained samples were incubated in the dark at room temperature for 15 min. and were subsequently washed three times with autoclaved, filter-sterilized deionized water. The DAPI-stained membranes were then mounted onto a glass slide with non-fluorescent immersion oil and a cover slip and were frozen at −20 °C until examined, if not examined immediately. The slides were examined using a Nikon Eclipse Ci-L using a single pass filter under type HF (halogen-free) immersion oil. For each sample, a minimum of 200 cells and a minimum of 20 fields were counted. Microbial cell numbers in the original GW sample were calculated using the following equation:

$$\text{microorganisms (cells/mL)} = (\text{N} \times \text{A}t)/(\text{V}f \times \text{A}g) \tag{1}$$

where N is the number of cells counted, At is the effective area of the filter (mm^2), Ag is the area of the counting grid (mm^2) and Vf is the volume of sample filtered [21].

2.4. *Aqueous Speciation Modeling*

Modeling of groundwater from well #1 (E33-268) and well #2 (E33-360) at a given pH and Eh was computed using The Geochemist's Workbench®, GWB version 12.0.1 (Aqueous Solutions LLC, Champaign, IL, USA) [22]. The Minteq thermodynamic database built within GWB was selected for these calculations after it was updated with equilibrium constants by Sehmel, 1989 [23], for all reactions involving aqueous ferric/ferrous cyanides complexes.

2.5. *Statistical and Geospatial Analyses*

Statistical analyses were performed using R software [24]. To determine environmental and geochemical factors associated with bacterial leucine uptake and GW cyanide content, ^{3}H-Leu and CN$^-$ were correlated to all other collected variables using one-tailed Pearson Product-Moment Correlation tests (significance level at 0.05). To determine current and historical spatial distributions of CN$^-$ (Figure 1), we interpolated observed values across the sampling domain using kriging estimation. Kriging predicts the value at a given point in space as a function of data in the neighborhood of the point. Kriging parameters were derived from maximum likelihood estimation with the function 'likfit' in the 'geoR' package, and predicted values were generated with the 'krige.conv' function in 'geoR' [25]. Predicted values were plotted using the function 'image.plot' in the 'fields' package [26].

3. Results

Tritiated leucine uptake by the native BP-5 GW bacteria, incubated aerobically (~20 °C) in ambient (recently collected) GW, varied widely among the wells assayed, from <0.03% in wells #2 (E33-360) and #8 (E33-34) to >8% in well #1 (E33-268) (Table 2; all values based on an 8-h sampling point). In contrast, standing biomass in the well water samples, as indicated by DAPI staining, varied less (Table 2, Figure 2), but still spanned an order of magnitude, from ~1 × 10^{+5} cells/mL in well #8 to ~1 × 10^{+6} cells/mL in well #4 (E32-9). These values are as much as ~10 to ~20 times the cell numbers previously observed in GW collected from the Hanford formation in the Columbia River hyporheic zone of the Hanford 300 Area (Figure S1b), 5 × 10^{+4} to 6 × 10^{+5} cells/mL [14]). Although rates of ^{3}H-leu uptake, an indication of heterotrophic bacterial productivity, can be related to the corresponding bacteria cell numbers to calculate bacterial growth rates and doubling times [14], such analyses should be treated with some caution. First, the ^{3}H-leu assay is specific for heterotrophic bacteria, such that if significant autotrophic bacteria are present and included in the DAPI counts, then cell doubling times, calculated from values for leucine uptake per cell per unit time, will be underestimated [20]. Second, since our assays were carried out aerobically, any in situ bacterial productivity arising from facultative or strict anaerobes would likely be undetected, again underestimating in situ bacterial productivity. Third, since the DAPI stain is specific for nucleic acids, DAPI counts could include non-bacterial entities, such as viruses, DNA-containing membrane vesicles, and eukaryotes, again leading to underestimates in cell-specific ^{3}H-Leu uptake.

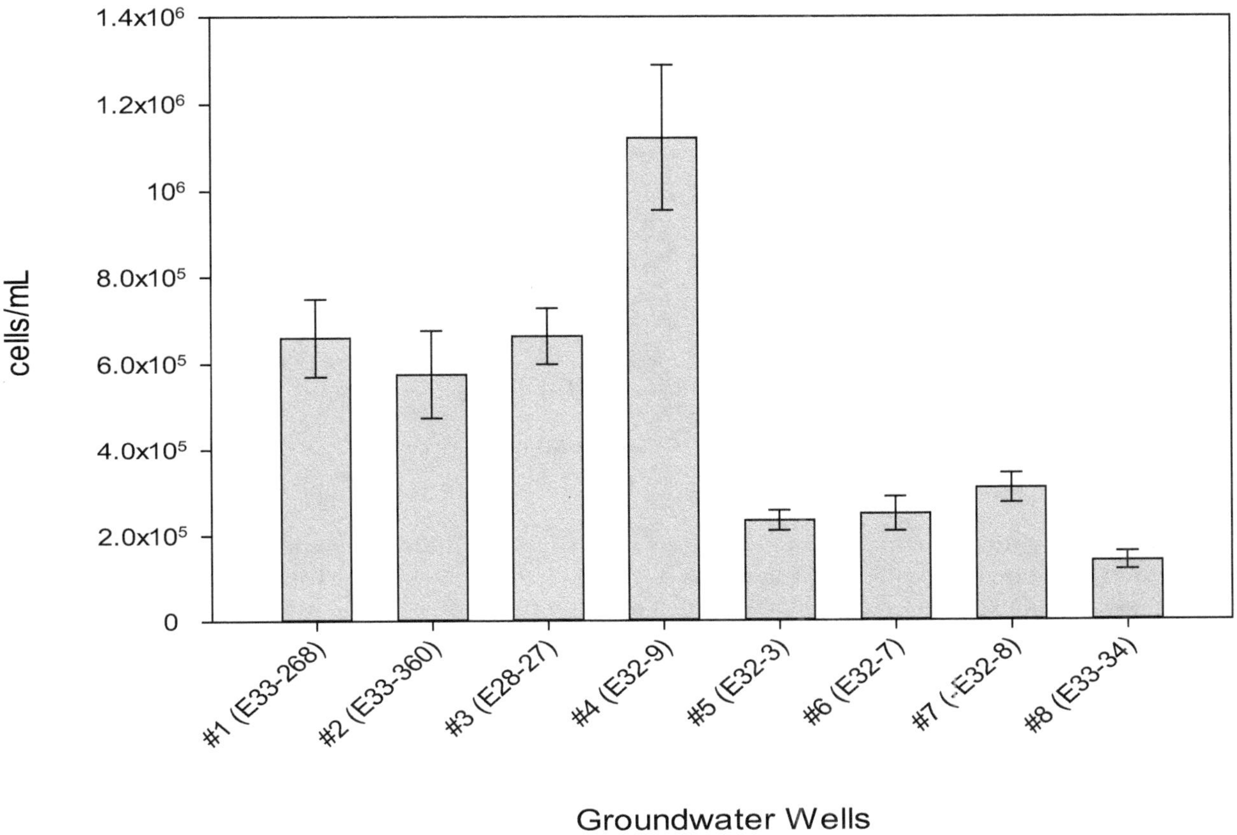

Figure 2. Microbial cell density (cells/mL) in groundwater samples, as determined by DAPI staining and epifluorescence microscopy.

We found ^{3}H-Leu uptake to be nearly linear from 0 to 8 h in all well-water samples (Figure 3a,b). However, in a subset of wells incubated for longer (~21 h), leucine uptake did not increase over the increased incubation time, except for well #1 (Figure 3a). These results suggest that in wells #3 (E28-27) and #4 (E32-9), where ^{3}H-Leu remained constant at ~3% and ~2%, respectively, either (1) some growth factor, such as inorganic phosphorous or organic carbon, became limiting over time or, less likely, (2) some component in the GW exerted a toxic effect that limited bacterial growth to the level seen at 8 h. In contrast, ^{3}H-Leu uptake in well #1 (the BP-5 extraction well for the 200 West Area pump-and-treatment facility) increased from ~8% at 8 h to ~24% at 21 h (Figure 3a), suggesting that neither factor considered above inhibited cell growth in this GW sample over this duration. In support of the first hypothesis, our analysis of this sample showed organic carbon to be 0.5 mg/L and inorganic phosphorous to be 4 μM (data not shown).

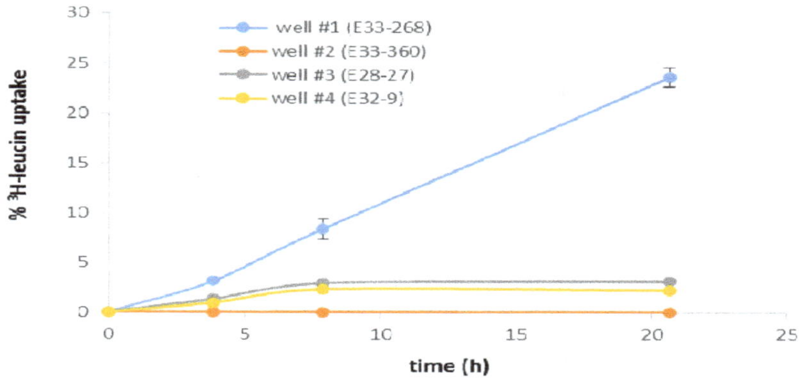

(a)

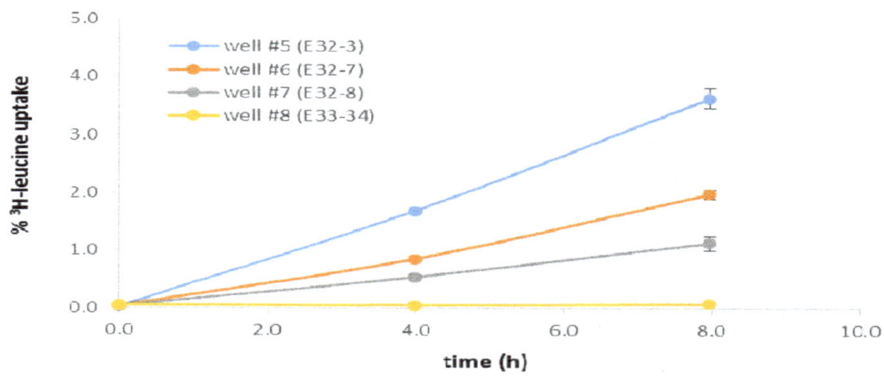

(b)

Figure 3. (**a**) Tritiated leucine uptake (as percentage of label added) in wells #1–#4, from 0 to 21 h. Results are based on duplicates, with standard deviation bars as shown. (**b**) Tritiated leucine uptake (as percentage of label added) in wells #5–#8, from 0 to 8 h. Results are based on duplicates, with standard deviation bars as shown.

Both the lowest and highest leucine uptake were in GW from wells with the highest most recently measured total cyanide concentrations (analysis on GW samples collected at the same time as the samples for biotic analysis), wells #1 and #2. Well #2, a recently (2014) installed monitoring well, had 0.01% 3H-Leu uptake at 8 h and contemporaneous total cyanide concentration 190 μg/L. Well #1, the pump-and-treat extraction well, had ~8% ^{3}H-Leu uptake at 8 h and a contemporaneous total cyanide concentration 297 μg/L. Total cell numbers, as indicated by DAPI staining, were similar in these wells (~6 × 10^{+5} cells/mL in well #2 and ~7 × 10^{+5} cells/mL in well #1). Likewise, the proportion of living cells to total cells, as indicated by live/dead staining were similar in both wells (data not shown). In addition to the aforementioned organic carbon and inorganic phosphorus measured in well #1, it is possible that the GW extraction process may increase the proportion of viable heterotrophic, aerobic bacteria in the GW collected via some physical effect. In addition, the redox potential in the P&T well (#1, +16 mV) was lower than in the monitoring well (#2, + 304 mV), possibly indicating greater electron-consuming bacterial activity in the P&T well. The other well showing low ^{3}H-Leu uptake (0.03%), well #8 (E33-34), also had the lowest total cell numbers (1.4 × 10^{+5} cells/mL, Table 2 and Figure 2). The most recent total CN^{-} measurement in this well was 100 μg/L, and this well had a historical maximum total CN^{-} concentration of 558 μg/L, the highest historical value seen in this set of wells (Table 2).

Aqueous speciation modeling of the most recent GW chemistry data from well #1 showed that strong-acid dissociable iron-cyanide species accounted for 93% of cyanide species, while free cyanide accounted for 6.5% (Table 3). In contrast, in well #2, only 77% of CN^{-} species were SAD, while 22% were free cyanide (Table 4). Weak-acid dissociable (WAD) species were 0.21% and 0.40% in wells

#1 and #2, respectively. Iron speciation in the two wells indicated that 88% and 78% of the Fe in wells #1 and #2, respectively, was cyanide-associated (Table S2). However, these values are for total (unfiltered) Fe, rather than filtered (aqueous) Fe. Likewise, CN^- values are on unfiltered samples. Given that $CN_{(aq)}$ and $Fe_{(aq)}$ values are not available for these wells (Table S1), we cannot determine how much of the Fe and CN^- may have been either colloidal or particulate. However, all of the other wells examined had recent Fe values that were at or near the Fe detection limit (Table S1), suggesting that $Fe_{(aq)}$ and $CN^-_{(aq)}$ may be lower than the total values used here for aqueous speciation analysis. Also, at the approximate time that well #2 (E33-360) was constructed (2014), the Eh measurement was −157 mV, significantly lower than the Eh value reported in Table 1 (+304 mV). Re-analysis of aqueous speciation for the given well chemistry data set using this much lower, reducing Eh indicated that lowering the redox potential increases the proportion of SAD (86%) and lowers the amount of free cyanide in solution (14% at the historical low Eh vs. 22% at the most recent, much higher Eh).

Statistical analyses indicated that ^{3}H-Leu uptake was significantly correlated with contemporaneous chemistry data for Fe (Pearson's $r = 0.76$, $P = 0.01$) and Eh (Pearson's $r = -0.68$, $P = 0.03$), and also with the historical maximums for manganese (Mn) (Pearson's $r = 0.63$, $P = 0.046$) (Table S3a). However, no significant correlation was found with total CN^- (Pearson's $r = 0.47$, $P = 0.12$) (Table S3a). We also found that the most recent total cyanide data for the eight wells correlated strongly with Eh (Pearson's $r = 0.94$, $P < 0.001$), NO_3^- (Pearson's $r = 0.94$, $P < 0.001$), and specific conductance (Pearson's $r = 0.95$, $P < 0.001$) (Table S3b), suggesting chemical rather than biotic drivers of CN^- recent concentrations.

Table 3. Aqueous cyanide species for well #1 (E33-268), based on speciation modelling of well groundwater chemistry contemporaneous to the date of groundwater collection for this study. Strong-acid dissociable (SAD), weak-acid dissociable (WAD), and free cyanide account for 93.29, 0.21, and 6.49% of species, respectively.

Aqueous Species	Concentration (µM)	CN^-%
[1] $Fe(CN)_6^{4-}$	1.296	67.07
[1] $NaFe(CN)_6^{3-}$	0.464	24.04
[2] $HCN_{(aq)}$	0.723	6.24
[1] $KFe(CN)_6^{3-}$	0.038	1.95
[2] CN^-	0.030	0.26
[1] $Na_2Fe(CN)_6^{2-}$	0.005	0.25
[3] $Ni(CN)_4^{2-}$	0.004	0.13
[3] $Cu(CN)_2^-$	0.005	0.08

Notes: [1] Strong-acid dissociable (SAD). [2] Free cyanide. [3] Weak-acid dissociable (WAD).

Table 4. Aqueous cyanide species for well #2 (E33-360), based on speciation modelling of well groundwater chemistry contemporaneous to the date of groundwater collection for this study. Strong-acid dissociable (SAD), weak-acid dissociable (WAD), and free cyanide account for 77.24, 0.40, and 22.37% of species, respectively.

Aqueous Species	Concentration (µM)	CN^-%
[1] $Fe(CN)_6^{4-}$	0.64	52.76
[2] $HCN_{(aq)}$	1.56	21.44
[1] $NaFe(CN)_6^{3-}$	0.24	19.88
[1] $Fe(CN)_6^{3-}$	0.04	3.14
[1] $KFe(CN)_6^{3-}$	0.02	1.44
[2] CN^-	0.05	0.64
[3] $Ni(CN)_4^{2-}$	0.01	0.39
[1] $Na_2Fe(CN)_6^{2-}$	0.00	0.22

Notes: [1] Strong-acid dissociable (SAD). [2] Free cyanide. [3] Weak-acid dissociable (WAD).

4. Discussion

Our results do not implicate cyanide toxicity in limiting heterotrophic bacterial productivity in the cyanide-affected aquifer examined here. This result is consistent with other aquatic toxicity studies of ferrocyanide, ferric ferrocyanide, and other iron cyanide compounds, and with what is known about the chemical stability of these compounds. A literature review indicates multiple instances of free cyanide ($HCN + CN^-$) rendering bacterial toxicity [9–13]. However, except for a single study with a cyanobacterium and a *Pseudomonas*, both in pure culture (where free cyanide was not measured) [9], bacterial toxicity has not been observed at total cyanide concentrations measured in BP-5 (~1600 μg/L maximum). However, confirmation of our results would require additional BP-5 GW sampling and ^{3}H-Leu analysis in space and time, and over a larger range of CN^- contamination levels in the aquifer. Additionally, iron concentrations in the BP-5 operable unit have been shown to roughly track total cyanide concentrations [5]. Consequently, it is possible that CN^--derived iron from the originally disposed ferrocyanide could also exert negative microbial affects in the aquifer, if the iron cyanide complexes over time have dissociated, or if the cyanide has been subject to biodegradation or other degradative fates, due to iron toxicity, or to bacterial stress from reactive oxygen species resulting from reactions of dissolved ferrous iron. In one study [27], ferrous iron toxicity to *Streptococci* from the human oral cavity was found to be relieved by the addition of ferrocyanide.

In addition, alternate BP-5 GW microbial probing, such as anaerobic tritiated leucine uptake for anaerobic bacteria or carbon-14 labelled bicarbonate ($H^{14}CO_3^-$) incorporation for autotrophs, might provide a more complete assessment of the bacterial productivity in the BP-5 Operable Unit, and could provide additional information on factors limiting bacterial productivity. It is possible that the bacterial biomass in GW from some wells might be dominated by autotrophs, such that bacterial productivity would not be amenable to measurement via the ^{3}H-Leu assay, which is specific for heterotrophs [20]. In Hanford formation sediments from the Hanford 300 Area within the Columbia River hyporheic zone, autotrophic ammonia oxidizers were found to dominate the microbial community in the seasonal absence of river water influx [28]. Likewise, Lee et al. [29] found a variety of autotrophic hydrogen-oxidizing, nitrate-reducing bacteria in GW in the Hanford formation of the 300 Area, including autotrophic species of *Acidovorax*, *Pseudomonas*, and *Pelosinus* (published and unpublished data from Lee et al. [29]). In addition, although CN^- is primarily inhibitory to aerobic metabolic processes, there are reports of CN^- inhibition of anaerobic microbial processes, such as denitrification [10,11], so effects on anaerobic bacteria may be relevant.

5. Conclusions

We did not find a correlation between tritiated leucine uptake, as an indication of aerobic heterotrophic bacterial productivity, and total cyanide values for the wells examined (both in terms of recent values and historical high values). However, more biological data in space and time would be required to conclude with certainty that CN^- contamination in the aquifer is unlikely to inhibit microbially facilitated in situ attenuation and degradation of co-contaminants, under either bioremediation or monitored natural attenuation scenarios. In addition, data on aqueous (filtered) values for Fe and CN^- would give a better picture, since some or most of the CN^- measured in the GW may be colloidal or suspended, and therefore presumably of low toxicity. Our results also indicate that the ^{3}H-leucine assay, which was developed for marine systems [20] and which has been widely adopted for surficial freshwater studies, is a valid tool to assess productivity of groundwater bacteria in relation to geochemical factors, confirming recent attempts to adapt the technique to aquifers, both those that are influenced by surface water inputs [14,18] and those that are not (this study [19]).

Supplementary Materials

Figure S1: (a): Location of the Hanford Site in Washington State, U.S. (inset), along with groundwater plumes (as of 2016), Figure S1: (b): Location of the Hanford Site in Washington State, U.S. (inset), and the Hanford 200 and 300 Areas, Figure S2: (a): Generalized Hanford Site Stratigraphy, Figure S2: (b): Generalized Hanford Site stratigraphy, lithostratigraphy, and hydrostratigraphy, Figure S3: (a): Groundwater plumes of cyanide and other contaminants in the BP-5 Operable Unit, Figure S3: (b): Locations of the Hanford Site cyanide plumes, Figure S4: Locations of the groundwater wells sampled for this study, Table S1: Summary of data for the wells examined, based on results of this study and data from the Hanford Environmental Information System (HEIS), Table S2: Aqueous speciation modeling results for well #1 (E33-268) and well #2 (E33-360), Table S3a: ^{3}H-Leu uptake correlations, Table S3b: Recent groundwater [CN^{-}] correlations.

Author Contributions: Conceptualization, A.P. and B.L.; Methodology, A.P., E.G., O.Q., and S.B.; Formal Analysis, E.G. and O.Q.; Investigation, A.P., J.W., and S.B.; Resources, B.L.; Writing-Original Draft Preparation, A.P., J.W., E.G., O.Q., and S.B.; Visualization, E.G.; Supervision, A.P.; Project Administration, A.P.; Funding Acquisition, A.P.

Acknowledgments: The authors thank Vicky Freedman and Chris Brown for supporting this project. We also thank Elsa Cordova, Danielle Saunders, and Michelle Snyder for assistance in sample acquisition and processing, and we thank Alicia Gorton for guidance in accessing the HEIS database. We also thank Kevin Rosso for providing a domestic well-water sample for methodological testing and Christopher (Kitt) Bagwell for reviewing a draft of the manuscript.

References

1. Gephart, R.E. A short history of waste management at the Hanford Site. *Phys. Chem. Earth.* **2010**, *35*, 298–306. [CrossRef]
2. Hoitink, D.J.; Burk, K.W.; Ramsdell, J.V., Jr.; Shaw, W.J. *Hanford Site Climatological Summary 2004 with Historical Data*; Pacific Northwest National Laboratory: Richland, WA, USA, 2005; pp. 1–382.
3. Fayer, M.J.; Keller, J.M. *Recharge Data Package for Hanford Single-Shell Tank Waste Management Areas*; PNNL-16688; Pacific Northwest National Laboratory: Richland, WA, USA, 2007.
4. *Hanford Groundwater Monitoring Report for 2016*; DOE/RL-2016-67; CH2M Hill Plateau Remediation Company: Richland, WA, USA, 2017.
5. *Remedial Investigation Report for the 200-BP-5 Groundwater Operable Unit*; DOE/RL-2009-127; CH2M Hill Plateau Remediation Company: Richland, WA, USA, 2015; pp. 1–424.
6. Cash, R.J.; Meacham, J.E.; Lilga, M.A.; Babad, H. *Resolution of the Hanford Site Ferrocyanide Safety Issue*; HNF-SA-3126-FP; DE&S Hanford, Inc., Pacific Northwest National Laboratory, Babad Technical Services: Richland, WA, USA, 1997; pp. 1–10.
7. What Are EPA's Drinking Water Regulations for Cyanide? Available online: https://safewater.zendesk.com/hc/en-us/articles/212077077-4-What-are-EPA-s-drinking-water-regulations-for-cyanide- (accessed on 10 August 2018).
8. Dzombak, D.; Ghosh, R.; Young, T. Physical-Chemical Properties and Reactivity of Cyanide in Water and Soil. In *Cyanide in Water and Soil: Chemistry, Risk, and Management*; Dzombak, D., Ghosh, R., Young, T., Eds.; Taylor & Francis: Boca Raton, FL, USA, 2006; pp. 57–92. ISBN 9781566706667.
9. Bringmann, G.; Kuehn, R. Comparative results of the damaging effects of water pollutants against bacteria (*Pseudomonas putida*) and blue algae (*Microcystis aeruginosa*). *GWF Wasser/Abwasser* **1976**, *117*, 410–413. (In Germany)
10. Kapoor, V.; Elk, M.; Li, X.; Santo Domingo, J.W. Inhibitory effect of cyanide on wastewater nitrification determined using SOUR and RNA-based gene-specific assays. *Lett. Appl. Microbiol.* **2016**, *63*, 155–161. [CrossRef] [PubMed]
11. Kim, Y.M.; Cho, H.U.; Lee, D.S.; Park, D.; Park, J.M. Comparative study of free cyanide inhibition on nitrification and denitrification in batch and continuous flow systems. *Desalin.* **2011**, *279*, 439–444. [CrossRef]
12. Kim, Y.M.; Cho, H.U.; Lee, D.S.; Park, D.; Park, J.M. Influence of operational parameters on nitrogen removal efficiency and microbial communities in a full-scale activated sludge process. *Water Res.* **2011**, *45*, 5785–5795. [CrossRef] [PubMed]
13. Kim, Y.M.; Lee, D.S.; Park, C.; Park, D.; Park, J.M. Effects of free cyanide on microbial communities and biological carbon and nitrogen removal performance in the industrial activated sludge process. *Water Res.* **2010**, *45*, 1267–1279. [CrossRef] [PubMed]
14. Konopka, A.; Plymale, A.E.; Carvajal, D.A.; Lin, X.; McKinley, J.P. Environmental controls on the activity of aquifer microbial communities in the 300 area of the Hanford Site. *Microb. Ecol.* **2013**, *66*, 889–896. [CrossRef] [PubMed]

15. Gensemer, R.W.; DeForest, D.K.; Stenhouse, A.J.; Higgins, C.J.; Cardwell, R.D. Aquatic toxicity of cyanide. In *Cyanide in Water and Soil: Chemisty, Risk, and Management*; Dzombak, D., Ghosh, R., Young, T., Eds.; Taylor & Francis: Boca Raton, FL, USA, 2006; pp. 251–284. ISBN 9781566706667.
16. Bååth, E.; Pettersson, M.; Söderberg, K.H. Adaptation of a rapid and economical microcentrifugation method to measure thymidine and leucine incorporation by soil bacteria. *Soil Biol. Biochem.* **2001**, *33*, 1571–1574. [CrossRef]
17. Demoling, F.; Figueroa, D.; Bååth, E. Comparison of factors limiting bacterial growth in different soils. *Soil Biol. Biochem.* **2007**, *39*, 2485–2495. [CrossRef]
18. Velasco Ayuso, S.; López-Archilla, A.; Montes, C.; Guerrero, M.C. Microbial activities in a coastal, sandy aquifer system (Doñana Natural Protected Area, SW Spain). *Geomicrobiol. J.* **2010**, *27*, 409–423. [CrossRef]
19. Wilhartitz, I.C.; Kirschner, A.K.T.; Stadler, H.; Herndl, G.J.; Dietzel, M.; Latal, C.; Mach, R.L.; Farnleitner, A.H. Heterotrophic prokaryotic production in ultra-oligotrophic alpine karst aquifers and ecological implications. *FEMS Microbiol. Ecol.* **2009**, *68*, 287–299. [CrossRef] [PubMed]
20. Kirchman, D. Measuring bacterial biomass production and growth rates from leucine incorporation in natural aquatic environments. *Methods Microbiol.* **2001**, *30*, 227–237.
21. Kepner, R.L.; Pratt, J.R. Use of fluorochromes for direct enumeration of total bacteria in environmental samples: past and present. *Microbiol. Rev.* **1994**, *58*, 603–615. [PubMed]
22. *The Geochemist's Workbench*; 12.0.1; Aqueous Solutions LLC: Champaign, IL, USA, 2015.
23. Sehmel, G.A. *Cyanide and Antimony Thermodynamic Database for the Aqueous Species and Solids for the EPA-Minteq Geochemical Code*; PNNL-6835; Pacific Northwest National Laboratory: Richland, WA, USA, 1989; pp. 1–224.
24. The R Project for Statistical Computing. Available online: https://www.r-project.org/ (accessed on 10 August 2018).
25. Ribeiro, P.J., Jr.; Diggle, P.J. geoR: A package for geostatistical analysis. *R News* **2001**, *1*, 14–18.
26. Furrer, R.; Nychka, D.; Sain, S. *Fields: Tools for Spatial Data*; 6.11. R Foundation, 2009. Available online: http://citeseerx.ist.psu.edu/viewdoc/summary?doi=10.1.1.304.1555 (accessed on 10 August 2018).
27. Dunning, J.C.; Ma, Y.; Marquis, R.E. Anaerobic killing of oral streptococci by reduced, transition metal cations. *Appl. Environ. Microbiol.* **1998**, *64*, 27–33. [PubMed]
28. Graham, E.B.; Crump, A.R.; Resch, C.T.; Fansler, S.; Arntzen, E.; Kennedy, D.W.; Fredrickson, J.K.; Stegen, J.C. Deterministic influences exceed dispersal effects on hydrologically-connected microbiomes. *Environ. Microbiol.* **2017**, *19*, 1552–1567. [CrossRef] [PubMed]
29. Lee, J.-H.; Fredrickson, J.K.; Plymale, A.E.; Dohnalkova, A.C.; Resch, C.T.; McKinley, J.P.; Shi, L. An Autotrophic H_2-oxidizing, Nitrate-Respiring, Tc(VII)-Reducing *Acidovorax* sp. Isolated from a Subsurface Oxic-Anoxic Transition Zone. *Environ. Microbiol. Rep.* **2015**, *7*, 395–403. [CrossRef] [PubMed]

9

Assessing Decadal Trends of a Nitrate-Contaminated Shallow Aquifer in Western Nebraska using Groundwater Isotopes, Age-Dating and Monitoring

Martin J. Wells [1], Troy E. Gilmore [1,2,*], Aaron R. Mittelstet [1], Daniel Snow [3] and Steven S. Sibray [2]

[1] Department of Biological Systems Engineering, University of Nebraska-Lincoln, Lincoln, NE 68583, USA; martin.wells@huskers.unl.edu (M.J.W.); amittelstet2@unl.edu (A.R.M.)

[2] Conservation and Survey Division, School of Natural Resources, University of Nebraska-Lincoln, Lincoln, NE 68583, USA; ssibray1@unl.edu

[3] Nebraska Water Sciences Laboratory, Nebraska Water Center, Lincoln, NE 68583, USA; dsnow1@unl.edu

* Correspondence: gilmore@unl.edu

Abstract: Shallow aquifers are prone to nitrate contamination worldwide. In western Nebraska, high groundwater nitrate concentrations ([NO_3^-]) have resulted in the exploration of new groundwater and nitrogen management regulations in the North Platte Natural Resources District (NPNRD). A small region of NPNRD ("Dutch Flats") was the focus of intensive groundwater sampling by the United States Geological Survey from 1995 to 1999. Nearly two decades later, notable shifts have occurred in variables related to groundwater recharge and [NO_3^-], including irrigation methods. The objective of this study was to evaluate how changes in these variables, in part due to regulatory changes, have impacted nitrate-contaminated groundwater in the Dutch Flats area. Groundwater samples were collected to assess changes in: (1) recharge rates; (2) biogeochemical processes; and (3) [NO_3^-]. Groundwater age increased in 63% of wells and estimated recharge rates were lower for 88% of wells sampled (n = 8). However, mean age and recharge rate estimated in 2016 (19.3 years; R = 0.35 m/year) did not differ significantly from mean values determined in 1998 (15.6 years; R = 0.50 m/year). δ^{15}N-NO_3^- (n = 14) and dissolved oxygen data indicate no major changes in biogeochemical processes. Available long-term data suggest a downward trend in normalized [NO_3^-] from 1998 to 2016, and lower [NO_3^-] was observed in 60% of wells sampled in both years (n = 87), but median values were not significantly different. Collectively, results suggest the groundwater system is responding to environmental variables to a degree that is detectable (e.g., trends in [NO_3^-]), although more time and/or substantial changes may be required before it is possible to detect significantly different mean recharge.

Keywords: groundwater nitrate; groundwater age; groundwater transit time; groundwater recharge rates; non-point source pollution; groundwater monitoring; isotopes; ^{3}H/^{3}He; surface irrigation; center pivot irrigation

---

1. Introduction

Elevated groundwater nitrate concentrations ([NO_3^-]) in shallow aquifers are often linked to a combination of high groundwater recharge rates and intensive agricultural land use [1–6]. Greater recharge rates in areas with intense nitrogen fertilizer loading generally lead to higher [NO_3^-] in groundwater. For example, the central Wisconsin sand-plains region requires additional water and fertilizer inputs to sustain healthy crop yields, with irrigated agriculture having a governing influence on groundwater [NO_3^-] [7,8]. Similarly, high [NO_3^-] have been observed in groundwater in Nebraska, especially beneath areas with sandy soils and/or sand and gravel aquifers [9–13].

Growing concerns over changes in the state's water quality and quantity led to the creation of what are now 23 Natural Resources Districts (NRD) across Nebraska. Established in 1972, NRDs develop management plans and regulations to protect groundwater [14–16]. Regulations aimed at decreasing [NO_3^-] in groundwater have shown some potential for success [4,17], though the exact impacts are not always clear [13,18]. Due to the tendency of nitrate to be transported with recharge water, agricultural water management (i.e., irrigation technology and practices) and groundwater [NO_3^-] are likely to have a direct relationship. In some areas, water allocations and/or moratoriums on new well drilling can incentivize greater irrigation efficiency, which has been found to decrease groundwater [NO_3^-] [12,19–21]. For instance, replacing furrow irrigated fields with sprinkler systems (i.e., center pivot) is one method believed to reduce [NO_3^-] leaching to groundwater [12,19]. Such changes in irrigation practice, driven in part by regulatory changes and economic drivers, have occurred in the Dutch Flats area of western Nebraska.

Groundwater age-dating has been used widely to determine historical trends in groundwater [NO_3^-] [17,18,22–25]. The United States Geological Survey's (USGS) National Water-Quality Assessment (NAWQA) program has emphasized the importance of implementing groundwater age-dating to evaluate long-term trends in groundwater characteristics and its contaminants [24]. Relatively few studies, however, have used groundwater age-dating to directly evaluate the impact of water and/or nutrient management regulations, or major shifts in irrigation management, on groundwater recharge rates and nitrate contamination. Visser et al. [17] used $^3H/^3He$ age-dating in the Netherlands to examine impacts of legislation aimed at decreasing groundwater [NO_3^-] in areas characterized by alluvial sand and gravel deposits. Groundwater age-dating showed trend reversal in groundwater nitrate, with old groundwater increasing in [NO_3^-], and young groundwater decreasing in [NO_3^-]. Further, groundwater age-dating provides a method for evaluating impacts of land use change on groundwater quality [26–28]. As a result, coupling groundwater [NO_3^-] trends with apparent age may be useful in assessing how changing groundwater recharge and quality respond to irrigation management changes, in the context of other environmental variables (e.g., precipitation).

Within North Platte Natural Resources District (NPNRD) in western Nebraska, irrigation canals provide a source of artificial recharge [11,29–33]. Locations of highest and lowest recharge potential in canals were captured using capacitively coupled and direct-current resistivity methods to profile lithology of two major canals in NPNRD [32]. Estimates suggest canals leak between 40% and 50% of their water within this region [34]. The Interstate Canal, with a water right of 44.5 m^3/s, operates during irrigation season and is the largest canal delivering water to the region [35]. Other large canals in the region include the Mitchell-Gering and Tri-State Canals. An extensive analysis of the relationship between surface irrigation and groundwater quantity and quality in this area was provided by Böhlke et al. [11].

Böhlke et al. [11] also summarized USGS reports from a five-year study beginning in the mid-1990s [30,31]. The investigation was conducted from 1995 to 1999 (referred to in this text as the 1990s study) in the Dutch Flats area, a region comprising roughly four percent of NPNRD. Crop production in this area historically depends on surface water with low [NO_3^-] (e.g., [NO_3^-] < 0.06 mg N L^{-1} in 1997) delivered via canals for irrigation supply. Groundwater age estimates ($^3H/^3He$), isotopes, nitrate, and other analytes were used to evaluate trends in groundwater recharge and nitrate contamination. For example, groundwater recharge rates and temporal changes in [NO_3^-] demonstrated the influence of canal seepage on nearby wells. Wells far from canals were more influenced by local irrigation practices. Relatively young groundwater ages (mean = 8.8 years) indicated that recharge was occurring from more than just regional precipitation (i.e., groundwater would be expected to reside in the aquifer much longer if recharge rates based on precipitation were assumed). As a result, Böhlke et al. [11] theorized that groundwater residence times and [NO_3^-] may be impacted if recharge from canals and/or irrigation were significantly reduced. Further, if groundwater residence times were to increase, then potential for biogeochemical activity such as denitrification might also increase, resulting in a decrease in groundwater [NO_3^-].

Since the 1990s USGS study, several variables related to groundwater recharge have changed in the extensively sampled Dutch Flats area. For example, a shift in irrigation practice and canal management have been noted in the region [36], with the largest changes in irrigation practice occurring during approximately 2000–2003. The timing of these changes relative to the USGS study, combined with the relatively young groundwater ages in the aquifer, provides a unique opportunity to evaluate the potential impact of changing water management on the overall timescale of groundwater movement through the aquifer, and subsequent impacts on groundwater quantity (recharge rate) and quality ([NO_3^-]). Other variables we considered were annual precipitation, volume of water diverted into the Interstate Canal, planted corn area, and fertilizer loads.

In this study, we evaluated how changes to water resources management, with respect to numerous underlying variables, have affected leaching and groundwater transport of nitrate nitrogen. More specifically, the objective of this study was to compare the composition of recently collected groundwater samples to those reported by Böhlke et al. [11] for changes in: (1) groundwater recharge rates; (2) biogeochemical processes (i.e., denitrification) affecting [NO_3^-]; and (3) groundwater [NO_3^-] in the Dutch Flats area.

2. Materials and Methods

2.1. Site Description

The study area is within NPNRD in western Nebraska (Figure 1), where climate is classified as semi-arid [37]. Climate data retrieved from Western Regional Airport in Scottsbluff, Nebraska display long-term average annual rainfall and snow of 390 mm and 1021 mm, respectively (1908–2016) [38]. The average annual maximum and minimum temperatures from 1908 to 2016 were 17.6 °C and 1.1 °C, respectively. Growing season rainfall is typically insufficient to support high crop yields; therefore, irrigation is used extensively with 86% previously estimated to originate from surface water [39]. In 2002, a moratorium was implemented to restrict drilling of additional irrigation wells in NPNRD. The state legislature passed Legislative Bill 962 in 2004, allowing the district to declare areas either fully or over-appropriated and led to development of an integrated management plan intended to protect both groundwater and surface water. Regulations on water and soil include groundwater allocations and flow meters on wells in over-appropriated areas, requirements for irrigators using chemigation systems, well registration, and irrigation runoff controls.

This study is focused within the Dutch Flats area in NPNRD [11,30,31,40]. The study area is in the North Platte River Valley, along the Nebraska-Wyoming border [39]. The Dutch Flats area is about 540 km^2 and located in Scotts Bluff and Sioux Counties (Figure 1). Approximately 48% of the study area is in Scotts Bluff County, while 52% is in Sioux County. Based on the 2011 National Land Cover Database (NLCD), 53.5% of the Dutch Flats area is agriculture, while Scotts Bluff and Sioux Counties are 47.0 and 4.3, respectively [41]. Due to similarities in land use between Dutch Flats area and Scotts Bluff County, Scotts Bluff County was used as a proxy when data could not be determined directly for the study area. While surface water is the most common source for irrigation in this region, accessible groundwater offers alternative methods. Irrigation withdrawal estimates in Scotts Bluff County suggest surface water has remained the dominant source of irrigation water, ranging from 84.4% to 98.6% from 1985 to 2010 [42].

An extensive monitoring well network in NPNRD has been used to measure and record changing groundwater levels and [NO_3^-] over several decades. The Dutch Flats area varies in both vadose and saturated zone thickness, and is characterized as a sand and gravel alluvial aquifer, with limited areas of silt and clay [39] (Figure 1). The alluvial aquifer is underlain by the Brule Formation, made up of siltstone, mudstone, volcanic ash beds, gravel, and fine-grained sand. Groundwater for irrigation is typically pumped from Quaternary-aged alluvial deposits or water-bearing units of the Brule Formation. The direction of groundwater flow is generally southeast from canals toward the North Platte River, though flow in some locations is redirected by what is referred to as the Brule High [30].

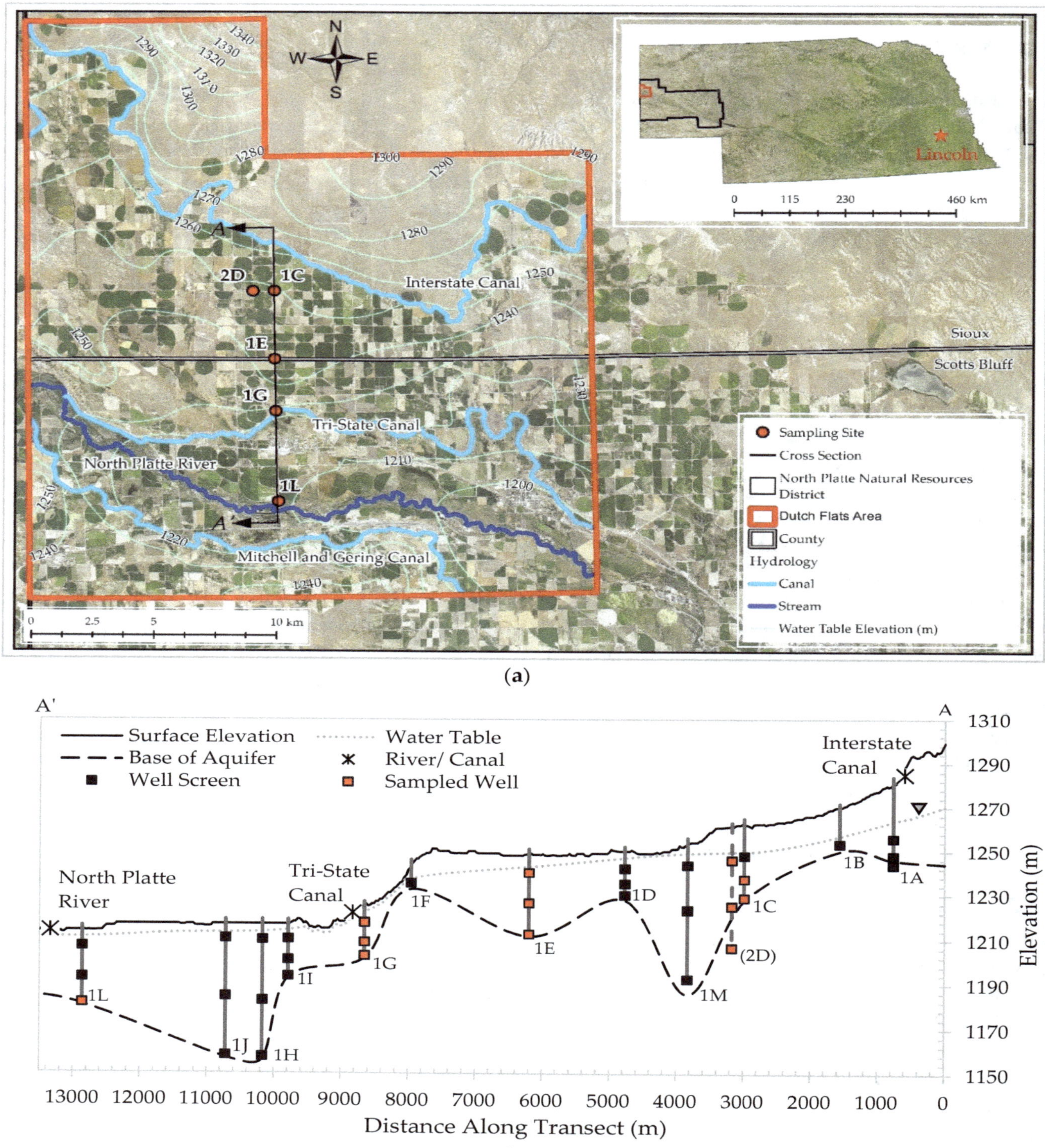

Figure 1. Site description: (**a**) location of Dutch Flats area within Nebraska's North Platte Natural Resources District, including water table elevation contours [43]; and (**b**) representative cross-section along well transect with mid-screen elevations at each well nest. Elevations were derived from previous Dutch Flats studies [11,31]. Red well screens indicate locations where the current study collected groundwater samples. Site 2D, shown with a dashed line and in parenthesis, is situated behind 1C, or into the page, and is located above the base of aquifer in its respective location. A small feedlot is directly adjacent to Well 1G.

2.2. Sample Sites

Within the Dutch Flats area, five well nests were selected for sampling in 2016. Wells were selected based on completeness of data from the previous study [11], so direct comparisons could be made to our results. Samples for noble gases, tritium, nitrogen and oxygen isotopes of nitrate, nitrate nitrogen, ammonium nitrogen, and dissolved organic carbon (DOC) analysis were collected following standard sampling procedures. Groundwater parameters logged and recorded at these five well nests were pH, temperature (°C), dissolved oxygen (mg O_2 L^{-1}), percent saturation of dissolved oxygen, specific conductivity (µS cm^{-1}), and total dissolved gas (g L^{-1}). Each well was measured for depth to groundwater and depth to well bottom relative to surface elevation. At least one well bore volume was pumped from each well prior to sampling (Geotech SS Geosub Controller and Pump). Groundwater parameters were monitored by a Hydrolab MS5 Multiparameter Sonde. After parameters stabilized, pump speed was decreased, and samples were collected. Sampling occurred three times over the course of 2016: spring, summer, and fall. Spring sampling included collection of groundwater from shallow, intermediate, and deep wells from nest 1G (Figure 1). Groundwater beneath irrigated fields, and away from major canals, was collected in summer for assessing spatial patterns in groundwater. Shallow, intermediate, and deep samples were collected from Well Nests 2D and 1E, intermediate and deep samples from nest 1C, and deep samples at Well 1L. To capture temporal influences, Well Nest 1G (Figure 1) was sampled again at a shallow and intermediate depth during fall.

2.3. $^3H/^3He$ Sampling and Noble Gas Modeling

After rinsing with sample water, 0.9 m copper refrigeration tubes for noble gas analysis were filled and crimped [44]. Samples for analysis of tritium was collected in 0.5-liter HDPE plastic bottles. Noble gases (Ar, He, Kr, Ne, and Xe) were analyzed via mass spectrometry at the University of Utah's Dissolved Gas Service Center in Salt Lake City, UT. Tritium activities were determined using the helium ingrowth method [45].

Excess air and recharge temperatures were modeled using iNoble Version 2.2 workbook developed by the International Atomic Energy Agency (IAEA). We assumed a terrigenic $^4He/^3He$ (R_{terr}) of 2.88×10^{-8}, and ^3H half-life of 12.32 years [46]. Apparent groundwater age was calculated from

$$\tau = \lambda^{-1} \ln\left(1 + \frac{^3He_{trit}}{^3H}\right) \tag{1}$$

where τ is the apparent groundwater age in years, $^3He_{trit}$ is the modeled tritiogenic helium, and ^3H is tritium at the time of sampling, determined from helium ingrowth. The decay constant, λ, is a function of the half-life of tritium and is determined as $\lambda = \frac{\ln 2}{12.32 \text{ years}}$.

2.4. Recharge Estimates

Groundwater recharge, defined as the rate at which water moves vertically across the water table, was calculated for each well. Shallow wells typically have screen lengths of 6.1 m with the midpoint originally designed to be located at the water table. Intermediate and deep wells both have screens of approximately 1.5 m. To maintain consistency, calculations were performed using methods and assumptions made by Böhlke et al. [11]. Groundwater age was assumed to follow a linear gradient as a slug of water traveling from the water table to the midpoint between the upper and lower well screen (Equation (2)). For shallow wells, recharge was determined at the midpoint between the bottom screen and water table. Recharge was calculated as follows:

$$R = \frac{z\theta}{\tau} \tag{2}$$

where R is recharge in m/year, θ is porosity, z is depth below water table to the screen midpoint (m), and τ is the groundwater age, in years, determined via apparent $^3H/^3He$ ages. Porosity was assumed to be 0.35. Intermediate and deep wells are screened below the water table. Therefore, z for these calculations was the distance from the water table to the midpoint of the screen. Again, recharge was estimated using Equation (2). Böhlke et al. [11] also used an exponential equation [47] to estimate recharge rates of intermediate wells, though this equation was not applied to shallow or deep wells (Equation (3)).

$$R = \frac{z\theta}{\tau} \ln\left(\frac{L}{L-z}\right) \tag{3}$$

where L is unconfined aquifer saturated thickness (m), and z is the distance from the water table to the screened midpoint (m). Applying this equation near the water table (shallow wells) or bottom of the aquifer (deep wells) may lead to uncertainties associated with groundwater mixing. While the simplistic modeling of Equations (2) and (3) fit the data reasonably well, it is acknowledged uncertainties related to $^3H/^3He$ age-dating and aquifer heterogeneity could affect calculated recharge rates. For instance, uncertainty in $^3H/^3He$ ages can be relatively large on a percentage basis [48]. Further, aquifer heterogeneity, namely porosity, directly influence Equations (2) and (3) calculations. However, assumptions by Böhlke et al. [11] were maintained in both studies, yielding similar calculated recharge uncertainties. The appropriate applications and limitations of these equations have been addressed in previous literature [49,50].

2.5. Nitrate Isotopes, Nitrate, Ammonium, and DOC Concentrations

Samples for isotope analysis were collected in 1-liter HDPE plastic bottles, placed on ice immediately after collection, frozen within 48 h, and analyzed at the University of Nebraska's Water Science Laboratory. The oxygen isotope composition of nitrate was measured according to methods described in Chang et al. [51] and Silva et al. [52]. A measured volume of sample containing 0.25 mg NO_3-N was then treated with 1 M barium chloride to precipitate sulfate and phosphate. The solution was filtered, passed through a cation exchange column to remove excess Ba^{2+}, and then through an anion exchange column to concentrate nitrate. Nitrate was eluted using 3 M hydrochloric acid, neutralized with Ag_2O, filtered to remove the AgCl precipitate, and then dried to produce purified $AgNO_3$. The $AgNO_3$ was dissolved in 1 mL of reagent water and 100 μL (25 μg N) aliquots were transferred to three silver cups and dried for analysis of oxygen isotope composition using high temperature pyrolysis on nickelized graphite in a closed tube to produce carbon monoxide (CO) on a Eurovector EA coupled to an Isoprime continuous flow isotope ratio mass spectrometer. The final result was averaged from the triplicate instrumental results and converted to the standard oxygen isotope reference (VSMOW = 0.00‰).

A reagent grade potassium nitrate (KNO_3) was used as a working standard, and reference sucrose oxygen isotope standards were analyzed with every sample batch (up to 20 samples) both for calibration and for drift correction. USGS 34 and USGS 35 reference standards are analyzed at least monthly to compare and convert working standards to a $\delta^{18}O$ isotope value with respect to Vienna Standard Mean Ocean Water (VSMOW). The 1σ measured analytical precision of $\delta^{18}O$-NO_3^- is ±0.5‰ for solutions of KNO_3 standard processed through the entire procedure. In addition to triplicate instrumental average measurement, laboratory duplicates were carried through the preparation process and analyzed at a rate of 5%.

The nitrogen isotope composition of nitrate ($\delta^{15}N$-NO_3^-) was measured according to methods previously described [53,54]. Ammonia-N was quantified after addition of MgO on a steam distillation line and titrated with standardized sulfuric acid [55]. Nitrate was then reduced to ammonia with Devarda's alloy, distilled separately into a boric acid indicator solution, and then quantified titrimetrically with standardized sulfuric acid. Distillates were acidified with sulfuric acid immediately after titration and evaporated to 1 to 2 mL on a hot plate. They were then reacted with lithium hypobromite on a high-vacuum preparation line and the ammonium quantitatively reduced to nitrogen gas, purified by passage through two liquid nitrogen cryotraps and a 400 °C copper oven, and collected in a gas sample bulb. Atmospheric nitrogen standards were prepared on the same high-vacuum preparation line. Ultrapure tank nitrogen was used as the working standard and was calibrated against the atmospheric nitrogen standard. All nitrogen isotope measurements were performed on either a Micromass OPTIMA or a GVI Isoprime dual inlet stable isotope ratio mass spectrometer (IRMS). The $\delta^{15}N\text{-}NO_3^-$ of the sample was measured and expressed relative to the atmospheric standard expressed in parts per thousand (‰). Quality control was monitored through the analysis of replicate standards to determine the accuracy and repeatability of the method.

The nitrogen ($^{15}N/^{14}N$) and oxygen isotope ($^{18}O/^{16}O$) composition of nitrate was expressed as the difference (‰) of the sample ratio relative to each international standard ratio using Equation (4).

$$\delta(‰) = \frac{(Ratio)_{Sample} - (Ratio)_{Standard}}{(Ratio)_{Standard}} \times 1000 \tag{4}$$

Nitrate concentration was determined for the five well nests selected for detailed sampling in 2016 using the Cd-reduction method [56] on a Seal AQ2 autoanalyzer. Dissolved organic carbon (DOC) samples were collected in 40 mL glass vials and preserved with sulfuric acid. Each sample was field filtered through a 0.45-micron filter attached to a syringe. Samples were analyzed following heated persulfate SM 5310 protocol using an OI Analytical model 1010 TOC Analyzer [56].

2.6. Evaluating Long-Term Trends in Nitrate Concentrations

In addition to nitrate evaluated in 1998 and this current study, long-term data collected and/or maintained by the NPNRD and Nebraska Agricultural Contaminant Database [57] were used for further analysis. These samples were collected with a low-volume pump after monitored parameters (temperature, pH, specific conductivity, DO, and total dissolved solids) stabilized. Samples were placed on ice and preserved with sulfuric acid prior to analysis at Midwest Laboratories, Inc in Omaha, NE, USA [58]. Sporadically sampled data from 1998 to 2016 were normalized about the maximum and minimum nitrate value over the sampling period, as

$$x' = \frac{x - \min(x)}{\max(x) - \min(x)} \tag{5}$$

where x' is the normalized nitrate value, x is the observed (or average if multiple samples were collected from a well within the same year) $[NO_3^-]$ for a specific year, and $\max(x)$ and $\min(x)$ are the maximum and minimum respective $[NO_3^-]$ over all the sample years for each well.

3. Results and Discussion

Groundwater samples analyzed for $^3H/^3He$, NO_3^-, $\delta^{15}N\text{-}NO_3^-$, and $\delta^{18}O\text{-}NO_3^-$, among other groundwater parameters, were used to evaluate the hypothesis that changes in environmental variables since the previous study would: (1) decrease recharge rates; (2) increase biogeochemical activity; and (3) result in lower groundwater $[NO_3^-]$. Unless otherwise noted, data collected in 2016 were analyzed and compared to 1998 groundwater data collected by the USGS [31] in August 1998 and limited to the five well nests sampled in 2016.

3.1. Groundwater Age-Dating

An increase in groundwater age between the 1998 and 2016 studies would indicate reduced rates of water movement through the aquifer over that period. Apparent groundwater ages from five of eight samples (63%) collected in 2016 were greater than groundwater ages estimated from samples collected in 1998 (Table 1; Figure 2). Mean groundwater age in the sampled wells increased from 15.6 years in 1998 to 19.3 years in 2016, but the difference was not statistically significant ($p = 0.53$; two-sample t-test assuming unequal variances).

Recent depth to groundwater data (2017; $n = 162$) vary throughout Dutch Flats, ranging from less than 1 m to over 30 m, with a mean of 10.6 m ($\pm$10.3 m) and median of 7.7 m. Vadose zone thickness of wells sampled in August 2016 were between 1.2 and 13.3 m. Well nests constructed for the 1990s study (and re-sampled in this study) had screen intervals designed to intercept groundwater at the water table, mid-aquifer, and at or near the base of the unconfined aquifer. In Dutch Flats, shallow wells typically have 6.1 m screens and were designed originally with roughly 3 m above and below the water table. Long well screens across the water table can increase error in groundwater age, due to mixing of a range of groundwater ages, and because fluctuations in the water table can lead to a loss of tritium-derived $^3He_{trit}$ escaping to the atmosphere. Among samples from shallow wells, apparent groundwater age sampled from Well 2D-S increased by approximately 2.5 years, or 93%, while 1E-S decreased by 0.5 years, or 9.3%.

Table 1. Apparent groundwater (GW) age from both 1998 and 2016 based on $^3H/^3He$ age estimates. Recharge rates were estimated with a linear equation in all cases, and with an exponential (Exp.) equation for intermediate wells.

Well ID	Böhlke et al. [11]				Current Study			
	Depth (m) *	GW Age (years)	Recharge—Linear (m/year)	Recharge—Exp. (m/year)	Depth (m)*	GW Age (years)	Recharge—Linear (m/year)	Recharge—Exp. (m/year)
1E-S	2.4	5.4	0.16	n.d.	1.9	4.9	0.13	n.d.
2D-S	4.9	2.8	0.61	n.d.	2.4	5.4	0.15	n.d.
1E-I	15.7	20.2	0.27	0.38	14.5	20.9	0.24	0.34
2D-I	25.8	11.3	0.80	1.2	22.6	20.5	0.39	0.56
1C-D	20.3	12.0	0.59	n.d.	19.4	12.0	0.57	n.d.
1E-D	29.4	31.5	0.33	n.d.	28.3	47.0	0.21	n.d.
1L-D	28.5	20.5	0.49	n.d.	30.3	20.2	0.53	n.d.
2D-D	44.3	20.9	0.74	n.d.	41.1	23.9	0.60	n.d.
Mean:		15.6	0.50			19.3	0.35	
Std. Dev.:		9.5	0.23			13.3	0.19	

Note: * Depth given as depth to mid-screen below water table; n.d., no data; S, Shallow well; I, Intermediate well; D, Deep well; Std. Dev., Standard Deviation.

Because of shorter (1.5 m) screened intervals, intermediate and deep wells may provide a better estimate of groundwater age and are subject to fewer uncertainties impacting shallow wells. Wells 1E-I and 2D-I had comparable groundwater ages of 20.9 and 20.5 years, respectively, in 2016. Well 1E-I groundwater age stayed similar between the two sampling periods (1998 = 20.2 years), while 2D-I increased by 81% (1998 = 11.3 years). Groundwater samples from 1G-I were collected in both the spring and fall of 2016 to explore temporal trends in groundwater age at a site near a canal. Although there is no direct comparison with the previous study, apparent groundwater age at the well found little variation in groundwater age, with spring and fall ages of just 5.9 and 5.3 years, respectively. Results from Well 1C-I displayed nearly modern groundwater age (i.e., age $\approx$ 0 years). Results from this well appear erroneous, and were excluded from the comparisons in Table 1, since it is unlikely modern groundwater would be observed 11 m below the water table.

Groundwater age of samples collected in 1998 from deep wells ranged from 12.0 to 31.5 years, while in 2016 apparent groundwater ages were between 12.0 to 47.0 years. The largest change in groundwater age was for groundwater sampled from Well 1E-D. While apparent groundwater age stayed similar in Well 1E shallow and intermediate depths, groundwater age from the deep well increased from 31.5 to 47.0 years, or 49%. This increase would suggest groundwater was nearly unaffected by recharge and water sampled in 1998 was essentially the same water collected in 2016. The groundwater age trend in Well Nests 2D and 1E is consistent with non-uniform recharge in the region, where screens at different depths in the aquifer are influenced by different recharge sources (i.e., localized irrigation and/or canals).

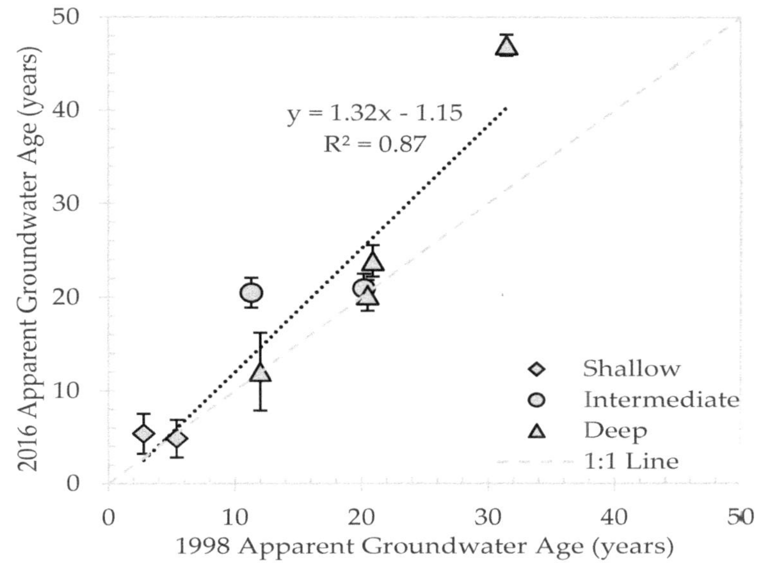

Figure 2. Apparent groundwater age determined in 2016 compared to apparent groundwater ages determined in 1998. Error bars are ±1σ from 2016 analysis using the International Atomic Energy Agency (IAEA) model.

3.2. Recharge Rates

Recharge rates (*R*) were estimated as a function of the vertical distance water travels below the water table, porosity, and apparent groundwater age (see Equations (2) and (3)). Figure 3 and Table 1 compare 1998 and 2016 recharge rates. Over nearly two decades, recharge rates decreased in each well, with exception to Well 1L, which had a minor increase. Recharge rates ranged from 0.16 to 0.80 m/year in 1998, with a mean rate of 0.50 m/year (±0.23). Rates in 2016 were not significantly different ($p = 0.19$; two-sample *t*-test assuming unequal variances) and varied between 0.13 and 0.60 m/year, averaging 0.35 m/year (±0.19). From 1998 to 2016, mean water depth increased from 7.9 m to 9.4 m.

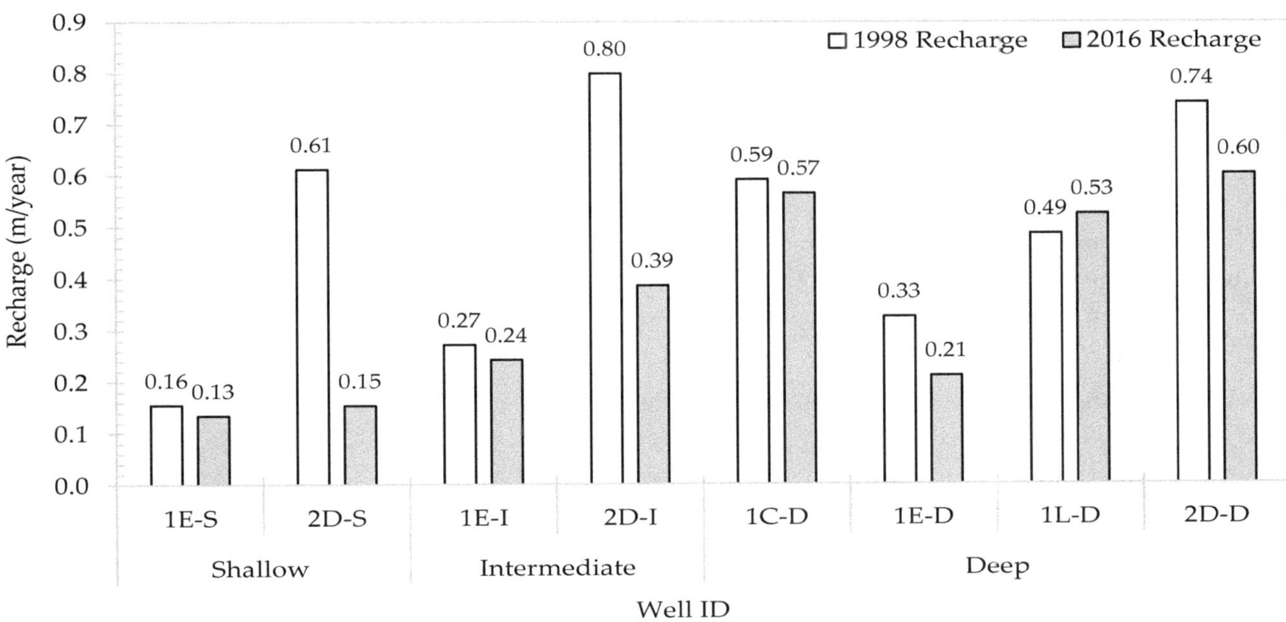

Figure 3. Comparison of 1998 and 2016 recharge rates categorized by shallow, intermediate, and deep well depths.

Shallow wells far from canals are believed to reflect localized recharge, while intermediate and deep wells are more likely to represent recharge sources from both localized irrigation and canal leakage [11]. Shallow wells had varying results, with a comparable recharge rate in Well 1E-S in both studies, and Well 2D-S less than 25% the 1998 recharge rate. The recharge rate in 1E-I, again, was similar between the two studies, while the 2016 recharge rate in 2D-I was less than half that of 1998. The 2016 mean recharge rate in deep wells was nearly double the mean from shallow wells. Deep wells, typically associated with greater groundwater ages, may take time before they reflect changes to environmental variables related to groundwater quantity. The lowest 1998 ($R = 0.33$ m/year) and 2016 ($R = 0.21$ m/year) recharge rates from deep wells were both in 1E-D, which is located far from the larger regional canals. The largest 1998 ($R = 0.74$ m/year) and 2016 ($R = 0.60$ m/year) deep well recharge rates were in Well 2D-D.

3.3. Nitrate Analysis

A combination of data collected in the 1990 and 2016 studies, in addition to long-term groundwater monitoring, provided two approaches to analyze nitrate concentration trends:

1. Comparison of data collected in 1998 [11,31] to data collected in 2016 at the same well nests where groundwater age-dating was conducted (Table 2 and Figure 4); and
2. Analysis of a long-term dataset from much broader groundwater collection efforts in the Dutch Flats area, including sporadic sampling between 1998 and 2016, and intensive sampling in 1998, 1999, 2008, and 2016 (Table 3, Figures 5 and 6, and Appendix A (Table A1)).

Focusing first on comparisons from Approach (1), nitrate samples collected in 1998 varied from 1.4 mg N L^{-1} to 15.8 mg N L^{-1}, while 2016 ranged from 1.1 mg N L^{-1} to 46.8 mg N L^{-1} (Table 2), and 6 out of 14 samples collected in 2016 had lower [NO_3^-] compared to samples collected in 1998 (Figure 4). Apart from Well 2D-I (1.3 mg N L^{-1}) and 2D-D (1.4 mg N L^{-1}), groundwater from well nests sampled in 2016 had lower [NO_3^-] with greater depth in the aquifer. Concentrations of ammonium in groundwater were below detection (<0.1 NH_4-N mg L^{-1}) in samples from the eleven wells where $\delta^{18}O$-NO_3^- was also determined (see wells listed in Table 2).

Prior to irrigation season, the 2016 Well 1G-S spring nitrate was 46.8 mg N L^{-1}, while post-irrigation season was 22.1 mg N L^{-1}. This trend is consistent with canal leakage diluting concentrations near canals, as suggested by Böhlke et al. [11]. The August 1998 [NO_3^-] was 8.8 mg N L^{-1}. Proximity to a nearby cattle feedlot could be influencing the high 2016 [NO_3^-] observed in 1G-S. It is unknown the exact date operations began at this feedlot, though it is believed between 1998 and 1999, and has increased over the past two decades.

Nitrate concentrations from Well 1E-S were high in both studies (1998 = 15.8 mg N L^{-1}; 2016 = 45.2 mg N L^{-1}), though 2016 was uncharacteristically high. Well 1E-I (3.6 mg N L^{-1}) and 1E-D (3.1 mg N L^{-1}), which had screens approximately 12.7 m and 26.5 m below 1E-S, respectively, had much lower [NO_3^-]. In addition to the 1998 and 2016 values reported above, data collected by NPNRD during 1996–1998 ($n = 26$) were used for supplementary analysis of Well 1E-S. Monthly averages over this period display increasing [NO_3^-] starting in June, peaking in August, and declining thereafter. If fertilizer is applied around growing season, Well 1E-S displays a very short transport rate through the vadose and saturated zones, in that groundwater [NO_3^-] quickly reflect surface activities. This is supported by a young groundwater age, and possibly suggests there is a preferential pathway to this well screen.

Table 2. Nitrate nitrogen and nitrogen isotopic ratio of nitrate from samples collected in 1998 and compared to 2016 samples, at well nests where age-dating was also conducted. Samples collected in 2016 were mostly analyzed for $\delta^{18}O\text{-}NO_3^-$, as shown.

Well ID	Böhlke et al. [11]			Current Study			
	Date Sampled	$\delta^{15}N\text{-}NO_3^-$ (‰)	$[NO_3^-]$ (mg N L^{-1})	Date Sampled	$\delta^{15}N\text{-}NO_3^-$ (‰)	$[NO_3^-]$ (mg N L^{-1})	$\delta^{18}O\text{-}NO_3$ (‰)
1G-S	27 August 1998	2.4	8.8	18 April 2016	17.0	46.8	n.d.
1G-I	27 August 1998	n.d.	10.6	18 April 2016	2.6	7.2	n.d.
1G-D	27 August 1998	2.5	8.0	18 April 2016	3.3	6.2	n.d.
2D-S	27 August 1998	5.7	8.3	16 August 2016	0.5	9.7	−4.5
2D-I	27 August 1998	5.6	5.1	16 August 2016	2.2	1.3	−9.16
2D-D	27 August 1998	4.9	1.4	16 August 2016	−2.9	1.4	−6.96
1E-S	24 August 1998	2.9	15.8	16 August 2016	−1.6	45.2	−5.45
1E-I	24 August 1998	2.7	10.8	16 August 2016	−1.3	3.6	−6.6
1E-D	24 August 1998	4.1	2.5	16 August 2016	−3.7	3.1	−5.37
1L-D	25 August 1998	10.2	2.4	17 August 2016	1.1	1.1	−3.67
1C-I	27 August 1998	4.2	2.5	17 August 2016	4.9	5.3	−7.27
1C-D	24 August 1998	4.5	2.5	17 August 2016	−2.0	3.8	−8.38
1G-S	27 August 1998	2.4	8.8	12 October 2016	18.4	22.1	4.08
1G-I	27 August 1998	n.d.	10.6	12 October 2016	9.5	6.9	0.33

Note: n.d., no data; S, Shallow well; I, Intermediate well; D, Deep well.

Well 1G-I was sampled twice in 2016 to evaluate temporal trends in both groundwater age and $[NO_3^-]$ near a canal. Apparent groundwater age was similar (spring = 5.9 years; fall = 5.3 years). Interestingly, $[NO_3^-]$ in this well decreased from 46.8 to 22.1 mg N L^{-1} in 2016. Similarity in groundwater ages between spring and fall sampling suggest groundwater $[NO_3^-]$ in this well are not diluted from a large percentage of 2016 canal water. That is, a seasonal pattern in $[NO_3^-]$ was observed, but if a significant fraction of canal water that infiltrated during the 2016 growing season was arriving at the well screen by fall 2016, then the groundwater age from fall sampling should be much less than groundwater age from spring sampling. Apparently, the mass flux of water leaking from canals drives groundwater deeper into the aquifer during irrigation season and dilutes groundwater nitrate with older (pre-2016) canal water with low $[NO_3^-]$.

Based on two relatively small datasets (Table 2 and Figure 4), it is difficult to identify trends in groundwater nitrate due to large variations in concentrations, and additional sources of nitrogen influencing Well 1G. As a result, additional long-term nitrate data collected by NPNRD were analyzed, as described in Approach (2) at the beginning of this section. In total, 2918 nitrate samples were collected in the Dutch Flats area between 1979 and 2016. However, wells are not consistently sampled, making it difficult to compare overall annual medians from one year to the next. Thus, only data from wells with two or more samples collected between 1998 and 2016 were used (n = 987 samples from a total of 160 wells; Figure 5a). If multiple samples were collected from a well within the same year, annual concentrations were averaged. The annual median and mean normalized values (Equation (5)) were then used to evaluate groundwater $[NO_3^-]$ trends in Dutch Flats (Figure 5b). Both mean and median of normalized annual nitrate concentrations suggest a decrease in groundwater $[NO_3^-]$ from 1998 to 2016, with statistically significant regression slopes (p-values of 0.04 in both cases).

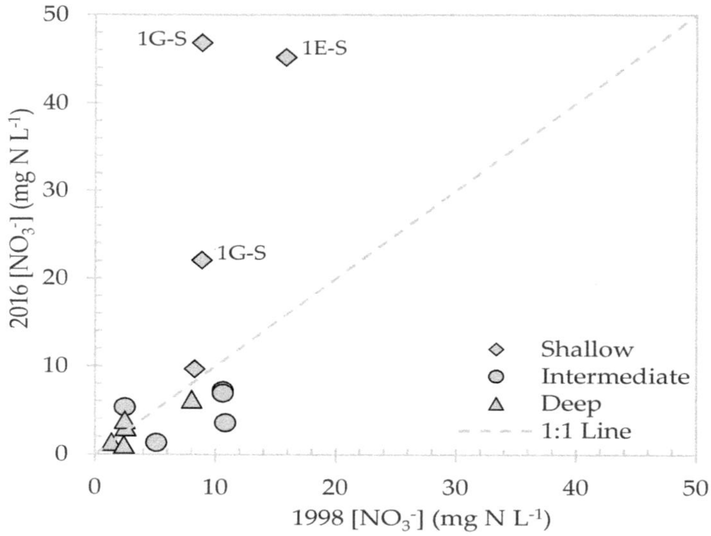

Figure 4. Comparison of nitrate concentrations from five well nests sampled in 2016 and 1998 (n = 14). Labels indicate the three wells with high 2016 [NO_3^-].

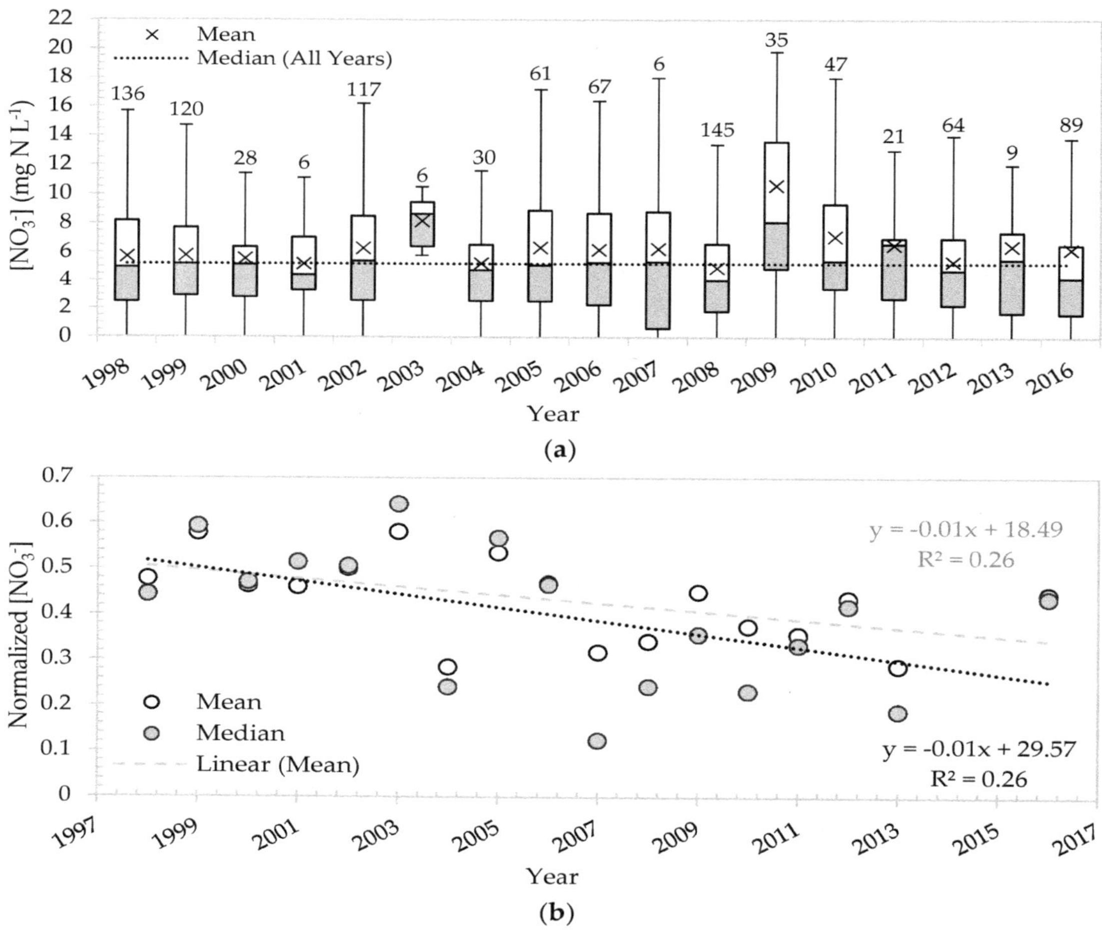

Figure 5. Nitrate data from 1998 to 2016 (n = 987) collected and/or maintained by the North Platte Natural Resources District, Nebraska Agricultural Contaminant Database, and the current study: (**a**) a Box-and-Whisker plot of all nitrate data, including the number of samples collected each year (referenced above the maxima) and long-term median of each annual median; and (**b**) mean and median of normalized annual Dutch Flats groundwater nitrate.

Given the suggested decrease in [NO_3^-] from the normalized data, we further explored trends in nitrate by evaluating three time periods. Time periods were selected based on the number of samples collected during a given year, as well as with respect to the time elapsed since the previous study. To characterize groundwater [NO_3^-] during the 1990s study, data from 1998 and 1999 were used as a base comparison. Two additional years, 2008 and 2016, were compared directly to [NO_3^-] in the late 1990s. These two years were selected because many samples were collected during 2008 and 2016. A total of 87 wells were sampled during all three time periods, and samples were further split based on screen depth (i.e., shallow, intermediate, and deep wells had n = 44, 16, and 27, respectively).

Overall, 52 out of 87 wells (60%) showed a decrease between 1998 and/or 1999 and 2016. The mean and median [NO_3^-] are reported in Table 3 for each well depth, and results from individual samples are shown in Appendix A and Figure 6. Since the data were not normally distributed and data transformation did not help, the Mann–Whitney test was used to determine if the median [NO_3^-] were different between the three periods (Table 3).

Table 3. The mean and median groundwater [NO_3^-] (mg N L^{-1}) for 1998 and/or 1999, 2008 and 2016. The calculated p-values are from Mann–Whitney tests comparing the medians between the three time periods.

Well Depth	Shallow (n = 44)			Intermediate (n = 16)			Deep (n = 27)		
	1990s *	**2008**	**2016**	**1990s ***	**2008**	**2016**	**1990s ***	**2008**	**2016**
Mean	7.2	6.2	8.0	6.6	5.6	5.1	4.0	3.5	3.5
Median	5.4	4.7	5.3	5.6	4.8	3.7	3.6	3.3	3.4
	Years		p-value	Years		p-value	Years		p-value
Mann–Whitney Test	1990s *–2008		0.15	1990s *–2008		0.66	1990s *–2008		0.68
	1990s *–2016		0.49	1990s *–2016		0.17	1990s *–2016		0.62
	2008–2016		0.70	2008–2016		0.38	2008–2016		0.94

Note: * Value shown is from 1998 or 1999, or the average from the two years.

The median [NO_3^-] for the 44 shallow wells decreased by 13% from 1998 to 2008, but increased by 13% from 2008 to 2016. These fluctuations may be attributed to the variation in precipitation, with high precipitation resulting in increased leaching rates. From 1996 to 1998 the average precipitation was 459 mm, compared to 303 mm and 497 mm from 2006 to 2008 and 2014 to 2016, respectively.

Though median concentrations were not significantly different (p-value 0.17; 5.6 versus 3.7 mg N L^{-1} for 1998 and 2016, respectively), 69% of the intermediate wells sampled had a reduction in [NO_3^-] from 1998 to 2008 and 75% had a reduction from 1998 to 2016. The median [NO_3^-] also decreased in the deep wells, but only by 8% from 1998 to 2008 and 6% from 1998 to 2016, and the differences were not statistically significant.

Although overall [NO_3^-] trends are decreasing in many individual wells, there is substantial variability and uncertainty in overall results, and a lack of statistical significance in median values. Other variables such as vadose zone depth, fertilizer application rates, percent cropland (specifically corn) should be considered. To develop a better understanding of the complexities of the system, a statistical model coupled with an increase in predictor variables may help explain the large fluctuations in nitrate trends. Continued long-term monitoring of groundwater [NO_3^-], perhaps with a sampling scheduled optimized by well characteristics and/or apparent vulnerability to groundwater nitrate contamination, will be critical for future studies.

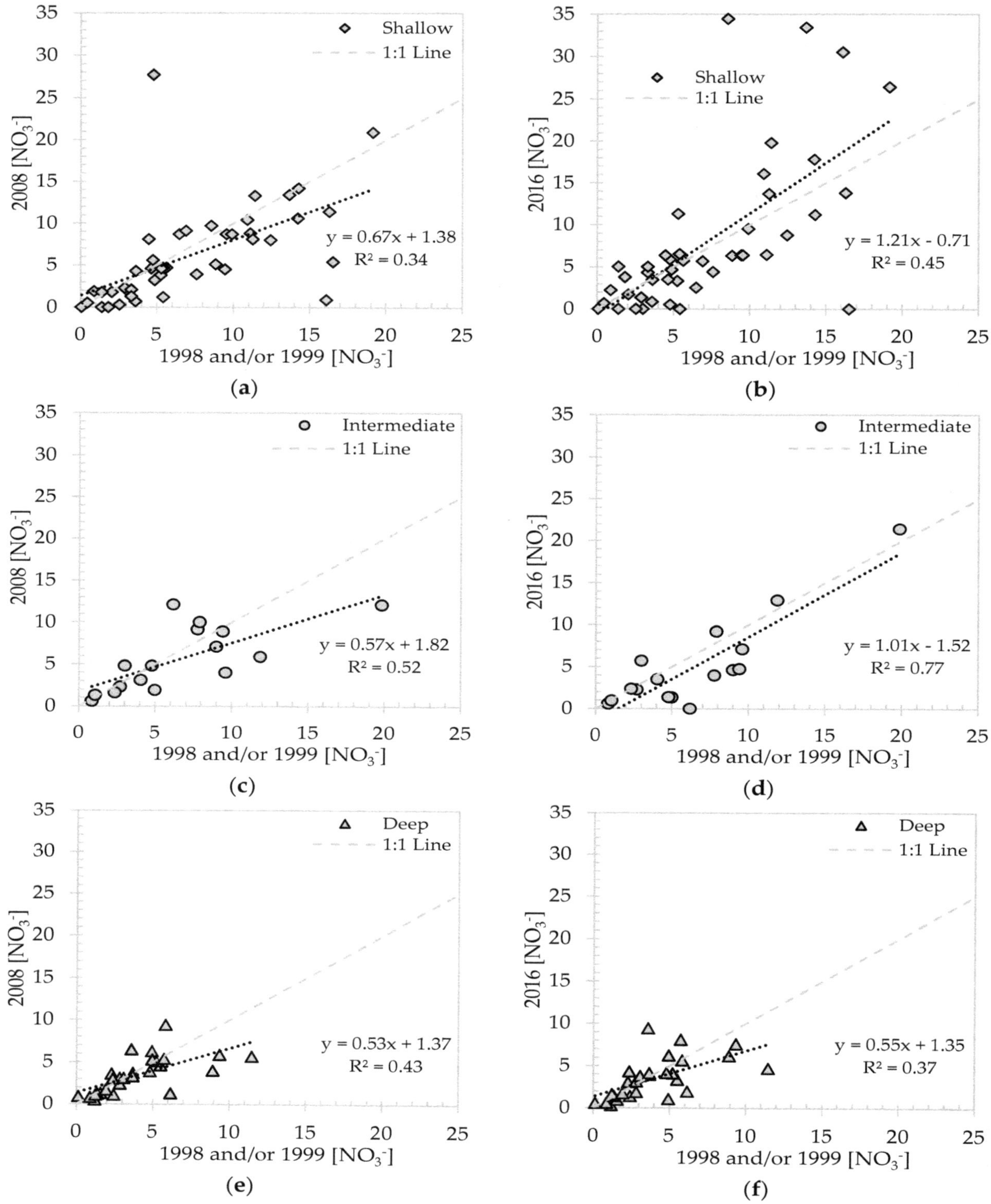

Figure 6. Comparison of groundwater $[NO_3^-]$ (mg N L^{-1}) from samples collected in 1998 and/or 1999 in shallow, intermediate, and deep wells to: (**a**) 2008 $[NO_3^-]$ in shallow wells (n = 44); (**b**) 2016 $[NO_3^-]$ in shallow wells (n = 44); (**c**) 2008 $[NO_3^-]$ in intermediate depth wells (n = 16); (**d**) 2016 $[NO_3^-]$ in intermediate depth wells (n = 16); (**e**) 2008 $[NO_3^-]$ in deep wells (n = 27); and (**f**) 2016 $[NO_3^-]$ in deep wells (n = 27).

3.4. Sources of Nitrate

Nitrogen-15 isotopes of nitrate were analyzed to characterize nitrogen sources (Table 2 and Figure 7). Oxygen-18 isotopes of nitrate were also analyzed in 2016. Nitrogen, of both ^{14}N and ^{15}N, is fixed from the atmosphere to produce fertilizer. Plants favor $^{14}N\text{-}NO_3^-$ to $^{15}N\text{-}NO_3^-$ during assimilation, increasing $^{15}N\text{-}NO_3^-$ observed in groundwater. It is believed this process also results in slightly elevated levels of $^{18}O\text{-}NO_3^-$ [59]. With relatively constant atmospheric ratios of stable oxygen (^{18}O:^{16}O) and nitrogen isotopes (^{15}N:^{14}N), Equation 4 may identify enrichment or depletion of $^{18}O\text{-}NO_3^-$ and $^{15}N\text{-}NO_3^-$ in groundwater. The combination of $\delta^{15}N\text{-}NO_3^-$ and $\delta^{18}O\text{-}NO_3^-$ may be used to determine sources of nitrate and potential for denitrification to have affected nitrate [60,61]. Different signatures suggest nitrogen sources may be naturally occurring or associated with anthropogenic activities.

Well Nest 1G was sampled in both spring and fall, but only fall samples were collected for oxygen isotopes. Nitrogen isotopes were collected during each survey (n = 14), but only samples with an oxygen isotope counterpart were used in this analysis (n = 11). Recently collected $\delta^{18}O\text{-}NO_3^-$ varied between −9.2 and 4.1‰, while $\delta^{15}N\text{-}NO_3^-$ spanned from −3.7‰to 18.4‰. The majority of nitrate in Dutch Flats study area appears to be derived from nitrification of ammonium in fertilizer and precipitation. Values from the 1G shallow well ($\delta^{18}O\text{-}NO_3^-$ = 4.08‰, $\delta^{15}N\text{-}NO_3^-$ = 18.4‰) and intermediate well ($\delta^{18}O\text{-}NO_3^-$ = 0.33‰, $\delta^{15}N\text{-}NO_3^-$ = 9.52‰) had higher nitrogen isotope composition in comparison to the other nine samples. Isotopically heavy $\delta^{15}N\text{-}NO_3^-$ in Well 1G is consistent with nitrate from organic sources (i.e., manure), and is consistent with its proximity to an adjacent feedlot (Figure 7).

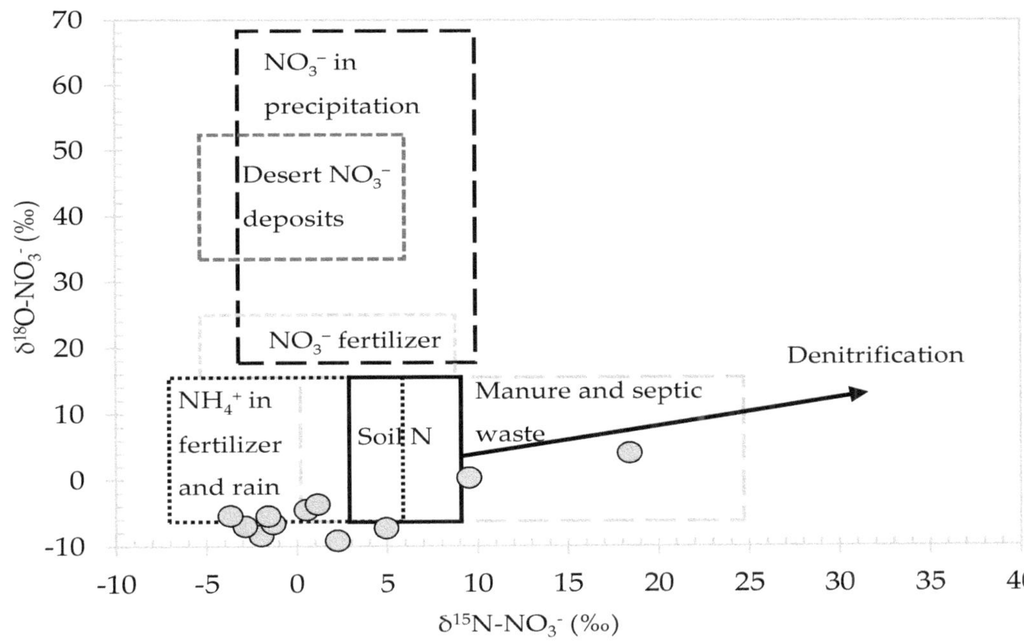

Figure 7. Determining sources of nitrogen in Dutch Flats area from oxygen and nitrogen isotopic ratios in nitrate. Figure labels are modified from Kendall [62].

3.5. Biogeochemical Processes

Nitrate isotope composition, dissolved oxygen (DO), and DOC were used to indicate changes in biogeochemical activity affecting nitrate, such as denitrification. Results from DO and DOC may be referenced in Appendix A (Figures A1 and A2). Nitrogen isotopes were collected and analyzed for their signatures in both 1998 and 2016 (Table 2). Samples from Well 1G-I were not available from the 1990s study. Denitrification could lead to increased $\delta^{15}N\text{-}NO_3^-$. However, data collected in 2016 generally displayed a leftward shift in δ^{15}N-NO_3^- compared to 1998. Overall, results do not suggest an increase in denitrification rates throughout the region (Figure 8). Values from 1998 ranged 2.4 to 10.2‰, while 2016 ranged from −3.7 to 18.4‰. Of the 14 samples collected in 2016 for δ^{15}N-NO_3^-, 12 samples had a 1998 counterpart. Eight samples show stable or even a slight decrease in δ^{15}N-NO_3^- (average change = −6.0‰), while two, excluding two outliers at 1G-S, increased (average change = 0.8‰).

Two nitrate isotope sample results stand out, both from Well 1G. From 1998 to 2016, δ^{15}N-NO_3^- in Well 1G-S increased from 2.4‰, to 17‰and 18.4‰in the spring and fall, respectively. It was noted during collection of groundwater in spring of 2016 that Well 1G-S had a yellow color, and spring-time sample collection occurred immediately after a large precipitation event. Thus, it is possible the organic nitrogen source of nitrate was a result of runoff. Increased nitrogen isotope composition in 1G-S indicates and increased influence of organic nitrogen sources, but could also suggest more prevalent microbiological activity coinciding with a decrease in dissolved oxygen (1998 = 5.6 mg O_2 L^{-1}, fall 2016 = 1.4 mg O_2 L^{-1}), although the groundwater was not strictly anoxic when samples in 2016. Dissolved organic carbon can serve as a mechanism driving microbial activity and denitrification, and within Well 1G-S, DOC increased from 1998 (2.9 mg C L^{-1}) to 2016 (10.2 mg C L^{-1}). Such increases indicate nitrogen from feedlot manure had little influence on samples collected in 1998, or yet to be in operation. Under anoxic conditions (DO < 0.5 mg O_2 L^{-1}), reduction of nitrate becomes favorable when concentrations are above 0.5 mg N L^{-1} [63]. It is possible recharge from high dissolved oxygen in canal water may prevent conditions from becoming anoxic at this site, and without canal leakage, denitrification could be higher.

Well 1L-D, located near North Platte River, had a water table approximately 2 m below the surface. This well had the largest 1998 (10.2‰) to 2016 (1.1‰) decrease in δ^{15}N-NO_3^-, although DO (1998 = 0.1 mg O_2 L^{-1}, 2016 = 0.4 mg O_2 L^{-1}) indicated consistent anoxic conditions. Slightly higher 1998 concentrations of NO_3-N and DOC, coupled with lower DO in 1998, may explain the larger δ^{15}N-NO_3^- value. From further examination of δ^{15}N-NO_3^- versus 1/[NO_3^-] and δ^{15}N-NO_3^- versus ln[NO_3^-], mixing analyses (Figure A3) were inconclusive as to whether mixing of high- and low-[NO_3^-] groundwater or denitrification were factors influencing groundwater nitrate in the Dutch Flats area [64].

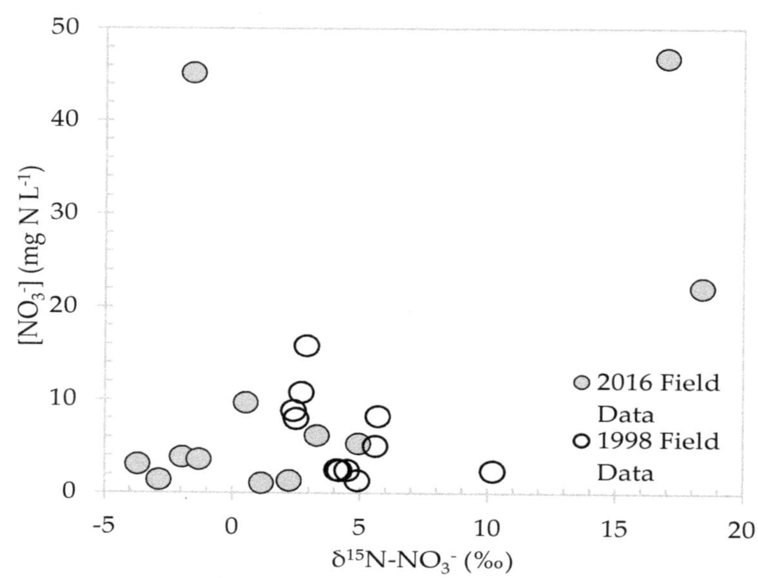

Figure 8. Comparison of [NO_3^-] and δ^{15}N-NO_3^- for 1998 and 2016 field data. Far right data points are from Well 1G-S and suggest organic nitrogen source.

3.6. Analysis of Other Relevant Environmental Variables

Potential changes in relevant environmental variables were evaluated for time periods prior to the 1998 and 2016 studies (i.e., before and after the USGS study). Variables associated with recharge (precipitation and Interstate Canal discharge) and nitrate (planted corn area and fertilizer loads) were analyzed for statistically significant differences between the two time periods. Precipitation records used were from the Western Regional Airport in Scottsbluff, NE [38]. Annual volume of water diverted into the Interstate Canal was from a gage station approximately 1.6 km downstream of Whalen Diversion Dam in Wyoming [65]. Planted corn area and fertilizer loads are both estimates for Scotts Bluff County [66,67]. Due to limited fertilizer application data, two time periods, each 13 years, were compared from 1987 to 1999, and 2000 to 2012. All other variables were compared over 17-year periods prior to and after the completion of the 1990s USGS study. Datasets were determined to follow a normal distribution, and a two-sample *t*-test was used to evaluate statistically significant differences between two time periods for each variable (Table 4).

Besides precipitation ($p = 0.11$), each environmental variable was determined to be significantly different when comparing the two time periods. Both Interstate Canal discharge and fertilizer loads exhibited a reduction, while planted corn area increased between the time periods. The inverse relationship between planted corn area and fertilizer loads is interesting and perhaps suggests an improvement in fertilizer application management, or possibly higher uncertainties associated with county level fertilizer estimates. The reduction in discharge from the canal may be attributed to the change in irrigation management practices.

Table 4. Summary of statistical analysis evaluating variables potentially influencing groundwater quantity and quality. *p*-values were determined from two-sample *t*-tests.

Variable	Mean (±std)		*p*-Value
	Year: 1983–1999	Year: 2000–2016	
Precipitation (mm)	431 (±97)	370 (±118)	0.11
Interstate Canal Discharge (km^3/year)	0.52 (±0.08)	0.44 (±0.08)	0.007 *
Planted Corn Area (hectares)	29,471 (±2568)	34,217 (±2608)	<0.001 *
	Year: 1987–1999	Year: 2000–2012	
Fertilizer Loads (kg)	11,503,061 (±1,150,187)	9,540,057 (±1,222,507)	<0.001 *

Note: * Statistically significant difference between two time periods ($\alpha = 0.05$).

3.7. Further Discussion

Since the previous 1990s study, numerous environmental variables related to groundwater nitrate contamination have undergone changes, including shifts in irrigation practices and canal management. Center pivot irrigated area has increased an estimated 270% from 1999 to 2017 within Dutch Flats. Much of the increase has occurred on fields previously irrigated by furrow systems, although there has likely been an increase in overall irrigated acres as well. Scotts Bluff County total irrigated area statistics from the National Agricultural Statistics Service [68,69] estimated an increase in irrigated area over a similar time period (1997 = 70,075 hectares; 2012 = 80,611 hectares). A significant difference in the means was found when comparing volumes diverted into the Interstate Canal from 1983–1999 to 2000–2016. While precipitation also displayed decreasing trends, statistical analysis did not find a significant difference over the same period. With more efficient irrigation methods, less precipitation, and decreased canal discharge, a decrease in recharge rates would be expected. Although seven of the eight wells did have decreased recharge rates, the mean recharge rates from the two studies were not significantly different. This may be attributed to insufficient time passing between the two sampling periods.

Groundwater denitrification still does not appear to be a major process affecting nitrate attenuation in Dutch Flats since the previous study. Nitrate data collected in 2016 and compared to a small subset of results from the previous study identified a large scatter in data. Three 2016 samples (two from the same well) had high nitrate concentrations, making it difficult to compare trends from the two small datasets. Additional analysis of long-term nitrate data collected by the NPNRD suggests [NO_3^-] have decreased in most wells between 1998 and 2016. Further, two variables related to nitrogen inputs were evaluated during this study: hectares of planted corn and estimated fertilizer loads. The two variables showed significant differences in their means; planted corn area has increased, while fertilizer loads have decreased since the previous study.

Decreased recharge rates were hypothesized to have reduced nitrate concentrations in the Dutch Flats area. However, trends from this study suggest that: (1) improved irrigation efficiency and changes to canal management have yet to significantly influence recharge rates in Dutch Flats; and (2) [NO_3^-] are currently decreasing as a result of a combination of variables, perhaps including improved nitrogen management practices. The second point is further suggested by a significant increase in planted corn area, yet significant decrease in estimated fertilizer application in Scotts Bluff County. In other words, it is possible other variables beyond irrigation practice and canal management are currently driving the decreasing trends in groundwater [NO_3^-], consistent with other studies that emphasize the need to improve both water and nitrogen management in agricultural production [12,13,70,71]. It should be noted, however, that, while this study was unable to detect a significant reduction in recharge rates, limitations and uncertainties associated with groundwater age-dating and recharge calculations may make it difficult to identify a small but meaningful decrease. If future research were performed in a similar manner and found decreased recharge rates compared to this study, an even greater reduction in groundwater [NO_3^-] is likely to be observed.

It is noteworthy that an estimated increase in irrigated area from 1997 to 2012 is being supplemented by a decrease in annual volume of water diverted into the Interstate Canal, with little indication of increased groundwater withdrawals. Without increased precipitation, these trends, nonetheless, serve as potential evidence of the extent at which irrigation efficiency has improved. Simply put, it is possible less water is being applied to fields regionally, even with increases in irrigated area.

An interesting dynamic to consider is the high leakage potential from canals in this region, and their association with diluting groundwater [NO_3^-]. If future efforts are made to improve irrigation efficiency through lining canals, less artificial recharge will be supplied to the region. Ultimately, this could result in a declining water table elevation, where it has been found artificial recharge is important in restoring aquifer storage and improving groundwater quality [72]. Further, it is unknown how the impact these water management improvements may have on groundwater [NO_3^-]. For instance, less nitrate might leach below the root zone with continued advancements in irrigation efficiency, however, less artificial recharge from low-[NO_3^-] canal water would be present to dilute groundwater [NO_3^-].

4. Conclusions

The study area, known locally as the Dutch Flats, has undergone changes in numerous variables influencing groundwater attributes, on timescales similar to those of groundwater movement through the aquifer. This study exemplifies the ability to use intensive snapshot sampling, coupled with long-term continuous data, to evaluate groundwater trends. Varying results in both recharge and [NO_3^-] from this study promote supplementary, and possibly more expansive investigations in the Dutch Flats area. It is possible more time is required to observe changes in groundwater recharge rates. Accounting for lag time through the vadose and saturated zones, some portions of the aquifer may have yet to reflect how changes in environmental variables will impact groundwater quantity and quality. Future resampling in the study area would be beneficial, though carefully-designed long-term monitoring and/or sampling from more wells would offer a better comparison to the more comprehensive 1990s survey. With a vast dataset of nitrate data available through the NPNRD, additional analysis could be beneficial in the identification of variables within Dutch Flats most strongly related to groundwater [NO_3^-]. Existence of the long-term nitrate dataset, coupled with groundwater age-dating to establish a range of lag times, makes for an opportune setting for additional analytics such as machine learning algorithms or classification techniques.

Author Contributions: Conceptualization, T.E.G.; Formal analysis, M.J.W., T.E.G. and A.R.M.; Funding acquisition, T.E.G., D.S. and S.S.S.; Investigation, M.J.W., T.E.G. and S.S.S.; Methodology, T.E.G. and D.S.; Project administration, T.E.G.; Resources, T.E.G.; Supervision, T.E.G.; Writing—original draft, M.J.W.; and Writing—review and editing, M.J.W., T.E.G., A.R.M., D.S. and S.S.S.

Acknowledgments: The authors acknowledge the North Platte Natural Resources District for providing technical assistance and resources, including long-term groundwater nitrate data accessed via the Quality-Assessed Agrichemical Contaminant Database for Nebraska Groundwater, and Mason Johnson (graduate research assistant) for his support in field sampling efforts. J.K. Böhlke (USGS) also provided insightful comments and suggestions during the project.

Appendix A

Table A1. Nitrate data from current study, North Platte Natural Resources District, and Nebraska Agricultural Contaminant Database. This dataset was used to create comparisons in Table 3 and Figure 6.

Well ID	1990s [NO_3^-] * (mg N L^{-1})	2008 [NO_3^-] (mg N L^{-1})	2016 [NO_3^-] (mg N L^{-1})
10A-S	1.4	0.0	0.0
10E-S	16.5	5.4	0.0
10K-S	7.6	3.9	4.4
10M-S	9.5	4.5	6.4
10N-S	3.0	2.1	0.0
1C-S	9.9	8.7	9.5
1E-S	13.7	13.4	33.5
1G-S	8.6	9.7	34.5
1H-S	11.3	8.1	13.7
1J-S	16.3	11.4	13.8
1M-S	10.9	10.4	16.1
1N-S	3.3	2.1	4.3
2E-S	4.9	3.2	4.7
2F-S	16.1	0.9	30.5
2J-S	12.4	8.0	8.8
3B-S	2.6	0.3	0.0
3C-S	1.8	0.0	3.8
3F-S	4.8	27.7	0.5
4A-S	0.5	0.5	0.7
5A-S	2.9	2.3	1.4
5B-S	0.9	1.9	2.2
5D-S	11.1	8.8	6.5
5F-S	4.7	4.6	3.5
5G-S	11.4	13.3	19.8
6C-S	4.8	5.6	5.9
6D-S	19.2	20.9	26.4

Table A1. *Cont.*

Well ID	1990s [NO_3^-] * (mg N L^{-1})	2008 [NO_3^-] (mg N L^{-1})	2016 [NO_3^-] (mg N L^{-1})
6E-S	8.9	5.1	6.3
6F-S	3.6	0.7	0.9
6G-S	2.1	1.8	1.7
6H-S	3.6	4.3	3.5
6M-S	4.5	8.1	6.4
6N-S	6.9	9.1	5.7
7A-S	5.3	3.8	3.3
7B-S	6.5	8.7	2.6
7C-S	14.3	14.2	11.2
7D-S	3.3	1.3	5.0
7H-S	14.2	10.6	17.8
8B-S	5.3	4.6	11.3
8C-S	5.6	4.7	5.7
8D-S	5.4	4.7	6.5
8E-S	0.1	0.0	0.0
8G-S	1.4	1.8	5.0
9D-S	5.4	1.2	0.0
9E-S	9.6	8.7	6.4
11D-M	2.7	2.3	2.2
1C-M	3.0	4.8	5.7
1E-M	9.0	7.1	4.6
1G-M	9.6	4.0	7.1
1H-M	4.0	3.1	3.5
1J-M	7.8	9.2	4.0
1M-M	11.9	5.9	12.9
2C-M	2.3	1.6	2.3
2D-M	5.0	1.9	1.3
2F-M	19.8	12.1	21.4
2J-M	9.4	8.9	4.7
2L-M	4.8	4.8	1.4
3C-M	1.1	1.3	0.9
3E-M	7.9	10.0	9.2
3F-M	6.2	12.1	0.0
5B-M	0.9	0.6	0.6
10A-D	3.6	6.5	9.4
10E-D	5.2	5.1	4.1
10K-D	5.0	5.3	6.2
10M-D	9.4	5.9	7.6
11A-D	4.8	3.9	4.2
1C-D	2.4	3.0	4.3
1E-D	2.3	3.6	3.1
1G-D	8.9	4.0	6.2
1L-D	4.9	6.3	1.1
2C-D	2.8	2.4	1.9
2D-D	1.4	1.2	1.4
2F-D	3.1	3.1	3.8
2L-D	1.6	1.3	1.0
3B-D	1.2	0.5	0.3
3C-D	1.2	1.1	1.6
3E-D	3.7	3.7	3.9
3F-D	11.5	5.7	4.7
5B-D	0.9	0.8	0.6
6G-D	6.2	1.3	2.0
6H-D	0.1	0.9	0.6
6M-D	2.4	1.1	1.4

Table A1. *Cont.*

Well ID	1990s $[NO_3^-]$ * ($mg\ N\ L^{-1}$)	2008 $[NO_3^-]$ ($mg\ N\ L^{-1}$)	2016 $[NO_3^-]$ ($mg\ N\ L^{-1}$)
7A-D	1.9	1.7	1.7
7C-D	5.8	9.4	5.7
7D-D	3.7	3.3	4.1
8D-D	5.7	5.4	8.1
9D-D	2.8	3.1	3.1
9E-D	5.5	4.6	3.4

Note: * Value shown is from 1998 or 1999, or the average from the two years.

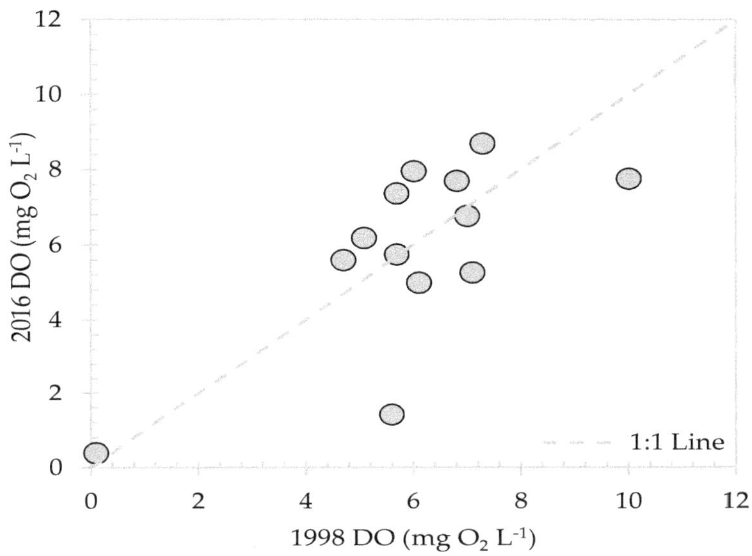

Figure A1. Comparison of groundwater dissolved oxygen (DO) in 2016 to 1998, where DO was mostly similar between both studies.

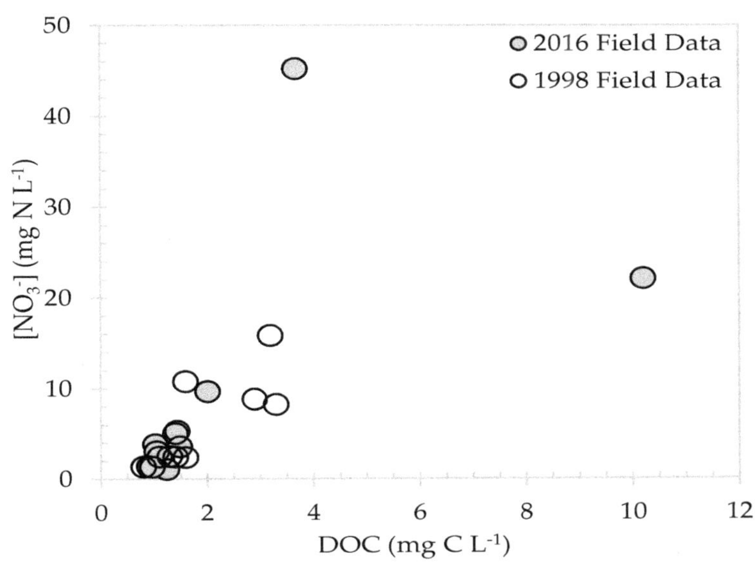

Figure A2. Comparison of $[NO_3^-]$ in 2016 and 1998 to dissolved organic carbon (DOC). Large $[NO_3^-]$ and DOC at Well 1G-S are consistent with isotopes indicating high organic nitrogen source.

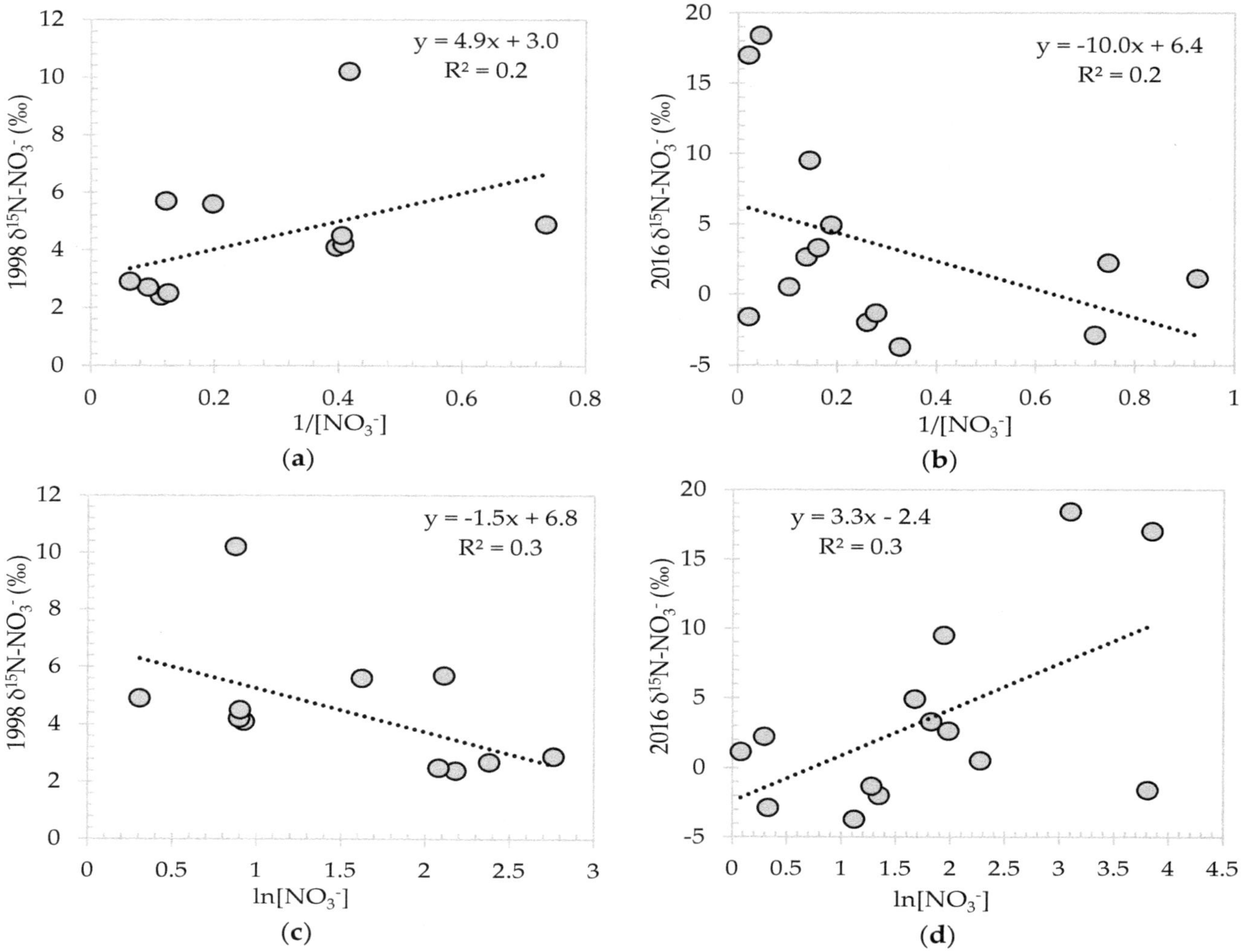

Figure A3. Evaluating whether $\delta^{15}N\text{-}NO_3^-$ suggests processes of mixing high- and low-$[NO_3]$ groundwater in: (**a**) 1998; and (**b**) 2016; or groundwater denitrification in: (**c**) 1998; and (**d**) 2016 [64]. Simple groundwater mixing nor denitrification were indicated from the composition of nitrogen isotopes throughout the Dutch Flats.

References

1. Almasri, M.N.; Kaluarachchi, J.J. Assessment and management of long-term nitrate pollution of ground water in agriculture-dominated watersheds. *J. Hydrol.* **2004**, *295*, 225–245. [CrossRef]
2. Burkart, M.R.; Stoner, J.D. Nitrate in aquifers beneath agricultural systems. *Water Sci. Technol.* **2007**, *56*, 59–69. [CrossRef] [PubMed]
3. Ritter, A.; Muñoz-Carpena, R.; Bosch, D.D.; Schaffer, B.; Potter, T.L. Agricultural land use and hydrology affect variability of shallow groundwater nitrate concentration in South Florida. *Hydrol. Process.* **2007**, *21*, 2464–2473. [CrossRef]
4. Derby, N.E.; Casey, F.X.M.; Knighton, R.E. Long-term observations of vadose zone and groundwater nitrate concentrations under irrigated agriculture. *Vadose Zone J.* **2009**, *8*, 290–300. [CrossRef]
5. Lockhart, K.M.; King, A.M.; Harter, T. Identifying sources of groundwater nitrate contamination in a large alluvial groundwater basin with highly diversified intensive agricultural production. *J. Contam. Hydrol.* **2013**, *151*, 140–154. [CrossRef] [PubMed]
6. Nolan, B.T.; Ruddy, B.C.; Hitt, K.J.; Helsel, D.R. Risk of nitrate in groundwaters of the united states—A national perspective. *Environ. Sci. Technol.* **1997**, *31*, 2229–2236. [CrossRef]

7. Saffigna, P.G.; Keeney, D.R.; Tanner, C.B. Nitrogen, chloride, and water balance with irrigated russet burbank potatoes in a sandy soil. *Agron. J.* **1977**, *69*, 251–257. [CrossRef]
8. Albertson, P.N. Agricultural Chemicals, Land Use, and Their Impacts on Stream and Ground Water Quality in the Little Plover River Watershed. Master's Thesis, University of Wisconsin—Stevens Point, Stevens Point, WI, USA, August 1998.
9. Piskin, R. Evaluation of nitrate content of ground water in Hall County, Nebraskaa. *Groundwater* **1973**, *11*, 4–13. [CrossRef]
10. Gosselin, D.C. *Bazile Triangle Groundwater Quality Study*; University of Nebraska: Lincoln, NE, USA, 1991; p. 29.
11. Böhlke, J.K.; Verstraeten, I.M.; Kraemer, T.F. Effects of surface-water irrigation on sources, fluxes, and residence times of water, nitrate, and uranium in an alluvial aquifer. *Appl. Geochem.* **2007**, *22*, 152–174. [CrossRef]
12. Exner, M.E.; Perea-Estrada, H.; Spalding, R.F. Long-term response of groundwater nitrate concentrations to management regulations in Nebraska's Central Platte Valley. *Sci. World J.* **2010**, *10*, 286–297. [CrossRef] [PubMed]
13. Exner, M.E.; Hirsh, A.J.; Spalding, R.F. Nebraska's groundwater legacy: Nitrate contamination beneath irrigated cropland. *Water Resour. Res.* **2014**, *50*, 4474–4489. [CrossRef] [PubMed]
14. Jenkins, W.J. *A History of Nebraska's Natural Resources Districts*; Hyer, R.B., Ed.; Nebraska Department of Natural Resources: Lincoln, NE, USA, 1975.
15. Exner, M.E.; Spalding, R.F. Groundwater quality and policy options in Nebraska. In *Groundwater Quality and Policy Options in Nebraska*; Smith, R., Ed.; Center for Applied Urban Research, University of Nebraska: Omaha, NE, USA, 1987.
16. Cash, D.W. Innovative natural resource management: Nebraska's model for linking science and decisionmaking. *Environ. Sci. Policy Sustain. Dev.* **2003**, *45*, 8–20. [CrossRef]
17. Visser, A.; Broers, H.P.; van der Grift, B.; Bierkens, M.F.P. Demonstrating trend reversal of groundwater quality in relation to time of recharge determined by $^3H/^3He$. *Environ. Pollut.* **2007**, *148*, 797–807. [CrossRef] [PubMed]
18. Wassenaar, L.I.; Hendry, M.J.; Harrington, N. Decadal geochemical and isotopic trends for nitrate in a transboundary aquifer and implications for agricultural beneficial management practices. *Environ. Sci. Technol.* **2006**, *40*, 4626–4632. [CrossRef] [PubMed]
19. Spalding, R.F.; Watts, D.G.; Schepers, J.S.; Burbach, M.E.; Exner, M.E.; Poreda, R.J.; Martin, G.E. Controlling nitrate leaching in irrigated agriculture. *J. Environ. Qual.* **2001**, *30*, 1184–1194. [CrossRef] [PubMed]
20. McMahon, P.B.; Plummer, L.N.; Böhlke, J.K.; Shapiro, S.D.; Hinkle, S.R. A comparison of recharge rates in aquifers of the United States based on groundwater-age data. *Hydrogeol. J.* **2011**, *19*, 779–800. [CrossRef]
21. Baudron, P.; Alonso-Sarría, F.; García-Aróstegui, J.L.; Cánovas-García, F.; Martínez-Vicente, D.; Moreno-Brotóns, J. Identifying the origin of groundwater samples in a multi-layer aquifer system with Random Forest classification. *J. Hydrol.* **2013**, *499*, 303–315. [CrossRef]
22. Johnston, C.T.; Cook, P.G.; Frape, S.K.; Plummer, L.N.; Busenberg, E.; Blackport, R. Ground water age and nitrate distribution within a glacial aquifer beneath a thick unsaturated zone. *Groundwater* **1998**, *36*, 171–180. [CrossRef]
23. Katz, B.G.; Chelette, A.R.; Pratt, T.R. Use of chemical and isotopic tracers to assess nitrate contamination and ground-water age, Woodville Karst Plain, USA. *J. Hydrol.* **2004**, *289*, 36–61. [CrossRef]
24. Rosen, M.R.; Lapham, W.W. Introduction to the U.S. Geological Survey National Water-Quality Assessment (NAWQA) of ground-water quality trends and comparison to other national programs. *J. Environ. Qual.* **2008**, *37*, 190–198. [CrossRef] [PubMed]
25. Puckett, L.J.; Tesoriero, A.J.; Dubrovsky, N.M. Nitrogen contamination of surficial aquifers—A growing legacy. *Environ. Sci. Technol.* **2011**, *45*, 839–844. [CrossRef] [PubMed]
26. Böhlke, J.K. Groundwater recharge and agricultural contamination. *Hydrogeol. J.* **2002**, *10*, 153–179. [CrossRef]
27. Moore, K.B.; Ekwurzel, B.; Esser, B.K.; Hudson, G.B.; Moran, J.E. Sources of groundwater nitrate revealed using residence time and isotope methods. *Appl. Geochem.* **2006**, *21*, 1016–1029. [CrossRef]
28. Carlson, M.A.; Lohse, K.A.; McIntosh, J.C.; McLain, J.E.T. Impacts of urbanization on groundwater quality and recharge in a semi-arid alluvial basin. *J. Hydrol.* **2011**, *409*, 196–211. [CrossRef]

29. Harvey, F.E.; Sibray, S.S. Delineating ground water recharge from leaking irrigation canals using water chemistry and isotopes. *Groundwater* **2001**, *39*, 408–421. [CrossRef]
30. Verstraeten, I.M.; Steele, G.V.; Cannia, J.C.; Hitch, D.E.; Scripter, K.G.; Böhlke, J.K.; Kraemer, T.F.; Stanton, J.S. *Interaction of Surface Water and Ground Water in the Dutch Flats Area, Western Nebraska, 1995–99*; Water-Resources Investigations Report; United States Geological Survey, United States Department of the Interior: Reston, VA, USA, 2001; p. 56.
31. Verstraeten, I.M.; Steele, G.V.; Cannia, J.C.; Böhlke, J.K.; Kraemer, T.E.; Hitch, D.E.; Wilson, K.E.; Carnes, A.E. *Selected Field and Analytical Methods and Analytical Results in the Dutch Flats Area, Western Nebraska, 1995–99*; United States Geological Survey: Reston, VA, USA, 2001; p. 53.
32. Ball, L.B.; Kress, W.H.; Steele, G.V.; Cannia, J.C.; Andersen, M.J. *Determination of Canal Leakage Potential Using Continuous Resistivity Profiling Techniques, Interstate and Tri-State Canals, Western Nebraska and Eastern Wyoming, 2004*; Scientific Investigations Report; United States Geological Survey, United States Department of the Interior: Reston, VA, USA, 2006; p. 59.
33. Hobza, C.M.; Andersen, M.J. *Quantifying Canal Leakage Rates Using A Mass-Balance Approach and Heat-Based Hydraulic Conductivity Estimates in Selected Irrigation Canals, Western Nebraska, 2007 through 2009*; Scientific Investigations Report; United States Geological Survey, United States Department of the Interior: Reston, VA, USA, 2010; p. 38.
34. Luckey, R.R.; Cannia, J.C. *Groundwater Flow Model of the Western Model Unit of the Nebraska Cooperative Hydrology Study (COHYST) Area*; Nebraska Department of Natural Resources: Lincoln, NE, USA, 2006; p. 63.
35. Nebraska Department of Natural Resources (NEDNR). *Fifty-fifth biennial report of the Department of Natural Resources*; Nebraska Department of Natural Resources (NEDNR): Lincoln, NE, USA, 2009; p. 675.
36. Cannia, J.C. (Aqua Geo Frameworks, Mitchell, NE, USA); Gilmore, T.E. (University of Nebraska, Lincoln, NE, USA). Personal communication, 2016.
37. Steele, G.V.; Cannia, J.C. *Reconnaissance of Surface-Water Quality in the North Platte Natural Resources District, Western Nebraska, 1993*; Water-Resources Investigations Report; United States Geological Survey, United States Department of the Interior: Reston, VA, USA, 1997.
38. NOAA Data Tools | Climate Data Online (CDO) | National Climatic Data Center (NCDC). Available online: https://www.ncdc.noaa.gov/cdo-web/datatools (accessed on 4 August 2017).
39. Verstraeten, I.M.; Sibray, S.S.; Cannia, J.C.; Tanner, D.Q. *Reconnaissance of Ground-Water Quality in the North Platte Natural Resources District, Western Nebraska, June–July 1991*; Water-Resources Investigations Report; United States Geological Survey, United States Department of the Interior: Reston, VA, USA, 1995; p. 114.
40. Babcock, H.M.; Visher, F.N.; Durum, W.H. *Ground-Water Conditions in the Dutch Flats Area, Scotts Bluff and Sioux Counties, Nebraska, With a Section on Chemical Quality of the Ground Water*; United States Geological Survey, United States Department of the Interior: Reston, VA, USA, 1951; p. 51.
41. Homer, C.G.; Dewitz, J.; Yang, L.; Jin, S.; Danielson, P.; Xian, G.Z.; Coulston, J.; Herold, N.; Wickham, J.; Megown, K. Completion of the 2011 National Land Cover Database for the conterminous United States—Representing a decade of land cover change information. *Photogramm. Eng. Remote Sens.* **2015**, *81*, 345–354.
42. Maupin, M.A.; Kenny, J.F.; Hutson, S.S.; Lovelace, J.K.; Barber, N.L.; Linsey, K.S. *Estimated Use of Water in the United States in 2010*; United States Geological Survey: Reston, VA, USA, 2014; ISBN 978-1-4113-3862-3.
43. *Conservation and Survey Division 1995 Water Table Contours*; Institute of Agriculture and Natural Resources, University of Nebraska-Lincoln: Lincoln, NE, USA, 2003.
44. Aeschbach-Hertig, W.; Solomon, D.K. Noble gas thermometry in groundwater hydrology. In *The Noble Gases as Geochemical Tracers*; Burnard, P., Ed.; Springer: Berlin, Germany, 2013; pp. 81–122. ISBN 978-3-642-28835-7.
45. Clarke, W.B.; Jenkins, W.J.; Top, Z. Determination of tritium by mass spectrometric measurement of ^{3}He. *Int. J. Appl. Radiat. Isot.* **1976**, *27*, 515–522. [CrossRef]
46. Lucas, L.L.; Unterweger, M.P. Comprehensive review and critical evaluation of the half-life of tritium. *J. Res. Natl. Inst. Stand. Technol.* **2000**, *105*, 541–549. [CrossRef] [PubMed]
47. Vogel, J.C. Investigation of groundwater flow with radiocarbon. In *Isotopes in Hydrology*; International Atomic Energy Agency: Vienna, Austria, 1967; pp. 355–369.
48. Gilmore, T.E.; Genereux, D.P.; Solomon, D.K.; Solder, J.E. Groundwater transit time distribution and mean from streambed sampling in an agricultural coastal plain watershed, North Carolina, USA. *Water Resour. Res.* **2016**, *52*, 2025–2044. [CrossRef]

49. Broers, H.P. The spatial distribution of groundwater age for different geohydrological situations in the Netherlands: Implications for groundwater quality monitoring at the regional scale. *J. Hydrol.* **2004**, *299*, 84–106. [CrossRef]
50. Solomon, D.K.; Cook, P.G.; Plummer, L.N. Models of groundwater ages and residence times. In *Use of Chlorofluorocarbons in Hydrology: A Guidebook*; Busenberg, E., Ed.; IAEA: Vienna, Austria, 2006; pp. 73–88.
51. Chang, C.C.; Langston, J.; Riggs, M.; Campbell, D.H.; Silva, S.R.; Kendall, C. A method for nitrate collection for δ^{15}N and δ^{18}O analysis from waters with low nitrate concentrations. *Can. J. Fish. Aquat. Sci.* **1999**, *56*, 1856–1864. [CrossRef]
52. Silva, S.R.; Kendall, C.; Wilkison, D.H.; Ziegler, A.C.; Chang, C.C.Y.; Avanzino, R.J. A new method for collection of nitrate from fresh water and the analysis of nitrogen and oxygen isotope ratios. *J. Hydrol.* **2000**, *228*, 22–36. [CrossRef]
53. Kreitler, C.W. Determining the Source of Nitrate in Ground Water by Nitrogen Isotope Studies. Ph.D. Dissertation, The University of Texas at Austin, Austin, TX, USA, August 1974; p. 57.
54. Gormly, J.R.; Spalding, R.F. Sources and concentrations of nitrate-nitrogen in ground water of the central platte region, Nebraskaa. *Groundwater* **1979**, *17*, 291–301. [CrossRef]
55. Bremner, J.M.; Keeney, D.R. Steam distillation methods for determination of ammonium, nitrate and nitrite. *Anal. Chim. Acta* **1965**, *32*, 485–495. [CrossRef]
56. American Public Health Association; American Water Works Association; Water Pollution Control Federation. Total Organic Carbon (TOC) Method #5310D, Wet-Oxidation. In *Standard Methods for the Examination of Water and Wastewater*; American Public Health Association; American Water Works Association; Water Pollution Control Federation: Washington, DC, USA, 1992.
57. NEDNR. University of Nebraska-Lincoln Quality-Assessed Agrichemical Contaminant Database for Nebraska Ground Water. Available online: https://clearinghouse.nebraska.gov/Clearinghouse.aspx (accessed on 22 September 2016).
58. Hudson, C. (North Platte Natural Resources District, Scottsbluff, NE, USA); Wells, M.J. (University of Nebraska, Lincoln, NE, USA). Personal communication, 2018.
59. Kendall, C.; Aravena, R. Nitrate isotopes in groundwater systems. In *Environmental Tracers in Subsurface Hydrology*; Cook, P.G., Herczeg, A.L., Eds.; Springer US: Boston, MA, USA, 2000; pp. 261–297. ISBN 978-1-4615-4557-6.
60. Amberger, A.; Schmidt, H.L. Natural isotope content of nitrate as an indicator of its origin. *Geochim. Cosmochim. Acta* **1987**, *51*, 2699–2705. [CrossRef]
61. Durka, W.; Schulze, E.-D.; Gebauer, G.; Voerkeliust, S. Effects of forest decline on uptake and leaching of deposited nitrate determined from ^{15}N and ^{18}O measurements. *Nature* **1994**, *372*, 765–767. [CrossRef]
62. Kendall, C. Tracing nitrogen sources and cycling in catchments. In *Isotope Tracers in Catchment Hydrology*; Elsevier: Amsterdam, The Netherlands, 1998; pp. 519–576. ISBN 978-0-444-81546-0.
63. McMahon, P.B.; Chapelle, F.H. Redox processes and water quality of selected principal aquifer systems. *Groundwater* **2008**, *46*, 259–271. [CrossRef] [PubMed]
64. Mariotti, A.; Landreau, A.; Simon, B. ^{15}N isotope biogeochemistry and natural denitrification process in groundwater: Application to the chalk aquifer of northern France. *Geochim. Cosmochim. Acta* **1988**, *52*, 1869–1878. [CrossRef]
65. USBR Hydromet: Archive Data Access. Available online: https://www.usbr.gov/gp/hydromet/hydromet_arcread.html (accessed on 22 May 2018).
66. Brakebill, J.W.; Gronberg, J.M. *County-Level Estimates of Nitrogen and Phosphorus from Commercial Fertilizer for the Conterminous United States, 1987–2012*; United States Geological Survey: Reston, VA, USA, 2017.
67. NASS USDA/NASS QuickStats Ad-hoc Query Tool. Available online: https://quickstats.nass.usda.gov/ (accessed on 15 February 2018).
68. *2012 Census of Agriculture, Nebraska State and County Data*; United States Department of Agriculture, National Agricultural Statistics Service: Washington, DC, USA, 2014.
69. *1997 Census of Agriculture, Nebraska State and County Data*; United States Department of Agriculture, National Agricultural Statistics Service: Washington, DC, USA, 1999.
70. Green, C.T.; Liao, L.; Nolan, B.T.; Juckem, P.F.; Shope, C.L.; Tesoriero, A.J.; Jurgens, B.C. Regional variability of nitrate fluxes in the unsaturated zone and groundwater, Wisconsin, USA. *Water Resour. Res.* **2018**, *54*, 301–322. [CrossRef]

71. Mas-Pla, J.; Menció, A. Groundwater nitrate pollution and climate change: Learnings from a water balance-based analysis of several aquifers in a western Mediterranean region (Catalonia). *Environ. Sci. Pollut. Res.* **2018**, *25*, 1–19. [CrossRef] [PubMed]
72. Ma, L.; Spalding, R.F. Effects of artificial recharge on ground water quality and aquifer storage recovery. *J. Am. Water Resour. Assoc.* **1997**, *33*, 561–572. [CrossRef]

10

Potential Impact of In-Situ Oil Shale Exploitation on Aquifer System

Shuya Hu [1,2,3], Changlai Xiao [1,2,3], Xue Jiang [4] and Xiujuan Liang [1,2,3,*]

1 Key Laboratory of Groundwater Resources and Environment of Ministry of Education, Jilin University, Changchun 130021, China; 90shuya@sina.com (S.H.); xiaocl@jlu.edu.cn (C.X.)
2 College of Environmental and Resources, Jilin University, Changchun 130021, China
3 National-Local Joint Engineering Laboratory of In-situ Conversion, Drilling and Exploitation Technology for Oil Shale, Changchun 130021, China
4 College of Water Resources and Architectural Engineering, The Northeast Agricultural University, Harbin 150030, China; jxue1986@163.com
* Correspondence: xjliang@jlu.edu.cn

Abstract: The effects of heat on physical and hydraulic properties of oil shale were investigated. The porosity and water absorption of oil shale increased with increasing pyrolysis temperature. The porosity increased by 19.048% and water absorption increased by 0.76% when oil shale was heated to 500 °C. Thus, originally impermeable oil shale was converted to a permeable rock formation, facilitating interactions between surrounding groundwater and oil. Heated oil shale was immersed in water, which showed strong alkaline properties. The content of Ca^{2+} remained stable and a slight decrease in SO_4^{2-} content was observed. Hydrocarbon content in the water samples reached maximum concentration within three days.

Keywords: oil shale; in-situ exploitation; aquifer systems; immersion test

1. Introduction

With rapid socio-economic development, the demand for energy is growing. Conventional petroleum resources can no longer meet the rapidly increasing demand for oil. Consequently, development and utilization of unconventional petroleum resources is gaining attention. Shale oil plays an important role in the world's energy supply as it is one of the main unconventional oil resources [1]. Significant reserves of oil shale exist globally, amounting to nearly 10 trillion tonnes, which can be converted to nearly 0.045 trillion tons shale oil [2]. This is nearly four times the amount of current oil resources [3,4]. Presently, oil shale is produced by strip mining or underground mining, and crushed and sieved to the desired particle size. The particles are subject to dry distillation in furnaces to generate shale oil [5]. However, the traditional method of developing and utilizing oil shale is not economically efficient and is environmentally unfriendly [6,7]. In-situ oil shale mining does not require mining, transportation, and ore processing. Additionally, it can mine deep and high-strength oil shale mines [8,9]. In-situ shale oil extraction heats sections of vast oil shale fields in-situ, releasing shale oil and oil shale gas from the rock, which is pumped to the surface and converted to fuel [10]. Scientists and energy companies have researched oil shale exploration and development technologies extensively, domestically and globally [11–13]. Current in-situ oil shale mining technologies mainly include Shell's in-situ conversion process (Shell ICP), ExxonMobil Electrofrac in-situ shale oil extraction technology, Geothermic Fuels Cells Process (IEP GFC) [11,12], and convection heating technology of Taiyuan University of Technology among others [13]. Although in-situ oil shale mining offers several advantages, the physico-chemical properties of the shale layer can change after heating to high temperatures. Pyrolysis products such as oil, gas, and residue generated during the mining process can enter groundwater and affect the aquifer water quality [14]. In-situ exploitation of industrialized oil shale has not been extensively investigated and limited research has focused on the influence of in-situ mining on groundwater quality.

In this paper, changes in physical and hydraulic properties of oil shale after high-temperature pyrolysis were tested in the laboratory. We selected oil shale samples from the in-situ oil shale mining demonstration area in Fuyu City, Jilin Province, China. Immersion tests were conducted to analyze water quality indices generated by different oil shales due to water–rock interactions and assess the potential impact of in-situ recovery of oil shale on aquifer water quality.

2. Materials and Methods

2.1. Study Area and Samples

Samples were collected from the in-situ oil shale mining demonstration area in Fuyu City, Jilin Province, China. Fuyu City is located in Northwest Jilin Province at 125°0′–126°10′ E and 44°30′–44°44′ N. Rock formation at the site was cretaceous. Based on hydraulic properties, characteristics, and occurrence, the groundwater in the study area can be divided into three categories: quaternary unconsolidated rock pore water, pore fissure water in clastic rocks of upper cretaceous, and bedrock fissure water construction.

Porosity is the ratio of the total volume of pores in the rock to the volume of the rock. Porosity has significant influence on groundwater movement as the number and size of pores determine the ability of the rock to contain water. Rock water absorption mainly depends on the number of pores, cracks, pore sizes, and opening degrees. Rock water absorption is the ratio of the rock's volume under atmospheric pressure to absorb water and the dry rock mass. It takes the degree of rock fracture development into account [15]. Han et al. tested original samples of oil shale and residues obtained after heating using Fourier transform infrared spectroscopy. They found that the characteristic peak of organic matter began to decrease at 300 °C and disappeared at 500 °C [16], indicating that kerogen began to pyrolyze at 300 °C and completely decomposed at 500 °C. Therefore, we chose heating temperatures of 300 °C and 500 °C to investigate the changes in physical, hydraulic, and chemical properties of oil shale. Separate samples were collected for each test.

2.2. Porosity Determination

2.2.1. Determination of Oil Shale Sample Weight

Nine oil shale samples were obtained from the study area for porosity tests. Each sample had a diameter of 60 mm and height between 40 mm to 60 mm. Porosity test samples were divided into three groups: Pa, Pb, and Pc, with three samples per group. Pa and Pb groups were heated to 300 °C and 500 °C, respectively, in a muffle furnace, and cooled to 20 °C after heating. Samples in the Pc group were not heated. All sets of oil shale samples were placed in an oven for 8 h at 105 °C and their dry weights were determined. The bulk density was calculated accordingly to the formula:

$$d_s = \frac{d_w}{V} \tag{1}$$

where d_s represents the bulk density of oil shale, g/cm^3; d_w the dry weight of the sample, g; and V the sample volume, cm^3.

2.2.2. Determination of the Specific Gravity of Oil Shale Samples

Individual test samples were ground in a ceramic bowl and passed through a 0.2 mm sieve. Next, 3.0 g of the sample was transferred to a 50 mL pycnometer and 25 mL of distilled water was added slowly along the vial wall. The pycnometer was boiled in 100 °C water bath for 1 h to remove gases adsorbed in the pores of the powder. The mixture was removed from the pycnometer and distilled water was poured dropwise along the bottle wall using a straw, which was closed using a bottle stopper, and placed at the bottle mouth. Excess distilled water overflowed through the capillaries of the stopper. Afterwards, the pycnometer was cooled to room temperature and dried. The three groups of samples were simultaneously transferred to an oven heated to 105–110 °C to dry for 24 h. Dried samples were placed in a desiccator, cooled to room temperature and weighed. Subsequently, the weight of the pycnometer, oil shale powder, and water (G_1) was determined. A control setup where oil shale powder was not added to the pycnometer was used as a reference. Finally, the weights of the pycnometer and water (G_0) were determined using the formula:

$$\gamma_s = \frac{3}{G_0 + 3 - G_1} \times \rho_w \tag{2}$$

where γ_s represents the proportion of oil shale; and ρ_w the density of water, g/cm^3.

2.2.3. Porosity Calculation

Porosity was calculated using the bulk density and specific gravity of the sample, as described below:

$$n = \left(1 - \frac{d_s}{\gamma_s}\right) \times 100\% \tag{3}$$

2.3. Water Absorption

Nine samples were tested to ensure accuracy of the water absorption test. The free immersion method was utilized to saturate the samples. The samples were placed in a sink and water was added to cover one-fourth of the samples' height. After two hours, additional water was added to cover one-half of the samples' height. After two more hours, water was added to cover three-fourth of the samples' height. Finally, after additional two hours (total of six hours), water was added to completely cover the samples. The samples were subsequently soaked in water for 48 h. Finally, the samples were taken out of the water bath, and surface moisture was wiped off and weighed. Sample water absorption was calculated according to the following formula:

$$\omega_a = \frac{m_o - m_s}{m_s} \times 100\% \tag{4}$$

where w_a represents rock water absorption, %; m_o the specimen weight after water quality testing, g; and m_s the dry specimen mass, g.

2.4. Determination of Water Quality Indices in Oil Shale Soaking Liquid

Unheated oil shale samples and oil shale samples heated to 300 °C and 500° C were weighed into seven portions (100 g each) and placed in individual conical flasks. Then, 1000 mL distilled water was added to each sample and the bottle was sealed. Sample sets were soaked for different durations: 3, 6, 9, 15, 21, 33, and 69 days. Values of regular indices in the water were measured after the stipulated immersion duration to determine pH, carbonate, bicarbonate, sulfate, calcium, iron, and ammonium ions. According to China quality standard for ground water, the pH of each sample was measured using pHS-3C pH meter. CO_3^{2-}, HCO_3^-, and Ca^{2+} contents were determined using titration tests. The SO_4^{2-} content was measured by barium chloride titration using methyl orange as an indicator. Iron ions in water samples were analyzed using Shimadzu AA-6000CF Flame Atomic Absorption Spectrophotometer. Ammonium ion content was determined using Nessler's reagent spectrophotometry.

2.5. Determination of Organic Matter in Oil Shale Immersion Samples

Unlike the determination of conventional indexes, the water samples were acidified to pH < 2 to inhibit the activity of microorganisms before immersion. After immersion, 300 mL of each water sample was poured into a 1000 mL separatory funnel and 60 mL dichloromethane (Chromatography grade) was added. Additionally, a small amount of sodium chloride was added to reduce organic matter and extraction solvent was added to improve the extraction effect.

Thereafter, the separatory funnels were shaken for 20 min and left to settle. After stratification, the lower extract was filtered using a rotary evaporator with anhydrous sodium sulfate. Anhydrous sodium sulfate needed to be baked at 600 °C for 4 h to improve water absorption capacity. Rotary evaporation bottles had to be dried in advance to avoid water contamination of the samples.

Afterwards, the heating temperature of the evaporation bottles was set to 40 °C to enable evaporation of the extraction solution (methylene chloride) and increase organic matter concentration in the extract. After the volume of the extract had decreased to 1 mL, the rotary evaporator was switched off, and the concentrated solution was transferred to an Agilent bottle for organics measurement.

Finally, water samples, in which oil shale had been immersed for different durations, were tested to analyze the organic content of the oil shale samples heated to different temperatures. Organic components in the water samples were determined according to the results of the automatic match between the internal database of the Agilent 6890/5973 GC-MS analysis and peak positions of the substances (Tables 1 and 2). A capillary column HP-5MS (30 m 0.32 mm 0.25 µm) was used in the experiment. Chromatographic setup conditions were: inlet 300 °C, constant flow mode, column flow 2 mL/min, and split ratio 50:1.

Table 1. Composition-matching statistics of water samples with oil shale heated to 300 °C.

Time (min)	14.16	15.29	18.67	21.35	24.04	25.01	27.84
Component Chemical formula	$C_{14}H_{22}O$	$C_{19}H_{41}NO_3S$	$C_{18}H_{24}O$	$C_7H_5NS_2$	$C_{28}H_{38}O_{10}$	$C_{30}H_{42}C_{12}N_4O_3$	$C_{30}H_{50}$
Match probability	50.90%	52.90%	52.70%	65.50%	57.00%	50.20%	57.90%

Table 2. Composition-matching statistics of water samples with oil shale heated to 500 °C.

Time (min)	11.99	12.62	13.82	14.82	14.96	15.97	23.18
Component Chemical formula	$C_{13}H_{28}$	$C_{14}H_{30}$	$C_{15}H_{32}$	$C_{15}H_{32}$	$C_{15}H_{32}$	$C_{16}H_{34}$	$C_{18}H_{35}NO$
Match probability	77.50%	75.90%	73.80%	80.30%	73.90%	78.90%	78.60%

As shown in Tables 1 and 2, the match probability for alkanes was much higher than for the other components. Therefore, the content of hydrocarbons in the samples was determined using mixed standard samples of C_{10}–C_{20}.

Therefore, 0.2 mg/mL of mixed standard samples were diluted to concentrations of 0.2, 0.5, 1.0, 2.0, and 4.0 mg/L to draw a standard curve. Then, the mass spectra of the mixed samples at different concentrations were integrated to represent the content of the substance. The mass spectrum of the tested sample was integrated ensuring any minor peak of the tested alkanes was also included in the integration. Finally, C_{10}–C_{20} was calculated quantitatively from the area obtained after integration.

3. Results and Discussion

3.1. Impact of High-Temperature Pyrolysis on Physical and Hydraulic Properties of Oil Shale

3.1.1. Change in Porosity

Figure 1 shows the average porosity of oil shale at different heating temperatures. The average porosity of an oil shale sample before heating was 2.552%; the porosity of oil shale increased significantly during heating. The increase in porosity of oil shale with increasing temperature was mainly caused by pyrolysis. After the oil shale was heated, kerogen decomposed to generate oil and gas.

Precipitation of petroleum products results in a large number of pore fissures in oil shale [17]. The difference in porosity from room temperature to 300 °C was less than the difference from 300 °C to 500 °C. This is mainly because when oil shale is heated to 300 °C, only water and some organic matter with low melting point is precipitated from oil shale [17]. When the heating temperature was increased from 300 °C to 500 °C, high-molecular organic compounds in oil shale, such as aliphatic hydrocarbons or aromatic hydrocarbons, begin to crack, contributing significantly to the increase in porosity after oil and gas pyrolysis [6,14].

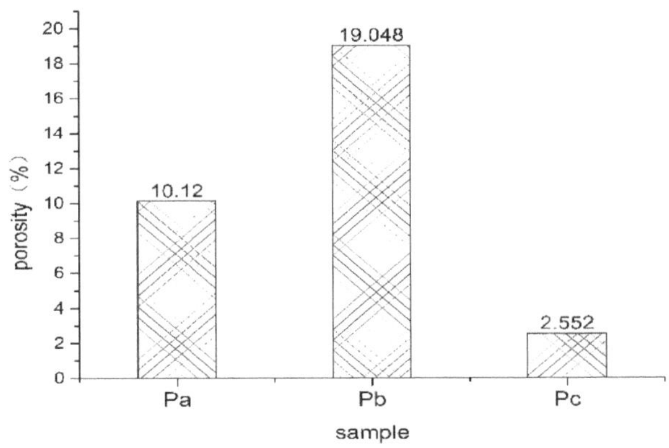

Figure 1. Average porosity of oil shale heated to different temperatures (Pa, Pb, Pc, respectively, represent oil shale heated to 300 °C, oil shale heated to 500 °C, and unheated oil shale).

3.1.2. Change in Water Absorption

Figure 2 shows the average water absorption of oil shale at different heating temperatures. As seen in Figure 2, oil shale water absorption increased significantly after heating. When heated to 300 °C, the water absorption of oil shale increased to 0.34%, which was five times higher than the water absorption at room temperature (0.06%). When heated to 500 °C, the water absorption of the oil shale increased to 0.76%, which was 12.5 times higher than the water absorption at room temperature. Water absorption rate of oil shale increased with pyrolysis temperature primarily due to the liberation of oil and gas products resulting in a large number of pore fractures within the oil shale. Higher numbers of pores contribute to higher water absorption. Water absorption and porosity change of oil shale after pyrolysis were synchronous; the change in water absorption from room temperature to 300 °C was less than the change in oil shale heated to 300 °C and 500 °C.

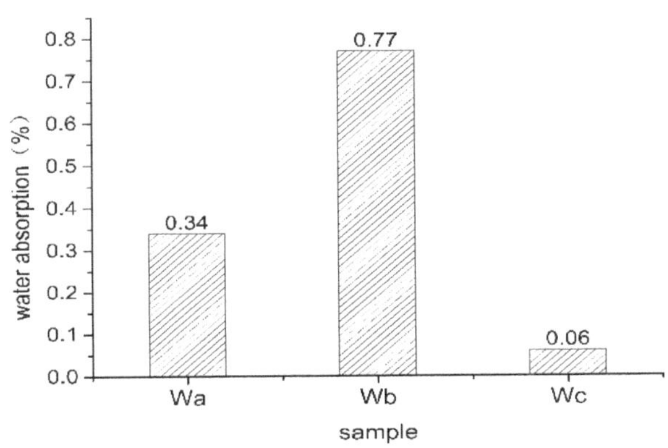

Figure 2. Average water absorption of oil shale at different heating temperatures (Wa, Wb, Wc, respectively, represent oil shale heated to 300 °C, oil shale heated to 500 °C, and unheated oil shale).

3.1.3. Analysis of Changes in Physical and Hydraulic Properties

Most fossil fuel chemical reactions such as pyrolysis, gasification, and combustion occur inside the pores of fuel particles [18–23]. When kerogen is stored in oil shale pyrolusite, pores arise inside the rock. Additionally, pyrolysis of generated oil and gas also causes expansion of pores. This process turns the originally dense, impermeable, or water-pervious oil shale into a more porous aquifer. In this paper, when the kerogen decomposed after heating to 500 °C, the porosity of oil shale increased to 19.048% and water absorption increased to 0.76%. Qiu et al. measured the permeability of oil shale in Meihekou City in Jilin Province and found that permeability increased from 1.62×10^{-3} μm^2 to 10.53×10^{-3} μm^2 after heating [19].

Thus, after thermal decomposition, physical and hydraulic properties of oil shale changed the impermeable oil shale to a more permeable rock formation. Further, the enlarged pores and fissures serve as conduit for water diversion, enabling the oil shale layer to come in hydraulic contact with the overlying sub-aquifer and underlying confined aquifer which was consistent with the research on the pyrolysis process of Huadian oil shale [21,22]. This enables water in the adjacent aquifer to enter the oil shale layer.

3.2. *Effect of Pyrolysis on Chemical Properties of Oil Shale*

3.2.1. Changes in pH and Ion Concentrations

Figure 3 shows the change in pH over immersion time. It was evident that the pH of the oil shale reached its maximum when it was heated to 300 °C. As the immersion time increased, the pH value increased but the rate of increase decreased over time. The change in pH of water samples with oil shale heated to 500 °C was similar to the pH change of water samples with oil shale heated to 300 °C. The pH change of water samples with unheated oil shale was the least; pH increased with time, but the rate was very slow and almost unchanged; the average pH was 7.72.

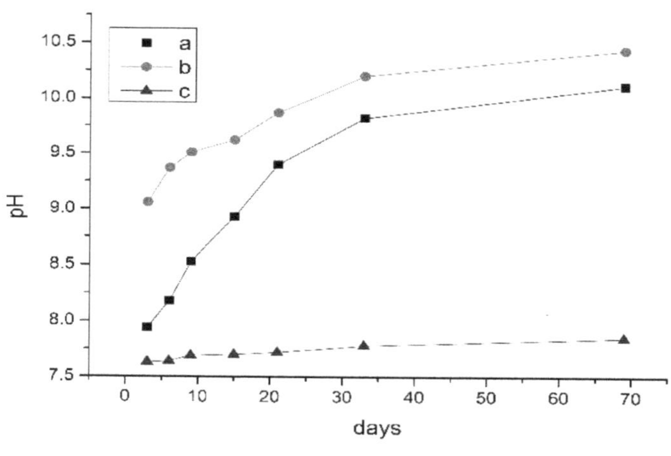

Figure 3. Change in pH over immersion time (a, b, and c, respectively, represent the pH of the liquids with oil shale heated to 300 °C, oil shale heated to 500 °C, and unheated oil shale).

The pH of the liquid with oil shale heated to 300 °C was the most alkaline and carbonate (Figure 4). Higher concentrations of $CO_3{}^{2-}$ promoted hydrolysis reactions and increased the OH^-. $CO_3{}^{2-}$ was not detected in the liquids with unheated oil shale and oil shale heated to 500 °C; however, the $HCO_3{}^-$ content of both liquids were very high, with an average content of 227.45 mg/L and 242.50 mg/L, respectively.

When the oil shale was heated to 300 °C, the thermal motion of carbonate minerals increased, which caused the crystal to break down and the material to change from crystalline to liquid phase. In this molten state, the carbonate minerals ionized to produce carbonate. When the oil shale was heated to 500 °C, most of the carbonates had been decomposed after the oil shale calcined at 500 °C; however, since the porosity of the oil shale became larger after calcination, the remaining small amount of soluble carbonate was separated out more easily. In addition, carbon dioxide from the surrounding air could also enter the sample. Under the corresponding pH conditions obtained from the experiment, bicarbonate was predominant, and slightly higher than in unheated oil shale.

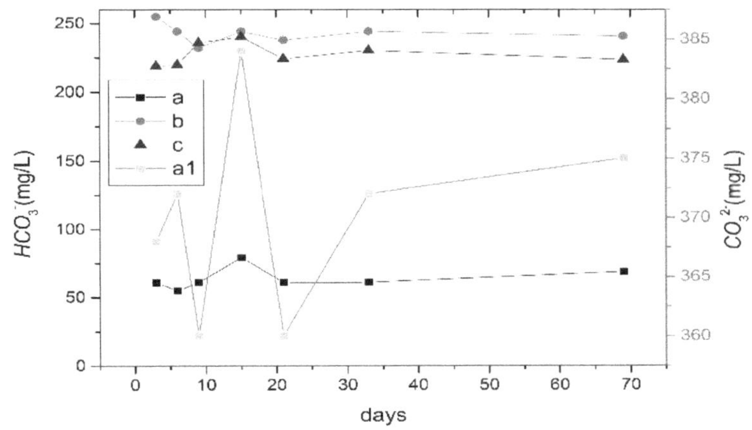

Figure 4. Concentration curves of CO_3^{2-} and HCO_3^- contents in sample liquids (a, b, and c, respectively, represent HCO_3^- content of liquids with oil shale heated to 300 °C, oil shale heated to 500 °C, and unheated oil shale; a1 represents the CO_3^{2-} content of liquids with oil shale heated to 300 °C).

The liquid with oil shale heated to 500 °C did not contain iron ions; the other two liquids contained iron ions. This can be attributed to the decomposition of iron-containing compounds at high temperatures [24]. Oxidation may occur in iron cement at high temperature. Ferric ion concentration curves of liquids with oil shale heated to 300 °C and unheated shale is shown in Figure 5.

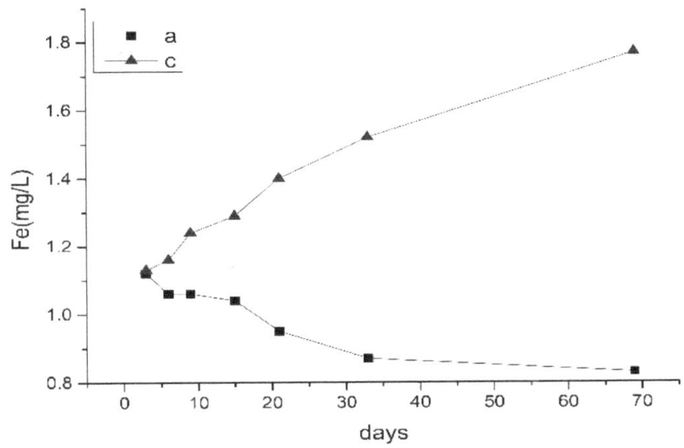

Figure 5. Ferric ion concentration curves of sample liquids (a and c, respectively, represent the ion concentration in the liquids with oil shale heated to 300 °C and unheated oil shale).

The liquid with unheated oil shale contained more iron ions than the liquid with oil shale heated to 300 °C. The concentration of iron ions in the liquid with unheated oil shale gradually built up over time. However, the concentration of iron ions in the liquid with oil shale heated to 300 °C decreased continuously with time. This is related to the pH of the soaking liquid and the type and content of iron-containing compounds in the oil shale. With continuous increase in pH, iron ions in the solution generate precipitates or ferrites, and decrease in concentration.

$$Fe^{3+} + 3OH^- = Fe(OH)_3 \downarrow$$

$$Fe(OH)_3 + 3OH^- = Fe(OH)_6{}^{3-}$$

Concentration of ammonium ions in the liquid with oil shale heated to 300 °C changed over time, as shown in Figure 6. When the oil shale was heated to 300 °C, organic compounds containing nitrogen decomposed into simpler nitrogenous organic matter. Under anaerobic conditions, the nitrogenous organic matter further decomposed into ammonium [23–25]. When the oil shale was heated to 500 °C, organic matters were decomposed and nitrogenous compounds turned into NOx and discharged. Hence, the liquid with oil shale heated to 500 °C did not contain ammonium ions.

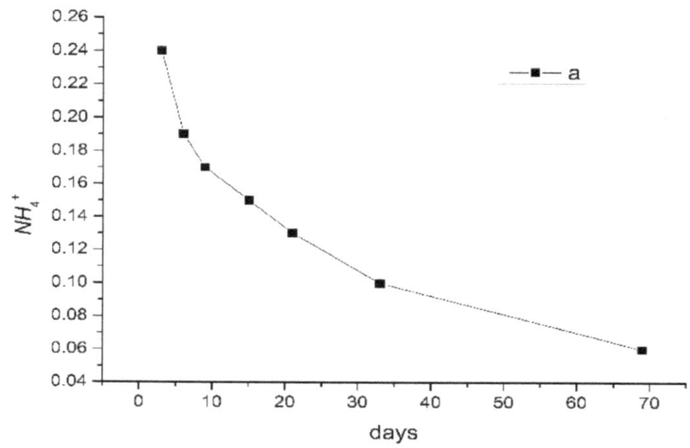

Figure 6. Ammonium ion concentration curve of liquids with oil shale heated to 300 °C.

Figure 7 shows that the content of Ca^{2+} of liquids with oil shale heated to 300 °C and 500 °C changed slightly as the mineral oil shale can only pyrolyze at temperatures over 500 °C [25].

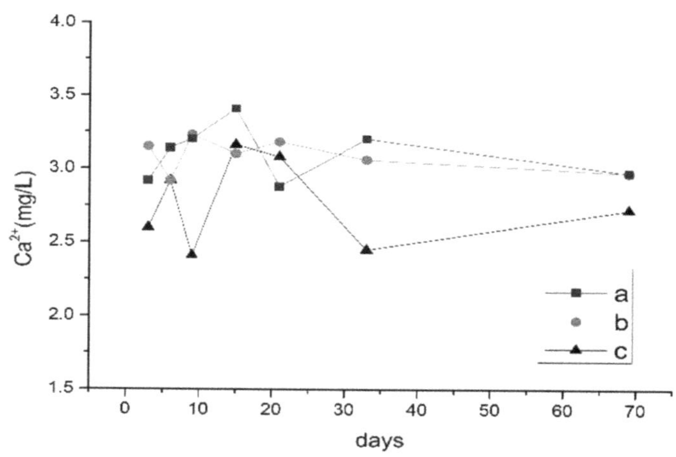

Figure 7. Ca^{2+} concentration curve of sample liquids (a, b, and c, respectively, represent the Ca^{2+} content in the liquids with oil shale heated to 300 °C, oil shale heated to 500 °C, and unheated oil shale).

Acid roots were measured in the three samples and Figure 8 shows their content curve over time. The concentration of $SO_4{}^{2-}$ in the liquid with oil shale heated to 300 °C was the lowest, with an average concentration of 77.26 mg/L.

Brandt found that kerogen decomposed into a mixture of oil, HC gas, and carbon-rich shale coke that adheres to shale particles (as well as CO_2, water vapor, and trace gases) [26]. Kerogen decomposition rates were dependent on temperature: 90% decomposition occurred within 5000 min at 370 °C and within 2 min at 500 °C [26]. The content of organic matter in the oil shale after 500 °C pyrolysis was less, the oil shale after pyrolysis at 300 °C was not completely pyrolyzed, and the organic matter content was higher. Sulfate and hydrocarbons reacted with sulfate thermochemical reduction (TSR) to reduce sulfate minerals to produce acid gases such as H_2S and CO_2 [27]. Therefore, the content of SO_4^{2-} in the water-soaked oil shale samples after 300 °C pyrolysis was the lowest. This also explains the high carbonate content in water-soaked oil shale after 300 °C pyrolysis.

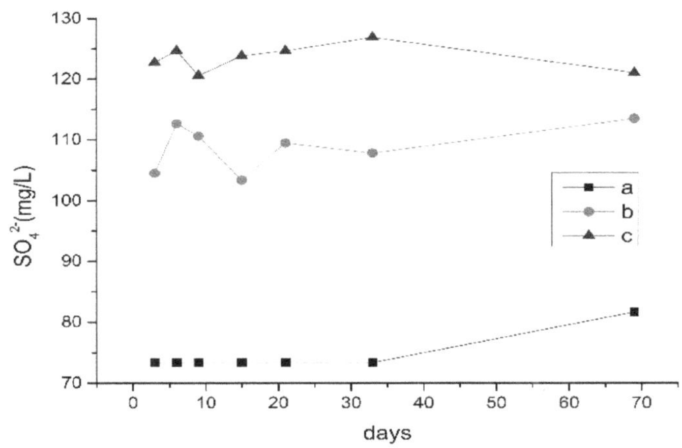

Figure 8. SO_4^{2-} concentration curves of sample liquids (a, b, c, respectively, represent liquids with oil shale heated to 300 °C, oil shale heated to 500 °C, and unheated oil shale).

3.2.2. Changes in Organic Content

Figure 9 shows the changes in C_{10}–C_{20} content in the tested samples. The content of C_{10}–C_{20} in the 300 °C sample was the highest; especially for C_{15}, C_{16}, and C_{17}, which were over 15%. When heated to 300 °C, kerogen in the oil shale was in the warm-up phase and at the beginning of decomposition. The rock samples in this stage are rich in organic matter. When heated to 500 °C, kerogen has been pyrolyzed and organic matter content in the samples has decreased.

Qiu [25] had also studied the effect of three treatment methods on organic matter content. The soaked oil shale solution heated to 300 °C had the largest organic matter content, followed by unheated oil shale. The soaked oil shale solution heated to 500 °C had the smallest organic matter content.

The composition of the samples did not change significantly over time. Hydrocarbon contents reached their maximum initial concentration within three days indicating that the C_{10}–C_{20} hydrocarbons release process was shorter than three days.

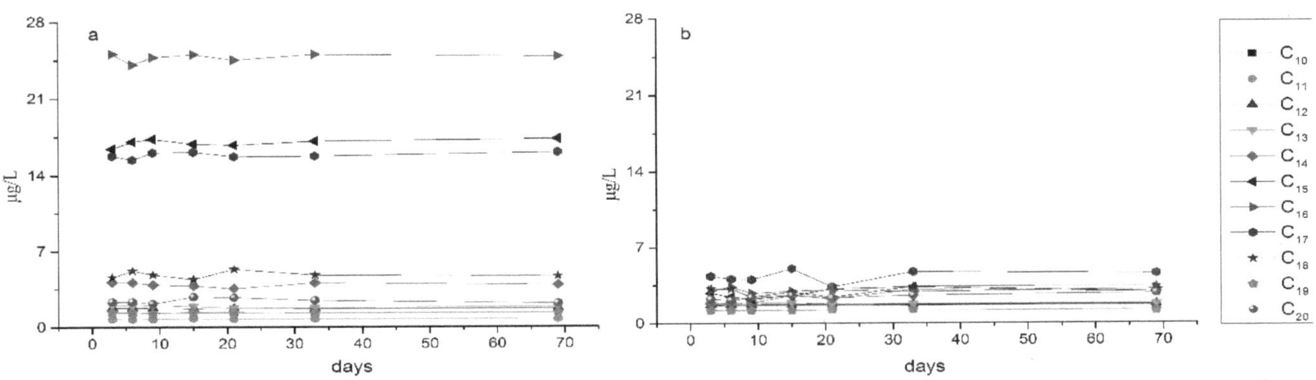

Figure 9. Content of C_{10}–C_{20} in liquid for (**a**) oil shale heated to 300 °C, and (**b**) oil shale heated to 500 °C.

3.2.3. Analysis of Water Quality Changes

Water is produced during retorting (up to 1.5% of the raw shale). Further, the retort water may contain up to 160 g/L of soluble organic carbon [28]. Both fracturing and pyrolysis during mining increases formation porosity and water absorption, and the dense oil shale layer becomes a weakly permeable or aquiferous layer. In this case, groundwater and hydraulic shale layers have a hydraulic connection and groundwater pollution can occur.

According to the Chinese "Sanitary Standards for Drinking Water" and "Quality Standards for Farmland Irrigation Water", water with pH higher than 8.5 should not be used for either drinking or farm irrigation. After oil shale pyrolyzed, the pH of soaking water increased with soaking time. After 60 days of soaking in oil shale and heating to 500 °C, the pH of the water sample exceeded 10 and showed strong alkalinity.

When the oil shale was heated to 500 °C, iron-containing compounds in the rock were decomposed and the organic matter containing nitrogen was also completely cracked. As a result, the soaked oil shale samples did not contain iron. However, iron and ammonium in the oil shale samples were not completely pyrolyzed at 300 °Ç and its content did not fulfill above standards.

Large numbers of hydrocarbons were present in the water sample after oil shale thermal cracking. The content of C_{10}–C_{20} in the soaked samples with oil shale heated to 300 °C was greater than in samples heated to 500 °C. Further, the hydrocarbons (C_{10}–C_{20}) were released in the water in less than three days. If a hydraulic connection between the oil shale layer and external aquifer exists, maximum initial concentration of hydrocarbons in the aquifer can be achieved within a short time, which pollutes the groundwater subsequently.

4. Conclusions

Changes in the physical and hydraulic properties of oil shale were tested to analyze the effect of in-situ oil shale mining. Further, changes in the chemical properties of water samples immersed with heated oil shale were tested to analyze the impact of in-situ mining on groundwater. Based on the results of the work, the following conclusions were drawn.

Porosity and water absorption of oil shale increased with increase in pyrolysis temperature. After kerogen was completely decomposed (after heating to 500 °C), the porosity of oil shale increased to 19.048%, which was 6.5 times greater than the porosity of unheated oil shale; corresponding water absorption of oil shale increased to 0.76%, 11.5 times greater than the water absorption of unheated oil shale. Permeability of rocks is a precondition for the transfer of elements and minerals from rocks into water; increase in porosity and water absorption promotes groundwater entry into the oil shale, facilitating entry of overlying and underlying water into the oil shale layer, and increase water–rock interaction.

After the oil shale was paralyzed, pH of the water samples increased with soaking time. When the oil shale sample was heated to 500 °C and soaked for 60 days, the pH of the water sample exceeded 10, exhibiting strong alkalinity.

Large amounts of hydrocarbon material were detected in the water samples. The content of C_{10}–C_{20} in the water with oil shale heated to 300 °C was greater than the 500 °C sample; after a short period of time (less than three days) the content of hydrocarbons in the water samples reached its maximum initial concentration.

In short, it also affects groundwater quality around the mining area and the damage comes from two aspects. On one hand, due to the leakage of production wells to underground rock formations during the mining process, organic components in groundwater increase, including volatile materials such as benzene, toluene, etc. Additionally, volume and quality of the groundwater changes, and pH is slightly increased and becomes alkaline. Therefore, this study on the impact of in-situ exploitation of oil shale on groundwater environment provides a reliable basis for groundwater protection.

Author Contributions: Shuya Hu and Xiujuan Liang conceived the manuscript; Changlai Xiao and Xue Jiang analyzed the data; Xue Jiang collected data; Shuya Hu and Xue Jiang wrote the paper.

Acknowledgments: The authors acknowledge the support of The National Natural Science Funds of China (41572216), The Water Resources Project of Jilin Province, China (0773-1441GNJL00390), The Natural Science Funds of Jilin Province, China (20140101164JC).

References

1. Wang, H.; Zhao, Q.; Liu, H. *Resource Distribution and Technical Progress of Oil Shale in China*; Petroleum Industry Press: Beijing, China, 2013.
2. Li, S.-Y.; Ma, Y.; Qian, J.-L. Global oil shale research, development and utilization today and an overview of three oil shale symposiums in 2011. *Sino-Glob. Energy* **2012**, *17*, 8–17. (In Chinese with English Abstract)
3. Guan, X.; Dong, L.-I.; Han, D.-Y. Development and utilization progress of foreign oil shale resources. *Contemp. Chem. Ind.* **2015**, *44*, 80–82. (In Chinese with English Abstract)
4. Sun, P.; Liu, Z.; Bai, Y.; Xu, Y.; Liu, R.; Meng, Q.; Hu, F. Accumulation stages and evolution characteristics of oil shale and coal in the Dunhua-Mishan fault zone, Northeast China. *Oil Shale* **2016**, *33*, 203–215. [CrossRef]
5. Chen, X.-F.; Gao, W.-J.; Zhao, J. Status of exploiting and utilization of oil shale in China. *Clean Coal Technol.* **2010**, *16*, 29–31.
6. Jalkanen, L.; Juhanoja, J. The effect of large anthropogenic particulate emissions on atmospheric aerosols, deposition and bioindicators in the eastern Gulf of Finland Region. *Sci. Total Environ.* **2000**, *262*, 123–136. [CrossRef]
7. Song, Y.; Liu, Z.; Meng, Q.; Xu, J.; Sun, P.; Cheng, L.; Zheng, G. Multiple controlling factors of the enrichment of organic matter in the upper cretaceous oil shale sequences of the Songliao basin, ne China: Implications from geochemical analyses. *Oil Shale* **2016**, *33*, 142–166. [CrossRef]
8. Zhong, S.; Tao, Y.; Li, X.; Li, T.; Zhang, F. Simulation and assessment of shale oil leakage during in situ oil shale mining. *Oil Shale* **2014**, *31*, 337–350.
9. McCarthy, H.E.; Cha, C.Y.; Bartel, W.J.; Burton, R.S. Development of the modified in situ oil-shale process. *Aiche J.* **1976**, *72*, 14–23.
10. Liu, D.-X.; Wang, H.-Y.; Zheng, D.-W.; Fang, C.-H.; Ge, Z.-X. World progress of oil shale in-situ exploitation methods. *Nat. Gas Ind.* **2009**, *25*, 128–132, (In Chinese with English Abstract).
11. Zheng, D.; Li, S.; Ma, G.; Wang, H. Autoclave pyrolysis experiments of Chinese Liushuhe oil shale to simulate in-situ underground thermal conversion. *Oil Shale* **2012**, *29*, 103–114. [CrossRef]
12. Maaten, B.; Loo, L.; Konist, A.; Nesumajev, D.; Pihu, T.; Külaots, I. Decomposition kinetics of American, Chinese and Estonian oil shales kerogen. *Oil Shale* **2016**, *33*, 167–183. [CrossRef]
13. Qian, J.; Yin, L. *Oil Shale—Petroleum Alternative*; Petroleum Industry Press: Beijing, China, 2010.
14. Brandt, A.R. Converting oil shale to liquid fuels: Energy Inputs and greenhouse gas emissions of the shell in situ conversion process. *Environ. Sci. Technol.* **2008**, *42*, 7489–7495. [CrossRef] [PubMed]
15. Tang, D. *Rock and Soil Engineering*; Geology Publishing House: Beijing, China, 1985. (In Chinese)
16. Zhou, K.; Sun, Y.-H.; Li, Q.; Guo, W.; Lyu, S.-D.; Han, J. Experimental research about thermogravimetric analysis and thermal physical properties of Nong'an oil shale. *Glob. Geology* **2016**, *35*, 1178–1184. (In Chinese)
17. Wang, W.; Ma, Y.; Li, S.; Shi, J.; Teng, J. Effect of temperature on the EPR properties of oil shale pyrolysates. *Energy Fuels* **2016**, *30*, 830–834. [CrossRef]
18. Decora, A.W.; Kerr, R.D. Processing use, and characterization of shale oil products. *Environ. Health Perspect.* **1979**, *30*, 217–223. [CrossRef] [PubMed]
19. Han, X.-X.; Jiang, X.-M.; Cui, Z.-G.; Yu, L.-J. Evolution of pore structure of oil shale particles during combustion. *Proc. Chin. Soc. Electr. Eng.* **2007**, *27*, 26–30.
20. Jiang, Q. *Study on Physical and Chemical Properties of Oil Shale in Huadian City*, Northeast Electric Power University: Jilin, China, 2006. (In Chinese)
21. Meng, Q. *Research on Petrologic and Geochemical Characteristics of Eocene Oil Shale and Its Enrichment Regularity, Huadian Basin*; Jilin University: Jilin, China, 2010. (In Chinese)
22. Bai, F.; Sun, Y.; Liu, Y.; Guo, M. Evaluation of the porous structure of Huadian oil shale during pyrolysis using multiple approaches. *Fuel* **2017**, *187*, 1–8. [CrossRef]
23. Xi, F.-Y.; Zhao, X.-E.; Cai, X. Study on kinetics of oxidation of iron sulfur compounds. *Appl. Chem. Ind.* **2017**, *5*, 825–828. (In Chinese)

24. Qiu, S. *Experimental Study on the Impacts of Oil Shale in-Situ Pyrolysis on Groundwater Hydrochemical Characteristics*; Jilin University: Jilin, China, 2016. (In Chinese)
25. Song, W.N.; Dong, Y.L.; Zhou, G.J.; Ding, H.X.; Li, Z. Research summarization of structure-constitute and application of oil shale. *J. Heilongjiang Hydraul. Eng.* **2010**, *3*, 20. (In Chinese)
26. Ying, H.-H.; Zhang, Z.-P.; Wu, Y.-B. Study on influence factors of acid solubility of fracturing proppant. *Liaoning Chem. Ind.* **2015**, *44*, 1052–1055, 1065. (In Chinese)
27. Zhu, G.-Y.; Zhang, S.-C.; Liang, Y.-B.; Li, J. Alteration of thermochemical sulfate reduction to hydrocarbons. *Acta Pet. Sin.* **2005**, *26*, 48–52.
28. Routson, R.C.; Wildung, R.E.; Bean, R.M. A Review of the environmental impact of ground disposal of oil shale wastes. *J. Environ. Qual.* **1979**, *8*, 14–19. [CrossRef]

11

Multivariate and Spatial Analysis of Physicochemical Parameters in an Irrigation District, Chihuahua, Mexico

Jesús Alejandro Prieto-Amparán [1], Beatriz Adriana Rocha-Gutiérrez [2], María de Lourdes Ballinas-Casarrubias [2], María Cecilia Valles-Aragón [3,*], María del Rosario Peralta-Pérez [2] and Alfredo Pinedo-Alvarez [1]

1. Facultad de Zootecnia y Ecología, Universidad Autónoma de Chihuahua, Periférico Francisco R. Almada, Km 1, Chihuahua, Chihuahua 31453, México; jesus_prieto06@hotmail.com (J.A.P.-A.); apinedoa@gmail.com (A.P.-A.)
2. Facultad de Ciencias Químicas, Universidad Autónoma de Chihuahua. Circuito Universitario s/n, Campus II, Chihuahua, Chihuahua 31125, México; rochafcq@gmail.com (B.A.R.-G.); lourdes.ballinas@gmail.com (M.d.L.B.-C.); mperalta@uach.mx (M.d.R.P.-P.)
3. Facultad de Ciencias Agrotecnológicas, Universidad Autónoma de Chihuahua, Av. Pascual Orozco s/n, Campus I, Chihuahua, Chihuahua 31200, México

* Correspondence: valles.cecilia@gmail.com

Abstract: Water quality is relevant due to the complexity of the interaction of physicochemical and biological parameters. The Irrigation District 005 (ID005) is one of the most important agricultural region in Chihuahua, México; for that reason, it was proposed to investigate the water quality of the site. Water samples were collected in two periods: Summer (S1) and Fall (S2). The samples were taken from 65 wells in S1, and 54 wells in S2. Physicochemical parameters (PhP) such as Arsenic (As), Temperature, Electrical Conductivity (EC), Oxide Reduction Potential (ORP), Hardness, pH, Total Dissolved Solids (TDS), and Turbidity were analyzed. The data were subjected to statistical principal component analysis (PCA), cluster analysis (CA) and spatial variability tests. In both seasons, the TDS exceeded the Mexican maximum permissible level (MPL) (35% S1, 39% S2). Turbidity exceeded the MPL in S1 (29%) and in S2 (12%). Arsenic was above the MPL for water of agricultural use in 9% (S1) and 13% (S2) of the wells. The PCA results suggested that most variations in water quality in S1 were due to As, pH and Temperature, followed by EC, TDS and Hardness; while in S2 to EC, TDS and Hardness, followed by As and pH.

Keywords: spatial distribution; PCA; CA; water quality

1. Introduction

Economic development, industrialization and urbanization, along with population growth, lead to an accelerated water consumption, which has generated concern of fresh water as a scarce resource [1,2]. Water quality is an important factor that affects human health and ecological systems [3]. In rural locations, groundwater is the support of agricultural activities and it is essential is an important standard for crop production and food security [4]. Due to, pollutants present in irrigation water can get accumulated in crops, causing serious clinical and physiological problems to humans when consumed it in large amounts in food [5,6]. In general, water quality for various applications is determined by its physical characteristics, chemical composition, biological parameters and uses [1,7]. These parameters reflect the inputs from natural sources, including atmosphere, soil and particular geological characteristics of each region, as well, as anthropogenic influence of various activities [8–10].

The evaluation of water quality in most countries has become a critical issue in recent years [2]. Water quality is subject to constant changes due to seasonal and climatic factors [8]. Likewise, spatial variations emphasize the need of water monitoring that provides a representative and reliable estimate [11]. Recently, several approaches have been used for water quality determination. Among them, it can be found methods based on modeling, monitoring or statistic techniques [12]. Modeling tools such as Soil and Water Assessment Tool (SWAT) or Agricultural Nonpoint have been employed to evaluate water quality at watershed scale. The commonly used statistic techniques for the monitoring of water quality include: Ordinary Least Square (OLS), Geographic Weighted Regression (GWR), among others. The monitoring techniques provide knowledge information for decision-making, regarding water quality [12]. However, in comparison to these approaches, multivariate techniques such as Principal Component Analysis (PCA) and Cluster Analysis (CA) could be used to analyze big water quality databases without losing important information [13–15].

Multivariate techniques and exploratory data analyses are appropriate for the data synthesis and its interpretation [16]. Classification, modeling and interpretation of the monitored data are the most important steps in the evaluation of water quality [17–19]. The application of multivariate statistical techniques, such as principal component analysis (PCA) or Cluster Analysis (CA), has significantly increased in recent years, especially for the analysis of environmental data and extracting significant information [20–22]. Additionally, these analyses have been reported as effective methods for the characterization and evaluation of water quality parameters [9]. PCA and CA are the most common multivariate statistical methods used in environmental studies [23].

The PCA is a mathematical technique used to reduce the dimensions of multivariate data and explain the correlation between a large number of variables observed by extracting a smaller number of new variables (i.e., the principal components or PC) [24–26]. The CA helps grouping objects (cases) based on homogeneity and heterogeneity between groups. The clusters characteristics are not known in advance but may be determined in the analysis. Such analysis benefits the interpretation of the data by pointing out associations among the studied variables [27,28]. The application of different multivariate statistical techniques aids in the interpretation of complex data matrices to better understand the water quality and ecological status of the studied systems [11,29]. It also allows the identification of possible factors or sources that influence water systems and offers a valuable tool for the reliable management of water resources, both in quantity and quality [21]. Previous research has shown that the use of multivariate analysis allows defining new variables that provide information on the variability of environmental data, as well as on the influence of each variable [30,31].

Furthermore, interpolation methods have been employed to map the spatial distribution of soil properties [32,33], heavy metals [34,35], population characteristics [36], precipitation [37,38], radioactive elements [39,40], among others. Data interpolation offers the advantage of projecting maps or continuous surfaces from discrete data [41]. Therefore, spatial interpolation techniques are essential to create a continuous (or predictable) surface from values of sampled points [37]. Interpolation is an efficient method to study the spatial allocation of elements, their inconsistency, reduce the error variance and execution costs [42]. The interpolation methods are useful to identifying contamination sources, assessing pollution trends and risks [43,44]. A growing number of studies have shown the need to determine the spatial distribution of pollutants. Spatial data helps to define areas where risks are higher and contribute in making decision to identify the locations where remediation efforts should be concentrated [45]. However, one of the characteristics of the spatial distribution of pollutants lies in their frequent spatial heterogeneity [46].

Few studies that combine multivariate techniques and interpolation methods have been completed [47,48] and many times, the studies are analyzed univariately. Therefore, the selection of PCA and CA methods was made to understand the multivariate relationship between parameters in this research, these techniques were used to compare the grouping analysis and the interpolations, to understood the spatial distribution of the parameters and even more the spatial distribution of the PCs. The objective of the present study was to analyze eight physicochemical parameters (PhP) in water samples from wells of the Irrigation District 005 (ID005) in Chihuahua, Mexico; perform a data analysis using multivariate techniques to evaluate the PhP contribution in water quality and apply interpolation methods to analyze the spatial variation of the PhP.

2. Materials and Methods

2.1. Research Area

The ID005 is located in the south-central region of the State of Chihuahua (Figure 1), among the geographical coordinates 105°40′ W–28°30′ N, 105°20′ W–28°30′ N, 105°40′ W–28°10′ N, 105°20′ W–28°10′ N. It has an average altitude of 1156 m above sea level. The predominant climate is semi-desert, with an average of annual rainfall of 350 mm [49]. The ID005 is composed by 10 irrigation modules, which are administered by local associations. The district is divided into two constituted irrigation units based on infrastructure characteristics, to facilitate water distribution [50]. Each unit is managed by a Limited Liability Corporation (Chihuahua, Mexico), that is integrated as follows: (1) The first unit called Conchos is composed by irrigation modules from 1 to 5 and module 12, which are mainly supplied by water from La Boquilla Dam; (2) The second unit, called San Pedro, is integrated by modules 6, 7, 8 and 9, which are supplied by water from the Francisco I. Madero dam, groundwater and, to a lesser extent, by water from the La Boquilla Dam [51].

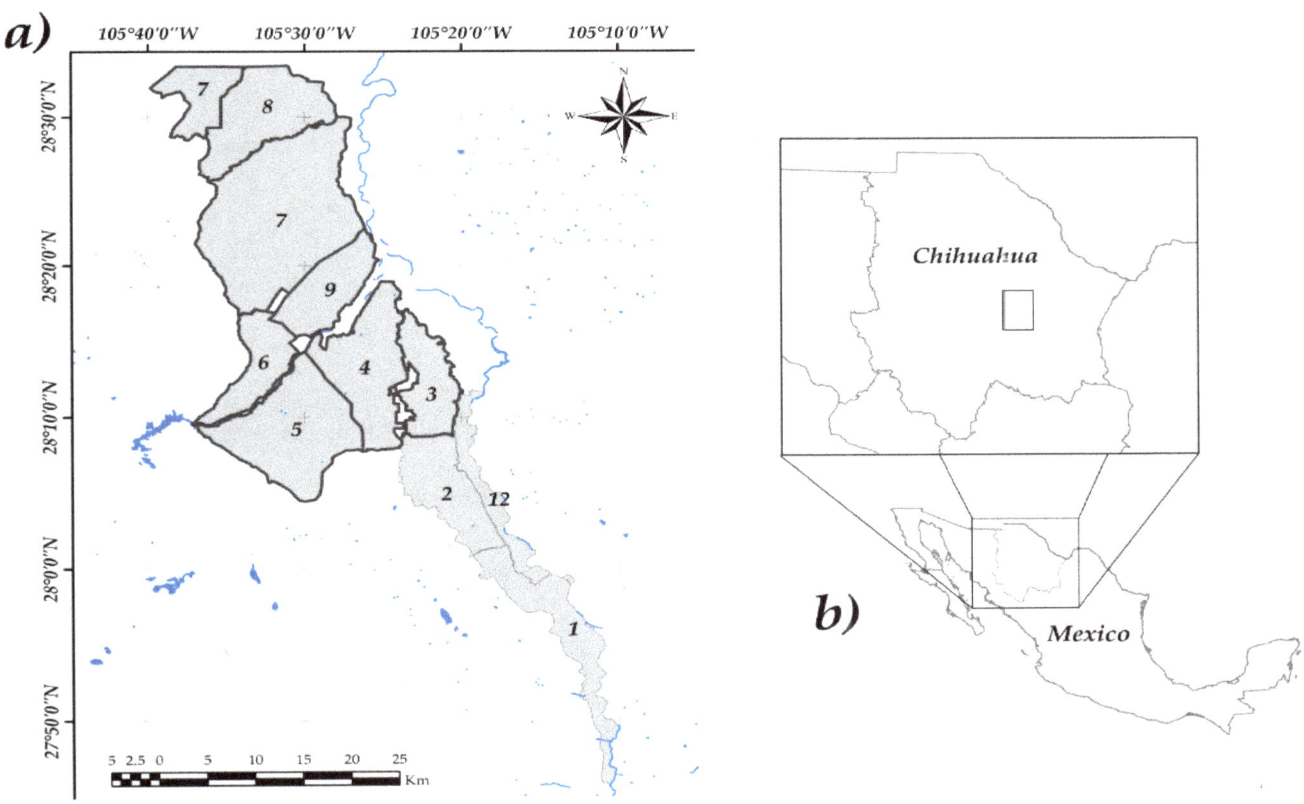

Figure 1. Geographical location of the study area: (**a**) Irrigation District 005 location ( ), boundary of the studied area (—), water bodies (—), bold numbers denote the module number, (**b**) Irrigation District 005 location in Chihuahua, Mexico.

2.2. Sampling

Two different sampling periods were performed in operating wells, on the studied area during 2016. The first sampling was performed in Summer (S1) and the second sampling in Fall (S2) (Figure 2), following the standard procedures of NOM-014-SSA1-1993 [52]. Two samples of 1 L were collected, one for PhP determination, and another for Arsenic (As) determination in which 2 mL of nitric acid (HNO_3) were added for its preservation. The samples were transported in coolers, taken to the laboratory and stored at 4 °C until their analysis. In the first period (S1), water samples were collected of 65 wells; while in the second period (S2), of 54 wells.

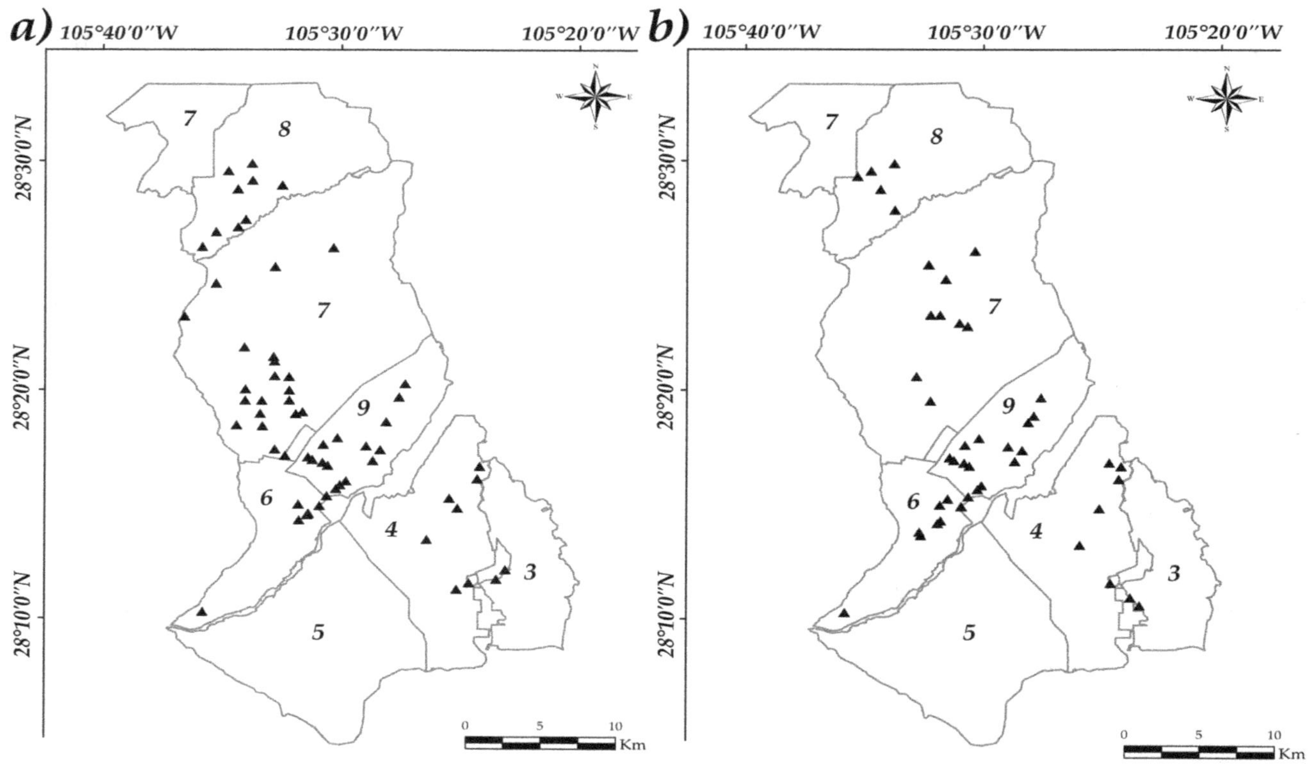

Figure 2. Sampling maps: (**a**) Summer (Sampling 1—**left**) and (**b**) Fall (Sampling 2—**right**). Studied modules (—), sampling points (▲), bold numbers denote the module number.

2.3. Physicochemical Parameters (PhP) Analysis

For As determination, the samples were filtered with 0.2 mm ash paper Whatman No. 41 (CTR Scientific, Chihuahua, Mexico). Subsequently, before the analysis, filtered with 0.45 µm Millipore filters (CTR Scientific, Chihuahua, Mexico). The As quantification was perform in an Atomic Absorption Spectrophotometer AAnalyst 700 (Perkin Elmer, Waltham, USA) to which the FIAS 100 Hydride Generator (Perkin Elmer, Waltham, USA) was coupled, following the NMX-AA-051-SCFI-2001 [53]. The detection limit of the equipment was 3.12 µg/L. The samples were analyzed in triplicate using the standard Trace Metals-Sand 1 Number CRM048 Sigma Aldrich (CTR Scientific, Chihuahua, Mexico) with a recovery percentage of 99%.

Moreover, different physicochemical parameters (PhP) were analyzed: Temperature, Electrical Conductivity (EC), Oxide Reduction Potential (ORP), Hardness, pH, Total Dissolved Solids (TDS) and Turbidity. These are listed in Table 1 along with their respective analytical method. All parameters were analyzed in triplicate. Temperature, pH and ORP were determined in situ, the rest in the laboratory.

Table 1. Physicochemical parameters in water, units, and analytical method.

Parameter	Unit	Analytical Method
As	mg/L	AAS Perkin Elmer Aanalist 700, coupled HG FIAS 100
Temp	°C	Potentiometer Hanna portable (in situ)
EC	µS/cm	Electrical conductivity meter CYBERSCAN
ORP	mV	Potentiometer Hanna portable (in situ)
Hardness	mg/L	Titration net NET (indicator)
pH	dimensionless	Potentiometer Hanna portable (in situ)
TDS	mg/L	Electrical conductivity meter CYBERSCAN
Turb	NTU	Electrical conductivity meter CYBERSCAN

As = Arsenic, EC = Electrical Conductivity, Turb = Turbidity, pH = pH, ORP = Oxide Reduction Potential, Temp = Temperature, Potentiometer Hanna portable (CTR Scientific, Chihuahua, Mexico), Electrical conductivity meter CYBERSCAN (FESTA, Chihuahua, Mexico), Titration net NET (indicator) (FESTA, Chihuahua, Mexico).

2.4. *Multivariate Statistical Methods*

Prior to the multivariate analysis, a Pearson correlation analysis was performed to understand the relationships among the PhP. To know the magnitude of the relationship between the parameters, the Pearson value is classified in 33 percentiles. The values of Pearson's linear correlation coefficient were classified as: Poor (0.0–0.3), Moderate (0.4–0.6) and Strong (0.7–1.0) [1,54]. Such analysis was performed in the SAS© 9.1.3 software [55].

The multivariate analysis of the data of the ID005 was realized by the PCA and CA methods [25,56]. The PCA is a method for pattern recognition that attempts to explain the variance of a large set of correlated variables (PhP); transforming the data set into a smaller set of independent (uncorrelated) principal components (PC). SAS© 9.1.3 software was used to describe these patterns. The PCA is a dimensionality reduction technique that helps to simplify the data and make it easier to visualize by looking for a PC set [57]. The PCs are orthogonal variables calculated by multiplying the original correlated variables with a list of coefficients that can be described as shown in Equation (1):

$$z_{ij} = a_{i1}x_{1j} + a_{i2}x_{2j} + \ldots + a_{im}x_{mj} \quad (1)$$

where: z = the component's coefficient, a = component weight, x = measured value of the variable, i = component number, j = sample number and m = number of variables.

The CA is an unsupervised pattern recognition technique that describes the structure of a data set [28]. The hierarchical grouping is the most common approach in which groups are formed sequentially, starting with the pair of most similar objects forming groups from the union of these objects. The Euclidean distance usually gives the similarity between two objects or groups of objects [58]. The resulting groups of objects should exhibit high internal homogeneity (within a group) and high external heterogeneity (among groups), where grouping is typically illustrated with a dendrogram [59]. The dendrogram provides a visual summary of the clustering processes, presenting an image of the groups and their proximity with a dramatic reduction in the dimensionality of the original data [60]. The CA was applied to classify the sampling sites by ascending cluster analysis with the Ward [61] criterion, using the determination coefficient R^2 as a measure of explanation of variation and pseudo T^2 served to confirm the number of groups [62]. It is possible to plot the pseudo values versus the number of clusters. If the values present a sudden change, the group value $n + 1$ that caused the change is a candidate for the number of groups to choose [63]. The CA was performed in the SAS© 9.1.3 software.

2.5. *Spatial Variability of the Physicochemical Parameters (PhP)*

The information of the PhP was used as input data to carry out an interpolation. To examine the spatial distribution of the studied variables, the interpolation method used was Inverse Distance Weight (IDW), available in ArcMap© 10.3 software (ESSRI, Redlands, CA, U.S.A.; https://www.

esri.com/en-us/home). The interpolation, through IDW has been widely used to map the spatial distribution of water elements [2,64,65]. The IDW method uses the existing values that are around the area to estimate the concentration of the non-sampled sites. The values of the closest observations will have a greater influence than those that are further away, i.e., the influence decreases with distance [66]. Equation (2) shows the algorithm for IDW.

$$Z(S_0) = \sum_{i=1}^{N} \lambda \times Z(S_i) \tag{2}$$

where: $Z(S_0)$ = value to be estimated in the place S_0, N = number of observations near to the place to estimate, λ = weight assigned to each observation to be used, decreases with distance, and $Z(S_i)$ = observed value of the place S_i.

3. Results

3.1. Analysis of Physicochemical Parameters (PhP)

Table 2 shows the results obtained from the As and the PhP analysis of water samples, the maximum permissible levels (MPL) established according to Mexican regulations for each parameter, and the percentage of samples exceeding the limits.

Table 2. Range of PhP concentrations, maximum permissible levels and percentage of samples that exceed it.

Parameter	Concentration Range S1	Concentration Range S2	MPL	Normative	Above MPL S1 (%)	Above MPL S2 (%)
As (mg/L)	ND–0.338	ND–0.576	0.100	[67]	9	13
Temp (°C)	22.1–30.1	22.8–27.5	-	Without regulation	-	-
EC (μS/cm)	13.8–1981.6	553.6–2600	-	Without regulation	-	-
ORP (mV)	85.6–267.7	98.1–306.3	-	Without regulation	-	-
Hardness (mg/L)	13.3–814	0–611	500	[52]	9	5
pH	7.5–9.6	7.3–9.0	6.0–9.0	[67]	1.5	0
TDS (mg/L)	0–990	0–932.3	500	[67]	35	39
Turb (NTU)	0–1000	0.2–519	10	[67]	29	12

As = Arsenic, EC = Electrical Conductivity, Turb = Turbidity, pH = pH, ORP = Oxide Reduction Potential, Temp = Temperature. MPL = Maximum permissible level.

In the two seasons, TDS was the parameter with the highest percentages of samples exceeding the MPL of the Mexican regulation (35% in S1, 39% in S2). Turbidity exceeded the MPL in S1 (29%) in more samples than S2 (12%). As concentrations were above the MPL of water for agricultural irrigation in 9% (Summer) and 13% (Fall) of the wells.

3.2. Multivariate Analysis

The correlation analysis reported the existence of significant positive and negative correlations ($p > 0.05$ and $p < 0.0001$) among the values of PhP from the first sampling (Table 3). As was positive moderately correlated with Turbidity and pH, and negative moderately correlated with hardness. EC was positive moderately correlated with TDS and Hardness; and negative moderately correlated with ORP. Regarding TDS was moderately correlated with Turbidity (negative) and Hardness (positive). Furthermore, pH and ORP were correlated positive strongly and negative moderately, with Temperature, respectively. The poor correlation between the other pairs of PhP indicates the presence of other variation sources.

In S2 (Table 4) it was observed that As was negative moderately correlated with Hardness while it was positive strongly correlated with pH. The EC was positive moderately correlated with Hardness and strongly correlated with TDS. Likewise, TDS was moderately negative correlated with Turbidity and ORP; and positive correlated with hardness.

Table 3. Pearson correlation among the PhP in the wells of the ID005, sampling 1 (S1).

	As	EC	TDS	Turb	Hardness	pH	ORP	Temp
As	1.00							
EC	0.07	1.00						
TDS	−0.17	0.625 **	1.00					
Turb	0.42	−0.01	−0.452 **	1.00				
Hardness	−0.477 **	0.493 **	0.586 **	−0.23	1.00			
pH	0.441 *	0.08	−0.04	0.17	−0.348 *	1.00		
ORP	−0.092 *	−0.44	−0.17	−0.18	−0.09	−0.398 *	1.00	
Temp	0.389 *	0.327 *	0.23	0.13	−0.17	0.827 **	−0.462 *	1.00

As = Arsenic, EC = Electrical Conductivity, Turb = Turbidity, pH = pH, ORP = Oxide Reduction Potential, Temp = Temperature, * = Significant $p > 0.05$, ** = Highly significant $p > 0.0001$.

Table 4. Pearson correlation among the PhP in the wells of the ID005, sampling 2 (S2).

	As	EC	TDS	Turb	Hardness	pH	ORP	Temp
As	1							
EC	−0.02	1						
TDS	0.08	0.89 **	1					
Turb	−0.14	−0.2	−0.55 **	1				
Hardness	−0.46 *	0.54 **	0.45 *	−0.15	1			
pH	0.77 **	−0.14	−0.03	−0.07	−0.60 **	1		
ORP	0.03	−0.33 *	−0.41 *	0.32 *	−0.13	−0.04	1	
Temp	−0.25	−0.19	−0.06	−0.21	0.04	−0.29 *	−0.04	1

As = Arsenic, EC = Electrical Conductivity, Turb = Turbidity, pH = pH, ORP = Oxide Reduction Potential, Temp = Temperature, * = Significant $p > 0.05$, ** = Highly significant $p > 0.0001$.

3.3. Principal Components Analysis (PCA)

The assumption that the parameters are linearly related was verified, then the PCA was realized to explore the relationships among the eight PhP. The first four PCs in S1 explained 87% of the variance (Table 5). In S1, PC1 contributed with 34% of the variance, PC2 with 30%, while PC3 and PC4 contributed with 12% and 9%, respectively. The dominant PhP in PC1 were As, pH and Temperature. Considering Table 3, there is a significant correlation between As and pH ($r = 0.44$, $p < 0.05$). In PC2, the coefficients that contributed the most were EC, TDS and Hardness. The parameters correlated were: EC and TDS ($r = 0.625$, $p < 0.0001$), EC and Hardness ($r = 0.493$, $p < 0.0001$) and TDS with Hardness ($r = 0.586$, $p < 0.0001$). The PC3 was influenced by Turbidity and PC4 by As and ORP.

Regarding S2, 86% of the variance was explained by considering four PCs (Table 5). The components contributed with 35%, 24%, 16% and 9% to PC1, PC2, PC3 and PC4, respectively. The PC1 was influenced by EC, TDS and Hardness with weak coefficients. These parameters strongly and moderately correlated as follows: EC and TDS ($r = 0.89$, $p < 0.0001$), EC and Hardness ($r = 0.54$, $p < 0.0001$), Hardness and TDS ($r = 0.45$, $p < 0.05$). The PC2 was influenced by As and pH, with moderate and highly correlated coefficients ($r = 0.77$, $p < 0.0001$). The PC3 explained the Turbidity and Temperature variability, with moderate to strong coefficients and PC4 was influenced by ORP. In Table 4 it was observed that there is a highly significant correlation of EC with TDS and Hardness, which indicates that these three components explain a large amount of variation in the study area.

The grouping of the sites is shown in the displacement plane of the first two PCs (Figure 3). The PhP for S1 and S2 were organized into 4 groups. Group 1: As, pH and Temperature; Group 2: EC, TDS and Hardness; Group 3: Turbidity; and Group 4: As and ORP. In regards to S2, Group 1 was composed of: Hardness, EC, TDS; Group 2: As and pH; Group 3: Turbidity and ORP; and Group 4: Temperature. Gebreyohannes et al. [68] determined in their area of study that TDS, Hardness and EC were positively associated and these were negatively associated with pH and Turbidity.

Table 5. Eigenvectors and eigenvalues of the PhP.

PhP	S1				S2			
	PC1	PC2	PC3	PC4	PC1	PC2	PC3	PC4
As	**0.441**	−0.11	0.108	**0.661**	−0.22	**0.56**	0.00	0.24
EC	0.066	**0.533**	0.312	0.261	**0.49**	0.21	0.31	0.07
TDS	−0.13	**0.544**	−0.19	0.318	**0.50**	0.32	0.03	0.17
Turb	0.305	−0.19	**0.727**	0.011	−0.28	−0.27	**0.53**	−0.31
Hardness	−0.33	**0.424**	0.263	−0.03	**0.47**	−0.23	0.20	0.16
pH	**0.518**	0.1	−0.34	−0.16	−0.28	**0.55**	−0.02	−0.03
ORP	−0.29	−0.33	−0.26	**0.604**	−0.29	−0.22	0.26	**0.87**
Temp	**0.48**	0.271	−0.27	−0.04	0.06	−0.26	**−0.72**	0.17
Eigenvalue	2.708	2.473	1.031	0.757	2.8	2.15	1.24	0.75
Variability	0.338	0.309	0.128	0.094	0.35	0.26	0.15	0.09
Cumulative	0.338	0.647	0.776	0.871	0.35	0.62	0.77	0.86

As = Arsenic, EC = Electrical Conductivity, Turb = Turbidity, pH = pH, ORP = Oxide Reduction Potential, Temp = Temperature, Bold letters indicate the dominant coefficients.

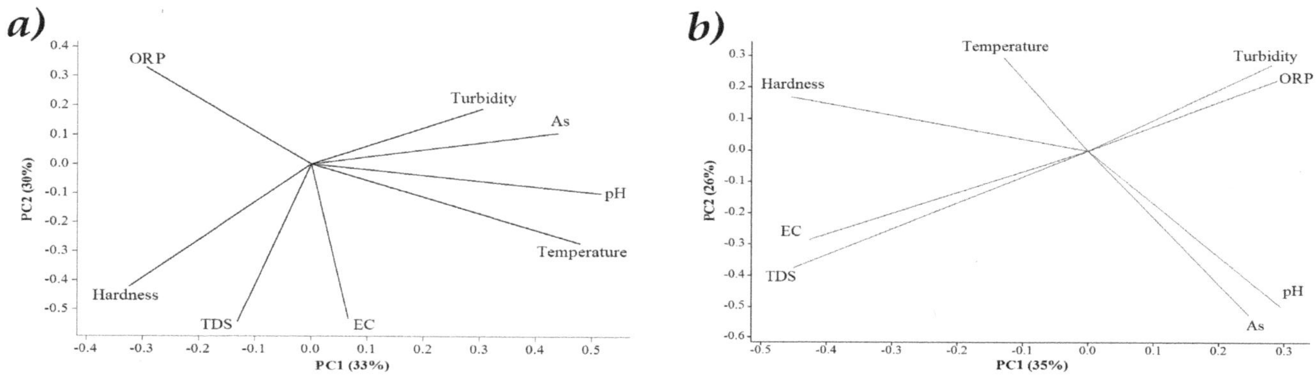

Figure 3. Plot S1 comparison PC1 vs. PC2 (**a**), Plot S2 comparison PC1 and PC2 (**b**).

The comparison plots of PC1 vs. PC2 in each sample indicate the displacement through components 1 and 2. These components, together explain more than 60% of the total variation. As, pH and Temperature in S1 move to the right side, indicating its dominance in PC1. While in S2, only As and pH remain dominant but in PC2. Furthermore, in S1 the variables with the greatest influence in PC2 were EC, TDS and Hardness and in S2 were also EC, TDS and but in PC1. The change in the station strongly influences the way in which the parameters are expressed in the well water, which explains the displacement of the parameters between stations.

3.4. Cluster Analysis (CA)

The definition of the number of groups was made considering the value of R^2 and the criterion of pseudo T^2. The value of R^2 indicated that, with four groups, up to 76% of the variability in S1 was explained, while in S2 with four groups 85% was explained. A line was drawn in both dendrograms to confirm the value of R^2. This line is imaginary and is traced in the dendrogram to support the definition of the groups [69]. The pseudo T^2 was useful to reaffirm the decision of the four groups, showing a value of 77.3 and 9.2, respectively [62]. The groups were significantly different based on the MANOVA test ($F = 25.65$, λ of Wilk's = 0.002, $p < 0.0001$).

In S1, Group 1 was made up of 9 wells; Group 2 was the largest with 38 wells; Group 3, the smallest with 7 wells and Group 4 included 9 wells (Figure 4). Each group was characterized with the average of the variables per group, presented in Table 6. In S1, Group 1 consisted of high values of As (0.098 mg/L), Turbidity (687.6 NTU), pH (8.2) and EC (1117.7 μS/cm), and low values of TDS and

Hardness (93.8 and 144.3 mg/L, respectively). Group 2 had moderate values with respect to almost all PhP, only with a low Turbidity value (3.9 NTU). Group 3 also presented moderate values in most of the PhP, with the exception of EC at low concentrations (15.3 μS/cm), and Turbidity at high concentrations (295.6 NTU). Lastly, Group 4 showed high values of TDS (883.9 mg/L), Hardness (497.4 mg/L), EC (1773.4 μS/cm) and Temperature (25.1 °C), while the lowest values corresponded to Turbidity (38.0 NTU). The remaining PhP had moderate values.

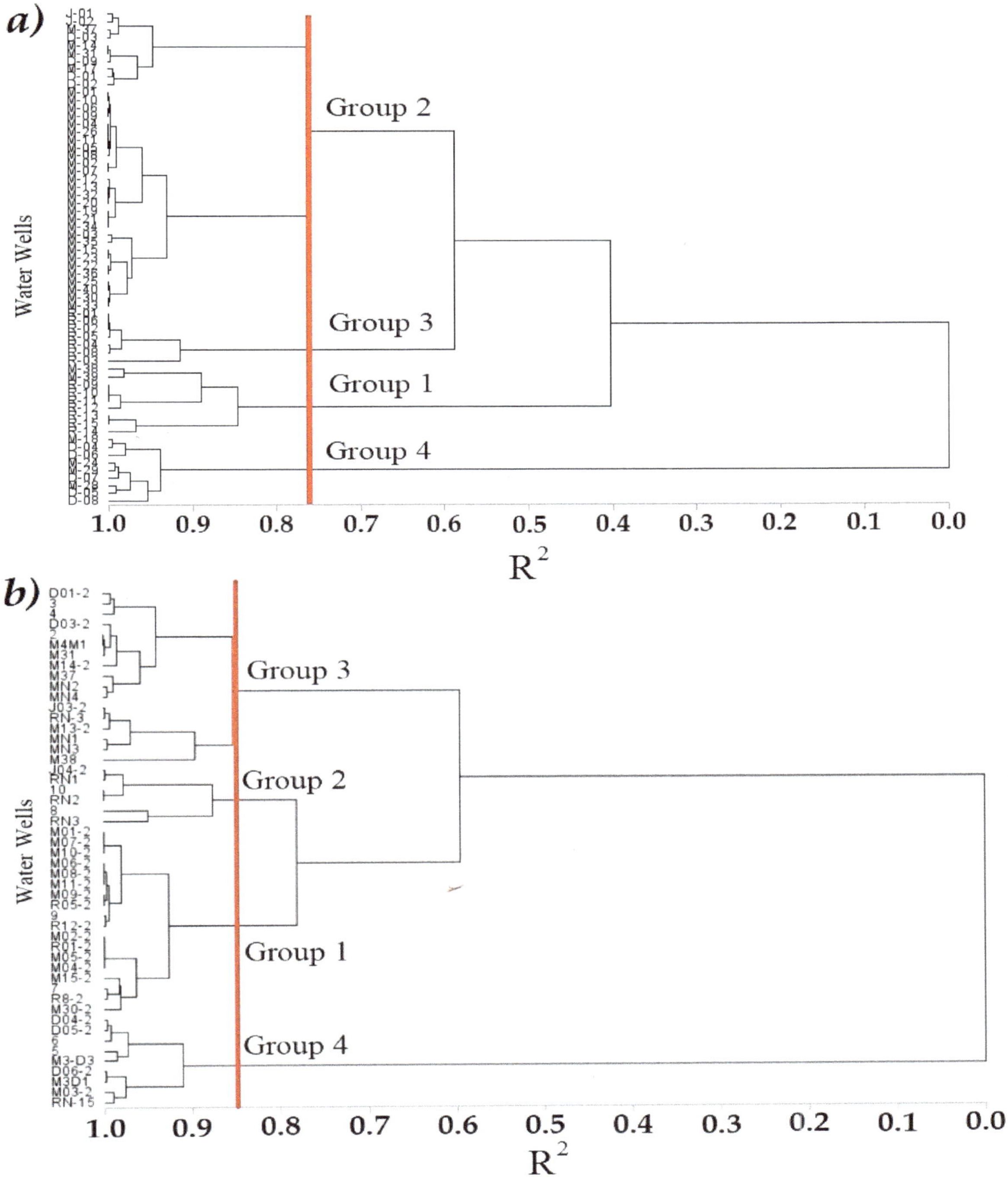

Figure 4. Dendogram. Grouping of sampling sites according to the PhP of the ID005 for S1 (**a**) and S2 (**b**).

Table 6. Average value of the PhP by groups.

G	S1								S2							
	As	EC	TDS	Turb	Hardness	pH	ORP	Temp	As	EC	TDS	Turb	Hardness	pH	ORP	Temp
1	0.098	1117.65	93.84	687.63	144.34	8.21	150.29	25.25	0.017	686.111	344.552	8.182	196.074	7.701	231.456	24.796
2	0.035	832.62	415.06	3.89	208.06	7.98	208.57	24.42	0.005	768.717	78.938	382.933	164.533	7.665	270.156	23.856
3	0.014	15.25	185.71	295.56	207.03	7.53	230.08	22.59	0.106	1102.486	563.371	5.753	157.647	8.024	210.418	24.170
4	0.008	1773.36	883.91	38.01	497.41	7.77	138.66	25.07	0.005	1779.137	885.278	5.228	384.678	7.525	195.219	24.203

G = Group, As = Arsenic, EC = Electrical Conductivity, Turb = Turbidity, pH = pH, ORP = Oxide Reduction Potential, Temp = Temperature.

In the S2, Group 1 was the largest with 18 wells; Group 2 was the smallest with 6 wells; Group 3 was comprised of 17 wells; and Group 4 of 9 wells (Figure 4). Group 1 was formed with high Turbidity values (196.1 NTU). Group 2 showed the highest values of Turbidity (164.5 NTU) and the lowest values of As (0.005 mg/L) and TDS (78.9 mg/L) among all groups. Group 3 showed high values of As (0.106 mg/L), EC (1.102.5 µS/cm), TDS (563.4 mg/L), Turbidity (157.6 NTU) and pH (8.02). Finally, Group 4 had the highest values of EC (1779.1 µS/cm), TDS (885.3 mg/L), Hardness (384.7 mg/L) and the lowest Turbidity (5.2 NTU), pH (7.3) and ORP (195.2 mV) values when compared to the other groups (Table 6).

The spatial distribution of S1 and S2 was observed by linking the database derived from the CA with the vector file of wells, using ArcMap 10.3©. In S1, the distribution of Group 1 (high values of As, Turbidity and pH) was homogeneous in the northern part of the ID005, located in modules 7 and 8. Group 2, with the highest number of wells (moderate values of all PhP), included modules 4, 7, 8, 9 and 3. Group 3 (high Turbidity), was located in module 6, showing a homogeneous spatial grouping. Group 4 (high TDS, Hardness and EC) was defined in modules 4, 8, 7 and 3, being the group with the greatest geographical dispersion.

In S2, Group 1 (high Turbidity) was homogeneously distributed between module 9 and 6. Group 2 (high Turbidity) was placed in module 6, only with one observation in module 8. Group 3, with high magnitudes of As, EC, TDS, Turbidity and pH, was presented in module 7, with some observations in modules 8 and 4. Group 4, with high magnitudes of EC, TDS and Turbidity, was distributed in the boundary between module 3 and 4 in the southwest portion of the ID005 (Figure 5).

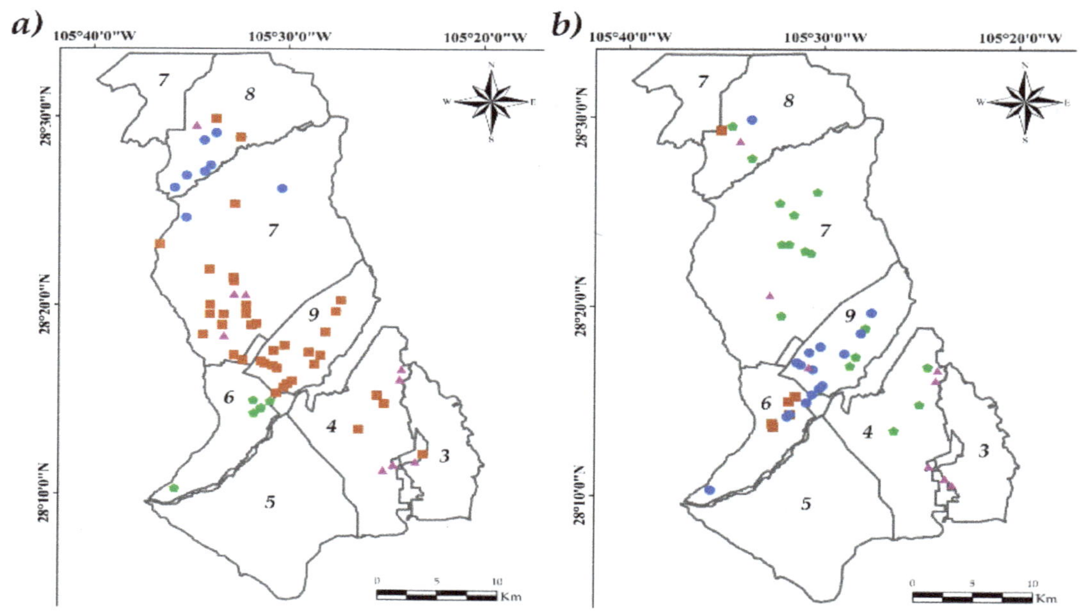

Figure 5. Spatial distribution of the groups in the IDDR005. S1 (**a**), S2 (**b**). Group 1 (●), Group 2 (■), Group 3 (⬟), Group 4 (▲), bold numbers denote the module number.

3.5. Spatial Variability of Physicochemical Parameters

The maps of the PhP are shown in Figures 6 and 7. The areas with low concentrations are colored in yellow, the blue colored areas represent moderate concentrations while the red colored areas represent high concentrations.

In S1, the PhP As, pH and Temperature showed a similar distribution where the highest and moderate concentrations are found in modules 6 and 7. The pattern of As (high concentration) may be due to a geological mineralization process [70], which seems to be present in these modules. Likewise, Hardness and TDS show a similar distribution. The highest concentrations are predominantly distributed in modules 3, 4 and 5. The EC shows a distribution pattern similar to Hardness and TDS but with some variations. The highest concentrations prevailed in modules 3, 4, 7 and 8.

In S2, As and pH showed a similar pattern with high concentrations in the northern part of module 7. The values of EC, TDS and Hardness showed a very similar spatial distribution in modules 8, 7, 4 and 3, at high concentrations. The Temperature and Turbidity PhP presented a similar pattern of high concentrations in module 6. Finally, ORP was the only variable that did not show a spatial behavior similar to the rest of the PhP.

The similarities in the spatial distribution among the PhP confirm the results of the multivariate analysis, where As-Turbidity, pH-Temperature, Hardness-TDS-EC and As-pH were grouped in S2, while TDS-EC-Turbidity were grouped in S1. In a study conducted by Li and Feng [71], similarities in the spatial distribution of the elements were found. In both, the S1 and S2 samplings, the spatial behavior for As, EC, Hardness, ORP, pH and TDS was similar. The PhP that varied were temperature and Turbidity. Temperature showed a greater variability in S1 compared to S2, which may be associated with the variation of the rest of the PhP.

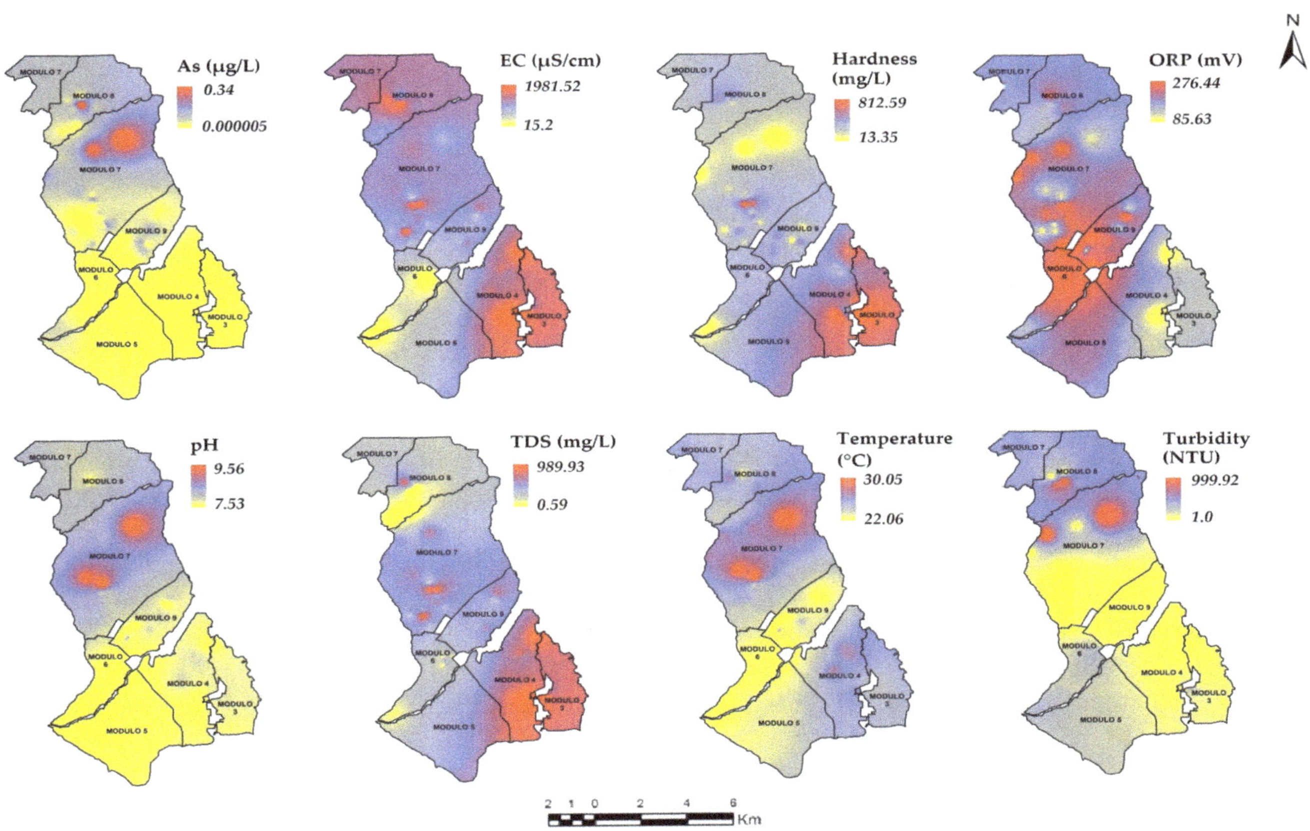

Figure 6. Spatial distribution of the PhP in the ID005 for S1. As = Arsenic, EC = Electrical Conductivity, pH = pH, ORP = Oxide Reduction Potential, studied modules (—).

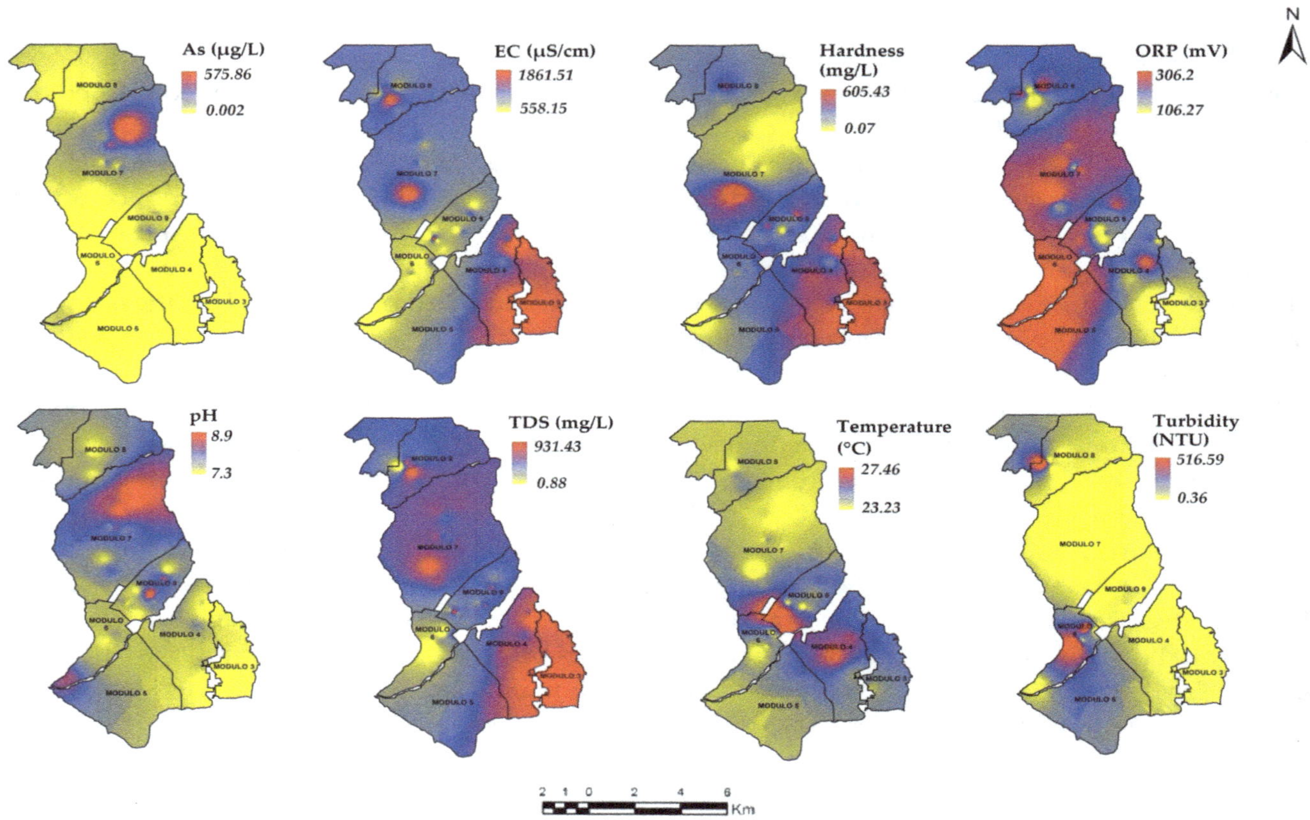

Figure 7. Spatial distribution of the PhP in the ID005 for S2, As = Arsenic, EC = Electrical Conductivity, pH = pH, ORP = Oxide Reduction Potential, studied modules (—).

Likewise, the coefficients of each PC of the PhP were used together with the geographical coordinates of the wells to generate the interpolation of the PCs (Figure 8). These interpolations spatially indicate the multivariate relationships among the PhP.

In S1, the interpolation of the PC1 coefficients (33% of the variability explained) indicated that in the yellow areas (negative coefficients) low concentrations of As existed (0.000005 mg/L). These areas registered values of pH between 7.5–7.8 and Temperatures between 22–23 °C. Conversely, the areas in red are those with high concentrations of As (0.33 mg/L), pH (9.5) and Temperatures of 30 °C. PC2 (30%), influenced by EC, TDS and Hardness, presented negative coefficients (areas in yellow), indicating the presence of low values of EC (13–16 μS/cm), TDS (20–100 mg/L) and Hardness (100–200 mg/L). PC3 (12%), influenced by Turbidity, depicted areas with negative coefficients indicating the presence of low values for this parameter (1.0 NTU). Meanwhile, positive coefficients corresponded to areas with high concentrations (900 NTU). Although Turbidity has the highest coefficient in the matrix of eigenvectors, Temperature also shows a similar behavior in the database (not shown). In this database, the negative coefficients correspond to zones with temperatures of 30 °C and positive coefficients to areas with temperatures of 24 °C. Finally, PC4 (9%) is influenced by As and ORP. The areas with negative coefficients correspond to low concentrations of As (0.000005 mg/L) and ORP (85–113 mV). The areas in red correspond to high concentrations of As (0.27–0.34 mg/L) and ORP (250 mV).

In the S2, PC1 (35%) represents the variability of EC, TDS and Hardness. In this component, the red areas represent EC values (1863 μS/cm), TDS (932.33 mg/L) and Hardness (594 mg/L). The concentrations in yellow are for EC (1078 μS/cm), TDS (573 mg/L) and Hardness (0 mg/L), which are distributed in module 6 and the northern part of 7. PC2 (26%) explains the variability of As and pH, where the high concentrations are distributed in module 7. In this module, the coefficients with positive value indicate As concentrations of 0.575 mg/L and pH of 8.9, while in modules 6 and 5, the concentrations are the lowest (As = 0 mg/L, pH = 7.4). The distribution of PC3 (15%) explains the variation of Turbidity and Temperature. In these zones (red color), the PC explains Turbidity concentrations of 519 NTU and Temperature 23.8 °C. Turbidity values of 0.43 NTU and Temperature of 27.5 °C are reported in the yellow zones. PC4 (9%) represents the variability of ORP. The zones in red tone indicate ORP concentrations of 306 mV and in yellow values of 106 mV.

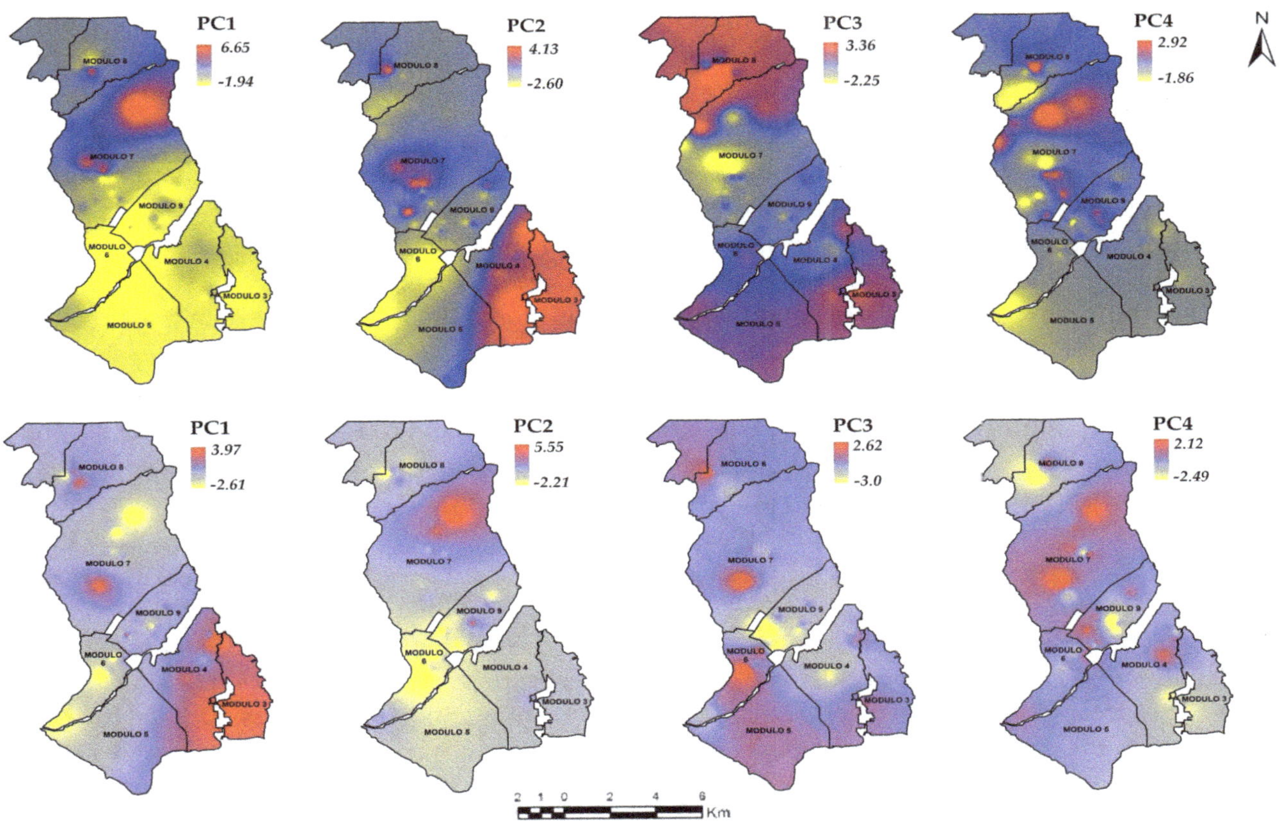

Figure 8. Spatial distribution of the PC scores of S1 (**top**) and S2 (**bottom**). PC = Principal component, studied modules (—).

4. Discussion

The multivariate techniques and the interpolation were useful to interpret the relationship between the PhP. The PCA has been previously used to examine and interpret the behavior of groundwater quality parameters [72–75]. The relationship between the PhP provided significant information on the possible sources of these parameters. In this study, four components were needed to explain the original data set. The four components showed that the behavior of the PhP in the wells was governed by more than one process or phenomenon.

According to Yidana et al. [76], in the analysis of dimensionality reduction of variables PC1 usually represents the most important mixture of processes in the study area. The PCA results suggest that most variations in water quality were found in S1 (summer) for As, pH and Temperature followed by EC, TDS and Hardness; and in S2 (Fall) for EC, TDS and Hardness, followed by As and pH. According to Bonte et al. [77], the increase in temperatures was associated with the As increase, which was shown in S1 where the main variables that explained the total variability were shown. The above was also demonstrated in the CA and the spatial interpolation of the individual parameters and the main components where high temperature zones show a spatial distribution similar to As.

The variables that were grouped in PC1 and PC2 of each sampling season had similar coefficients, which imply the existence of some similarities in the way they influence the groundwater concentration. It was observed that As, EC, TDS, Hardness and pH were shown in components 1 and 2 in both samples. This was consistent with the results of the interpolation, showing that the distribution of these parameters had a similar dispersion in the ID005. The interpolations of the PC coefficients are similar to the maps of the main components derived from the water sampling from the wells. Previous studies have shown similar results to improve the interpretation of PC [78,79]. These variables together accounted for more than 60% of the variability of the original data set. Also, it was observed that there was an exchange of these variables in PC1 and PC2. This behavior may have been caused the result of the different sampling seasons.

These relationships agree with the natural dynamics of water PhP. The pH is the main factor that controls the concentrations of soluble metals [80]. As well as, the arid climate leads to evaporation which can interfere in the concentration of As [9] and cause seasonal variations. It was observed that As concentrations higher than the MPL (0.1 μg/L) of water for agricultural irrigation established in the Mexican regulation [67], were presented in the northern area of the territory in both S1 and S2. The EC showed a significant correlation with parameters such as Hardness, TDS [10], which can be related to water salinity [9].

In wells that contain high amounts of As, the pH is also high. This was reaffirmed by the CA method and interpolation, where Group 1 showed the highest As and pH values for S1 and S2. The first two main components (PC1 and PC2) in both seasons (S1 and S2) showed similar variations. This same behavior was observed in wells where high concentrations of EC, TDS and Hardness were obtained, which was also observed by the CA and spatial interpolation.

For this study, the wells near the city showed the highest concentrations of EC, TDS and Hardness. The EC and TDS measurements for the S1 and S2 samples showed that the salinity is classified according to the Food and Agriculture Organization of the United Nations (FAO) as moderate (EC 700–3000 μS/cm, TDS 450–2000 mg/L), especially in the southern area of ID005 [81]. EC and high TDS limit the absorption of water by crops because of the salt that stays in the roots. Due to the difficult access to water, the growth rate of plants is reduced, which limits agricultural production [61]. The EC and the TDS in groundwater samples are significantly correlated with cations and anions (Ca^{2+}, K^{+}, Na^{+}, Cl^{+}, NO_3^{-} and SO_4^{2-}), which can be the result of ionic changes in the aquifer [9].

In the case of Hardness, it was within the MPL established by the Mexican regulation (500 mg/L) [52]. However, according to Gebreyohannes et al. [68], water with Hardness greater than 151 mg/L is classified as hard water. Considering these criteria, 75 and 77% of the samples (S1 and S2, respectively) were classified as hard water. This classification may indicate that there are deposits with high Mg^{2+} and Ca^{2+} contents [68]. Likewise, it is considered that hard water is not suitable for industrial and agricultural purposes [10].

The interpretation of the spatial behavior of the water quality in the studied area was possible when the scores derived from the PC were mapped. Previous studies have used geostatistical methods to map the scores resulting from PCA and used the resultant maps to predict the factors that may be impacting groundwater quality [82–84]. Based on the results of PCA and CA it was possible to understand the multivariate relationship of the set of parameters. In turn, with the application of the IDW interpolation technique on the scores of the PCA, it was possible to analyze the spatial variability. The combination of both methods was useful to examine patterns in common groups of parameters allowing to summarize the multiple relationship of variables on geographic regions to use in water quality analysis.

The PCA, the CA, the correlation coefficients and the interpolation were consistent with these interpretations. Although the results of the present study provided important conclusions regarding the origin of each PhP, more studies are needed to obtain a better understanding of the sources of the PhP and their concentrations.

5. Conclusions

The As is an element present in the ID005, and it is at levels that can cause a risk to agricultural production, mainly in the northern region. In addition, it is important to continue this investigation to determine the As traceability in the medium and to identify the risk of introducing this metalloid to the food chain by diet intake.

A slight issue was observed with indicators that affect the salinity of water. If such high levels persist, it can be detrimental to the optimal development of crops. Therefore, it will be necessary to look for alternatives to ameliorate this situation. Perhaps, starting with a continuous monitoring of wells in the ID005.

Multivariate statistical methods and spatial interpolation can be useful to identify locations of priority concern and potential sources of PhP, and to evaluate water quality from wells in an agricultural area. The multivariate geographic information system (GIS) approach showed the spatial relationships between the PhP (As, pH, EC, TDS and Hardness), proving to be convenient for the confirmation and refinement of PhP interpretations through the statistical results.

Author Contributions: J.A.P.-A.: sample collection, field analysis, geoinformatics and drafted the manuscript; B.A.R.-G.: carried out the physicochemical parameters in the water, evaluated the water quality in the area for irrigation and drafted the manuscript; M.d.L.B.-C.: carried out the water analysis for arsenic, data analysis and drafted the manuscript; M.C.V.-A.: conceived the study, designed of the studied area, sample collection, field analysis, processed analyzed the data, prepared the manuscript and designed the research; M.d.R.P.-P.: sample processing, statistical analysis and drafted the manuscript; A.P.-A.: contributed in the multivariate analysis and integration of geospatial information.

References

1. Kengnal, P.; Megeri, M.N.; Giriyappanavar, B.S.; Patil, R.R. Multivariate Analysis for the Water Quality Assessment in Rural and Urban Vicinity of Krishna River (India). *Asian J. Water Environ. Pollut.* **2015**, *12*, 73–80.
2. Varol, S.; Davraz, A. Evaluation of the groundwater quality with WQI (Water Quality Index) and multivariate analysis: A case study of the Tefenni Plain (Burdur/Turkey). *Environ. Earth. Sci.* **2015**, *73*, 1725–1744. [CrossRef]
3. Qadir, A.; Malik, R.N.; Husain, S.Z. Spatio-temporal variations in water quality of Nullah Aik-tributary of the River Chenab, Pakistan. *Environ. Monit. Assess.* **2008**, *140*, 43–59. [CrossRef] [PubMed]
4. Morris, B.L.; Lawrence, A.R.; Chilton, P.J.C.; Adams, B.; Calow, R.C.; Klinck, B.A. *Groundwater and its susceptibility to degradation: A global assessment of the problem and options for management*; United Nations Environment Programme: Nairobi, Kenya, 2003; Volume 3, p. 118. ISBN 92-807-2297-2.
5. Sharma, R.K.; Agrawal, M.; Marshall, F. Heavy metal contamination of soil and vegetables in suburban areas of Varanasi, India. *Ecotoxicol. Environ. Saf.* **2007**, *66*, 258–266. [CrossRef] [PubMed]
6. Khan, S.; Cao, Q.; Zheng, Y.M.; Huang, Y.Z.; Zhu, Y.G. Health risks of heavy metals in contaminated soils and food crops irrigated with wastewater in Beijing, China. *Environ. Pollut.* **2008**, *152*, 686–692. [CrossRef] [PubMed]
7. Gupta, I.; Dhage, S.; Kumar, R. Study of variations in water quality of Mumbai coast through multivariate analysis techniques. *Indian J. Mar. Sci.* **2009**, *38*, 170–177.
8. AlSuhaimi, A.O.; AlMohaimidi, K.M.; Momani, K.A. Preliminary assessment for physicochemical quality parameters of groundwater in Oqdus Area, Saudi Arabia. *J. Saudi Soc. Agric. Sci.* **2017**. [CrossRef]
9. Brahman, K.D.; Kazi, T.G.; Afridi, H.I.; Naseem, S.; Arain, S.S.; Ullah, N. Evaluation of high levels of fluoride, arsenic species and other physicochemical parameters in underground water of two sub districts of Tharparkar, Pakistan: A multivariate study. *Water. Res.* **2013**, *47*, 1005–1020. [CrossRef] [PubMed]
10. Patil, P.N.; Sawant, D.V.; Deshmukh, R.N. Physico-chemical parameters for testing of water-A review. *Int. J. Environ. Sci.* **2012**, *3*, 1194. [CrossRef]
11. Muangthong, S.; Shrestha, S. Assessment of surface water quality using multivariate statistical techniques: Case study of the Nampong River and Songkhram River, Thailand. *Environ. Model. Assess.* **2015**, *187*, 548. [CrossRef] [PubMed]

12. Giri, S.; Qiu, Z. Understanding the relationship of land uses and water quality in Twenty First Century: A review. *J. Environ. Manag.* **2016**, *173*, 41–48. [CrossRef] [PubMed]
13. Helena, B.; Pardo, R.; Vega, M.; Barrado, E.; Fernández, J.M.; Fernández, L. Temporal evolution of groundwater composition in an alluvial aquifer (Pisuerga river, Spain) by principal component analysis. *Water Res.* **2000**, *34*, 807–816. [CrossRef]
14. Singh, K.P.; Malik, A.; Sinha, S. Water quality assessment and apportionment of pollution sources of Gomti river (India) using multivariate statistical techniques: A case study. *Anal. Chim. Acta* **2005**, *538*, 355–374. [CrossRef]
15. Wang, Y.; Wang, P.; Bai, Y.; Tian, Z.; Li, J.; Shao, X.; Mustavish, L.F.; Li, B.L. Assessment of surface water quality via multivariate statistical techniques: A case study of the Songhua River Harbin region, China. *J. Hydro-Environ. Res.* **2013**, *7*, 30–40. [CrossRef]
16. Singh, K.P.; Malik, A.; Mohan, D.; Sinha, S. Multivariate statistical techniques for the evaluation of spatial and temporal variations in water quality of Gomti River (India)—A case study. *Water Res.* **2004**, *38*, 3980–3992. [CrossRef] [PubMed]
17. Zhao, Y.-F.; Shi, X.-Z.; Huang, B.; Dong-Sheng, Y.U.; Wang, H.-J.; Sun, W.-X.; Öboern, I.; Blombäck, K. Spatial Distribution of Heavy Metals in Agricultural Soils of an Industry-Based Peri-Urban Area in Wuxi, China. *Pedosphere* **2007**, *17*, 44–51. [CrossRef]
18. Brogna, D.; Michez, A.; Jacobs, S.; Dufrêne, M.; Vincke, C.; Dendoncker, N. Linking forest cover to water quality: A multivariate analysis of large monitoring datasets. *Water* **2017**, *9*, 176. [CrossRef]
19. Boyacioglu, H. Surface water quality assessment using factor analysis. *Water SA* **2006**, *32*, 389–393. [CrossRef]
20. Bhuiyan, M.A.; Rakib, M.A.; Dampare, S.B.; Ganyaglo, S.; Suzuki, S. Surface water quality assessment in the central part of Bangladesh using multivariate analysis. *KSCE. J. Civ. Eng.* **2011**, *15*, 995–1003. [CrossRef]
21. Batayneh, A.; Zumlot, T. Multivariate statistical approach to geochemical methods in water quality factor identification; application to the shallow aquifer system of the Yarmouk basin of North Jordan. *Res. J. Environ. Earth Sci.* **2012**, *4*, 756–768.
22. Oketola, A.A.; Adekolurejo, S.M.; Osibanjo, O. Water quality assessment of River Ogun using multivariate statistical techniques. *J. Environ. Prot.* **2013**, *4*, 466. [CrossRef]
23. Miranda, J.; Andrade, E.; López-Suárez, A.; Ledesma, R.; Cahill, T.A.; Wakabayashi, P.H. A receptor model for atmospheric aerosols from a southwestern site in Mexico City. *Atmos. Environ.* **1996**, *30*, 3471–3479. [CrossRef]
24. Jackson, B.B. *Multivariate Data Analysis: An Introduction*; Prentice Hall: Irwin, Homewood, IL, USA, 1983; pp. 154–196. ISBN 978-0256028485.
25. Wunderlin, D.A.; Diaz, M.P.; Ame, M.V.; Pesce, S.F.; Hued, A.C.; Bistoni, M.A. Pattern recognition techniques for the evolution of spatial and temporal variations in water quality. A case study: Suquia river basin (Cordoba-Argentina). *Water Res.* **2001**, *35*, 2881–2894. [CrossRef]
26. Loska, K.; Wiechuła, D. Application of principal component analysis for the estimation of source of heavy metal contamination in surface sediments from the Rybnik Reservoir. *Chemosphere* **2003**, *51*, 723–733. [CrossRef]
27. Vega, M.; Pardo, R.; Barrado, E.; Deban, L. Assessment of seasonal and polluting effects on the quality of river water by exploratory data analysis. *Water Res.* **1998**, *32*, 3581–3592. [CrossRef]
28. Al-Bassam, A.M. Evaluation of ground water quality in Al-Qassim area, Saudi Arabia, using cluster and factor analyses. *Kuwait J. Sci. Eng.* **2006**, *33*, 101–121.
29. Kazi, T.G.; Arain, M.B.; Jamali, M.K.; Jalbani, N.; Afridi, H.I.; Sarfraz, R.A.; Baig, J.A.; Shah, A.Q. Assessment of water quality of polluted lake using multivariate statistical techniques: A case study. *Ecotoxicol. Environ. Saf.* **2009**, *72*, 301–309. [CrossRef] [PubMed]
30. Chiu, W.C. Modelación y Simulación Para el Drenaje de Tierras, en la Planicie Aluvial del Estado de Tabasco, México. Ph.D. Thesis, Autonomous University of Nuevo León, Marin, Nuevo León, Mexico, 2000.
31. Medellín-Vázquez, J.J. Análisis de la Vegetación en un Gradiente Altitudinal Mediante Técnicas Multivariadas, en el Campo Santa María, Lampazos de Naranjo, Nuevo León y Candela Coahuila. Master Thesis, Autonomous University of Nuevo León, Linares, Nuevo León, México, 2003.
32. Villatoro, M.; Henríquez, C.; Sancho, F. Comparación de los interpoladores IDW y Kriging en la variación espacial de pH, Ca, CICE y P del suelo. *Agron. Costarric.* **2008**, *32*, 95–105.

33. Bhunia, G.S.; Shit, P.K.; Maiti, R. Comparison of GIS-based interpolation methods for spatial distribution of soil organic carbon (SOC). *J. Saudi Soc. Agric. Sci.* **2016**. [CrossRef]
34. Xie, Y.; Chen, T.B.; Lei, M.; Yang, J.; Guo, Q.J.; Song, B.; Zhou, X.Y. Spatial distribution of soil heavy metal pollution estimated by different interpolation methods: Accuracy and uncertainty analysis. *Chemosphere* **2011**, *82*, 468–476. [CrossRef] [PubMed]
35. Yan, W.; Mahmood, Q.; Peng, D.; Fu, W.; Chen, T.; Wang, Y.; Li, S.; Chen, J.; Liu, D. The spatial distribution pattern of heavy metals and risk assessment of moso bamboo forest soil around lead–zinc mine in Southeastern China. *Soil Till. Res.* **2015**, *153*, 120–130. [CrossRef]
36. Navarrete Álvarez, M. Modelos Geoestadísticos del Precio de la Vivienda: Aproximación al Conocimiento Intraurbano de la Ciudad de Madrid. Ph.D. Thesis, Autonomous University of Madrid, Madrid, Spain, 2012.
37. Wang, S.; Huang, G.H.; Lin, Q.G.; Li, Z.; Zhang, H.; Fan, Y.R. Comparison of interpolation methods for estimating spatial distribution of precipitation in Ontario, Canada. *Int. J. Climatol.* **2014**, *34*, 3745–3751. [CrossRef]
38. Núñez López, D.; Treviño Garza, E.J.; Reyes Gómez, V.M.; Muñoz Robles, C.A.; Aguirre Calderón, O.A.; Jiménez Pérez, J. Uso de modelos de regresión para interpolar espacialmente la precipitación media mensual en la cuenca del río Conchos. *Rev. Mex. Cienc. Agríc.* **2014**, *5*, 201–213.
39. Skeppström, K.; Olofsson, B. A prediction method for radon in groundwater using GIS and multivariate statistics. *Sci. Total Environ.* **2006**, *367*, 666–680. [CrossRef] [PubMed]
40. Maroju, S. Evaluation of five GIS-based interpolation techniques for estimating the radon concentration for unmeasured zip codes in the state of Ohio. Master Thesis, University of Toledo, Toledo, OH, USA, 2007.
41. Johnston, K.; Ver Hoef, J.M.; Krivoruchko, K.; Lucas, N. *Using ArcGIS Geostatistical Analyst*; Redlands ESRI: Redlands, CA, USA, 2001.
42. Behera, S.K.; Shukla, A.K. Spatial distribution of surface soil acidity, electrical conductivity, soil organic carbon content and exchangeable potassium, calcium and magnesium in some cropped acid soils of India. *Land Degrad. Dev.* **2015**, *26*, 71–79. [CrossRef]
43. Markus, J.; McBratney, A.B. A review of the contamination of soil with lead II. Spatial distribution and risk assessment of soil lead. *Environ. Int.* **2001**, *27*, 399–411. [CrossRef]
44. Rawlins, B.G.; Lark, R.M.; Webster, R.; O'Donnell, K.E. The use of soil survey data to determine the magnitude and extent of historic metal deposition related to atmospheric smelter emissions across Humberside, UK. *Environ. Pollut.* **2006**, *143*, 416–426. [CrossRef] [PubMed]
45. Maas, S.; Scheifler, R.; Benslama, M.; Crini, N.; Lucot, E.; Brahmia, Z.; Benyacoub, S.; Giraudoux, P. Spatial distribution of heavy metal concentrations in urban, suburban and agricultural soils in a Mediterranean city of Algeria. *Environ. Pollut.* **2010**, *158*, 2294–2301. [CrossRef] [PubMed]
46. Walter, A.M.; Christensen, S.; Simmelsgaard, S.E. Spatial correlation between weed species densities and soil properties. *Weed Res.* **2002**, *42*, 26–38. [CrossRef]
47. Chaoyang, W.E.I.; Cheng, W.A.N.G.; Linsheng, Y.A.N.G. Characterizing spatial distribution and sources of heavy metals in the soils from mining-smelting activities in Shuikoushan, Hunan Province, China. *J. Environ. Sci.* **2009**, *21*, 1230–1236.
48. Fu, S.; Wei, C.Y. Multivariate and spatial analysis of heavy metal sources and variations in a large old antimony mine, China. *J. Soil Sediments* **2013**, *13*, 106–116. [CrossRef]
49. Ortega-Gaucin, D.; Mejía Sáenz, E.; Palacios Vélez, E.; Rendón Pimentel, L.; Exebio García, A. Modelo de optimización de recursos para un distrito de riego. *Terra Latinoamericana* **2009**, *27*, 219–226.
50. Aguirre-Grijalva, E. Estudio de Factibilidad Técnica y Económica de 3 Métodos de Tecnificación del Riego en el Modulo 07 del Distrito 005. Master Thesis, Instituto Tecnológico de la Construcción, Chihuahua, Mexico, 2003.
51. Ortega-Gaucin, D. Reglas de operación para el sistema de presas del Distrito de Riego 005 Delicias, Chihuahua, México. *Ing. Agric. Biosist.* **2012**, *4*, 31–39. [CrossRef]
52. SSA (Secretaría de Salud) 1993. NOM-014-SSA1. Procedimientos Sanitarios para el Muestreo de Agua para Uso y Consumo Humano en Sistemas de Abastecimiento de Agua Públicos y Privados. Available online: http://www.salud.gob.mx/unidades/cdi/nom/014ssa13.html (accessed on 15 May 2017).
53. SCFI (Secretaría de Comercio y Fomento Industrial) 2001. NMX-AA-051-SCFI. Análisis de Agua Medición de Metales por Absorción Atómica en Aguas Naturales, Potables, Residuales y Residuales Tratadas—Método

de Prueba. Available online: http://www.economia-nmx.gob.mx/normas/nmx/2010/nmx-aa-051-scfi-2016.pdf (accessed on 15 May 2017).

54. Jothivenkatachalam, K.; Nithya, A.; Chandra, M.S. Correlation analysis of drinking water quality in and around Perur block of Coimbatore District, Tamil Nadu, India. *Rasayan J. Chem.* **2010**, *3*, 649–654.
55. SAS (Statistical Analysis Software) Institute. *SAS Software*; Version 9.1.3; SAS Inc.: Cary, NC, USA, 2006.
56. Simeonov, V.; Stratis, J.A.; Samara, C.; Zachariadis, G.; Voutsa, D.; Anthemidis, A. Assessment of the surface water quality in Northern Greece. *Water Res.* **2003**, *37*, 4119–4124. [CrossRef]
57. Jolliffe, I.T. *Principal Component Analysis*, 2nd ed.; John Wiley & Sons Inc.: Charlottesville, VA, USA, 2002; pp. 150–166. ISBN 978-0-387-22440-4.
58. Otto, M. Multivariate methods. In *Analytical Chemistry*; Kellner, R., Mermet, J.M., Otto, M., Widmer, H.M., Eds.; Wiley-VCH: Weinheim, Germany, 1998; p. 916. ISBN 3-527-28881-3.
59. Shrestha, S.; Kazama, F. Assessment of surface water quality using multivariate statistical techniques: A case study of the Fuji River Basin, Japan. *Environ. Model. Softw.* **2007**, *22*, 464–475. [CrossRef]
60. McKenna, J.E., Jr. An enhanced cluster analysis program with bootstrap significance testing for ecological community analysis. *Environ. Model. Softw.* **2003**, *18*, 205–220. [CrossRef]
61. Ward, J.H., Jr. Hierarchical grouping to optimize an objective function. *J. Am. Stat. Assoc.* **1963**, *58*, 236–244. [CrossRef]
62. Eder, B.K.; Davis, J.M.; Bloomfield, P. An automated classification scheme designed to better elucidate the dependence of ozone on meteorology. *J. Appl. Meteorol.* **1994**, *33*, 1182–1199. [CrossRef]
63. Neil, H.T. *Applied Multivariate Analysis*, 1st ed.; Springer: Board, NY, USA, 2002; p. 532. ISBN 0-387-95347-7.
64. Arslan, H.; Turan, N.A. Estimation of spatial distribution of heavy metals in groundwater using interpolation methods and multivariate statistical techniques; its suitability for drinking and irrigation purposes in the Middle Black Sea Region of Turkey. *Environ. Monit. Assess.* **2015**, *187*, 516. [CrossRef] [PubMed]
65. Ishaku, J.M.; Ankidawa, B.A.; Pwalas, A.J.D. Evaluation of Groundwater Quality Using Multivariate Statistical Techniques, in Dashen Area, North Eastern Nigeria. *Br. J. Appl. Sci. Technol.* **2016**, *14*. [CrossRef]
66. Moreno, J.A. *Sistemas y Análisis de la Información Geográfica. Manual de Autoaprendizaje con ArcGIS*, 2nd ed.; Ra-Ma: Madrid, Spain, 2008; p. 908. ISBN 978-84-7897-838-0. (In Spanish)
67. CONAGUA (Comisión Nacional del Agua) 2017. Ley Federal de Derechos Disposiciones Aplicables en Materia de Aguas Nacionales. Available online: https://www.gob.mx/cms/uploads/attachment/file/105138/Ley_Federal_de_Derechos.pdf (accessed on 21 June 2017).
68. Gebreyohannes, F.; Gebrekidan, A.; Hedera, A.; Estifanos, S. Investigations of physico-chemical parameters and its pollution implications of Elala River, Mekelle, Tigray, Ethiopia. *Momona Ethiop. J. Sci.* **2015**, *7*, 240–257. [CrossRef]
69. Güler, C.; Thyne, G.D.; McCray, J.E.; Turner, K.A. Evaluation of graphical and multivariate statistical methods for classification of water chemistry data. *Hydrogeol. J.* **2002**, *10*, 455–474. [CrossRef]
70. Bu, J.; Sun, Z.; Zhou, A.; Xu, Y.; Ma, R.; Wei, W.; Liu, M. Heavy Metals in Surface Soils in the Upper Reaches of the Heihe River, Northeastern Tibetan Plateau, China. *Int. J. Environ. Res. Public Health* **2016**, *13*, 247. [CrossRef] [PubMed]
71. Li, X.; Feng, L. Multivariate and geostatistical analyzes of metals in urban soil of Weinan industrial areas, Northwest of China. *Atmos. Environ.* **2012**, *47*, 58–65. [CrossRef]
72. Sánchez-Martos, F.; Jiménez-Espinosa, R.; Pulido-Bosch, A. Mapping groundwater quality variables using PCA and geostatistics: A case study of Bajo Andarax, Southeastern Spain. *Hydrol. Sci. J.* **2001**, *46*, 227–242. [CrossRef]
73. Liu, C.-W.; Lin, K.-H.; Kuo, Y.-M. Application of factor analysis in the assessment of groundwater quality in a blackfoot disease area in Taiwan. *Sci. Total. Environ.* **2003**, *313*, 77–89. [CrossRef]
74. Chapagain, S.K.; Pandey, V.P.; Shrestha, S.; Nakamura, T.; Kazama, F. Assessment of deep groundwater quality in Kathmandu Valley using multivariate statistical techniques. *Water Air Soil Pollut.* **2010**, *210*, 277–288. [CrossRef]
75. Belkhiri, L.; Boudoukha, A.; Mouni, L. A multivariate statistical analysis of groundwater chemistry data. *Int. J. Environ. Res.* **2011**, *5*, 537–544.
76. Yidana, S.M.; Banoeng-Yakubo, B.; Akabzaa, T.M. Analysis of groundwater quality using multivariate and spatial analyses in the Keta basin, Ghana. *J. Afr. Earth Sci.* **2010**, *58*, 220–234. [CrossRef]
77. Bonte, M.; van Breukelen, B.M.; Stuyfzand, P.J. Temperature-induced impacts on groundwater quality and

arsenic mobility in anoxic aquifer sediments used for both drinking water and shallow geothermal energy production. *Water Res.* **2013**, *47*, 5088–5100. [CrossRef] [PubMed]

78. Lu, A.; Wang, J.; Qin, X.; Wang, K.; Han, P.; Zhang, S. Multivariate and geostatistical analyses of the spatial distribution and origin of heavy metals in the agricultural soils in Shunyi, Beijing, China. *Sci. Total Environ.* **2012**, *425*, 66–74. [CrossRef] [PubMed]
79. Mueller, U.A.; Grunsky, E.C. Multivariate spatial analysis of lake sediment geochemical data; Melville Peninsula, Nunavut, Canada. *Appl. Geochem.* **2016**, *75*, 247–262. [CrossRef]
80. Acosta, J.A.; Faz, A.; Martínez-Martínez, S.; Zornoza, R.; Carmona, D.M.; Kabas, S. Multivariate statistical and GIS-based approach to evaluate heavy metals behavior in mine sites for future reclamation. *J. Geochem. Explor.* **2011**, *109*, 8–17. [CrossRef]
81. Ayers, R.S.; Westcot, D.W. *Water Quality for Agriculture*; Food and Agriculture Organization of the United Nations: Rome, Italy, 1994.
82. Kolsi, S.H.; Bouri, S.; Hachicha, W.; Dhia, H.B. Implementation and evaluation of multivariate analysis for groundwater hydrochemistry assessment in arid environments: A case study of Hajeb Elyoun–Jelma, Central Tunisia. *Environ. Earth Sci.* **2013**, *70*, 2215–2224. [CrossRef]
83. De Freitas Alves, S.M.; de Queiroz, D.M.; de Alcântara, G.R.; dos Reis, E.F. Spatial Variability of Physical and Chemical Attributes of Soil Using Techniques of Principal Component Analysis. *Biosci. J.* **2014**, *30*, 22–30.
84. Ha, H.; Olson, J.R.; Bian, L.; Rogerson, P.A. Analysis of heavy metal sources in soil using kriging interpolation on principal components. *Environ. Sci. Technol.* **2014**, *48*, 4999–5007. [CrossRef] [PubMed]

12

Comparison of Time Nonlocal Transport Models for Characterizing Non-Fickian Transport: From Mathematical Interpretation to Laboratory Application

Bingqing Lu [1], Yong Zhang [1,2,*], Chunmiao Zheng [3], Christopher T. Green [4], Charles O'Neill [5], Hong-Guang Sun [2] and Jiazhong Qian [6]

1 Department of Geological Sciences, University of Alabama, Tuscaloosa, AL 35487, USA; blu5@crimson.ua.edu
2 Department of Engineering Mechanics, Hohai University, Nanjing 210098, Jiangsu, China; shg@hhu.edu.cn
3 Guangdong Provincial Key Laboratory of Soil and Groundwater Pollution Control, School of Environmental Science & Engineering, Southern University of Science and Technology, Shenzhen 518055, Guangdong, China; zhengcm@sustc.edu.cn
4 U.S. Geological Survey, Menlo Park, CA 94025, USA; ctgreen@usgs.gov
5 Department of Aerospace Engineering and Mechanics, University of Alabama, Tuscaloosa, AL 35487, USA; croneill@eng.ua.edu
6 School of Resources and Environmental Engineering, Hefei University of Technology, Anhui 230009, Hefei, China; qianjiazhong@hfut.edu.cn
* Correspondence: yzhang264@ua.edu

Abstract: Non-Fickian diffusion has been increasingly documented in hydrology and modeled by promising time nonlocal transport models. While previous studies showed that most of the time nonlocal models are identical with correlated parameters, fundamental challenges remain in real-world applications regarding model selection and parameter definition. This study compared three popular time nonlocal transport models, including the multi-rate mass transfer (MRMT) model, the continuous time random walk (CTRW) framework, and the tempered time fractional advection–dispersion equation (tt-fADE), by focusing on their physical interpretation and feasibility in capturing non-Fickian transport. Mathematical comparison showed that these models have both related parameters defining the memory function and other basic-transport parameters (i.e., velocity v and dispersion coefficient D) with different hydrogeologic interpretations. Laboratory column transport experiments and field tracer tests were then conducted, providing data for model applicability evaluation. Laboratory and field experiments exhibited breakthrough curves with non-Fickian characteristics, which were better represented by the tt-fADE and CTRW models than the traditional advection–dispersion equation. The best-fit velocity and dispersion coefficient, however, differ significantly between the tt-fADE and CTRW. Fitting exercises further revealed that the observed late-time breakthrough curves were heavier than the MRMT solutions with no more than two mass-exchange rates and lighter than the MRMT solutions with power-law distributed mass-exchange rates. Therefore, the time nonlocal models, where some parameters are correlated and exchangeable and the others have different values, differ mainly in their quantification of pre-asymptotic transport dynamics. In all models tested above, the tt-fADE model is attractive, considering its small fitting error and the reasonable velocity close to the measured flow rate.

Keywords: time nonlocal transport model; Non-Fickian diffusion; breakthrough curve

1. Introduction

Non-Fickian or anomalous transport, where the plume variance grows nonlinearly in time, has been documented extensively for solute transport in heterogeneous aquifers [1–4], soils [5–7], and rivers [8,9]. Non-Fickian transport for dissolved contaminants can occur at all scales, varying from field scale [10] to micro- and nano-scale media [11,12]. Non-Fickian diffusion, which is characterized by slow diffusion (sometimes further classified as sub-diffusion), has long been believed to relate to sorption/desorption between mobile and immobile domains under an equilibrium assumption [13] or kinetic conditions [14,15], or mass exchange between flow regions with relatively high and low velocities [16]. Note here the term "diffusion" contains both molecular diffusion and mechanical dispersion (representing the local variation of advection from the mean velocity), and hence "non-Fickian diffusion" is used interchangeable with "non-Fickian transport" in this study.

To capture non-Fickian transport induced by solute retention, various transport models have been developed, starting from the standard advection–dispersion equation (ADE) with either equilibrium or kinetic sorption [17,18] and the two-domain or two-site models proposed originally in chemical engineering [19]. Time nonlocal transport models were then developed to capture solute retention in natural geologic media with intrinsic physical and chemical heterogeneity. There are at least three popular time nonlocal transport models, which are the multi-rate mass transfer (MRMT) model [16,20], the hydrologic version of the continuous time random walk (CTRW) developed by Berkowitz and colleagues (see the extensive review in [21] and the mathematical version of CTRW in [22]), and the tempered time fractional advection–dispersion equation (tt-fADE) model [23]. Some of these models have been compared theoretically. For example, mathematical similarity between the CTRW framework, the tt-fADE, and the MRMT model was explored by [24] and [21] (whose main conclusion is discussed below for clarification), and the numerical approximation of the MRMT model using CTRW schemes was developed by [25].

While previous studies showed that most of the time nonlocal models are identical with correlated parameters, fundamental challenges remain in real-world applications regarding model selection and parameter definition. Although various stochastic models have been developed for three decades in hydrology, they have not become routine modeling tools, due to many reasons. For example, given well-controlled laboratory transport experiments using heterogeneous sand columns and conservative tracers (which have been widely used to understand real-world diffusion and the resultant transport dynamics that are non-Fickian), a newcomer faces the challenge of model selection: which time nonlocal model should be used to capture the observed non-Fickian dynamics under specific flow/transport conditions, such as conservative tracer transport in saturated, heterogeneous columns repacked in the laboratory with a stable, relatively high water flow rate? To our best knowledge, there is, unfortunately, no literature providing such an answer for this simple question. The above time nonlocal transport models were originally built upon different physical theories and contain different quantities and types of parameters. A better understanding of the potential benefits and limitations of different time nonlocal transport models as applied to non-Fickian dynamics, therefore, is required before they can be reliably applied for real-world applications and attract new users with limited knowledge in stochasticity.

This study will systematically evaluate the above (MRMT, CTRW, and tt-fADE) theoretical treatments of non-Fickian transport of conservative solutes, with respect to their parameters and ability to represent non-Fickian dispersion, especially the late-time tailing behavior, which is typical for hydrological processes in heterogeneous geological media. Late-time dynamics of contaminant transport is also a major factor in many environmental issues, such as groundwater contamination remediation and aquifer vulnerability assessment. We will then apply all the modeling methods for laboratory column transport experiments and field tracer tests. We emphasize here that the application of a time nonlocal transport model to capture non-Fickian transport is not new. What is new in this study is the quantitative, practical comparison of nonlocal transport models given experimental data.

The rest of the work is organized as follows. In Section 2, we review the above time nonlocal transport models. In Section 3, the time nonlocal transport models are applied to non-Fickian transport in multidimensional heterogeneous porous media. The applicability of those time nonlocal models is checked using laboratory experiments and field tests. Our laboratory experiments, where a conservative tracer moves through heterogeneous lightweight expanded clay aggregate (Leca) beads, exhibit typical non-Fickian diffusive behavior with elongated late-time tails. The field tracer tests also exhibit apparent tailing behaviors. Section 4 checks and compares all the time nonlocal models for characterizing the observed non-Fickian dynamics. Further analyses are shown in Section 5, where we group the transport models and briefly discuss parameter uncertainty. Conclusions are finally drawn in Section 6.

2. Review and Evaluation of Time Nonlocal Transport Models

The core of the time nonlocal models is the appropriate definition of the memory function, which controls the distribution of waiting times for contaminants trapped by immobile zones. In this section, we focus on the theoretical background, especially the memory function, for each model and explore potential correlation of critical parameters in different models. For example, previous studies emphasized that the tt-fADE is a specific form of the CTRW framework since the tt-fADE assumes a truncated power-law memory function, which is one of the memory functions previously assumed by the CTRW framework [21]. Identical functionality was also pointed out for the MRMT and the CTRW framework (see Section 1).

2.1. Multi-Rate Mass Transfer Model

The MRMT model describes mass transfer between a mobile domain and any number of immobile domains with varying properties. The linear, multi-rate, first-order solute transport equations in the absence of sources/sinks can be written as [20]:

$$\frac{\partial C_m}{\partial t} + \sum_{j=1}^{n} \beta_j \frac{\partial C_{im,j}}{\partial t} = -\nabla \cdot [v\, C_m - D\, \nabla C_m], \tag{1}$$

$$\frac{\partial C_{im,j}}{\partial t} = \alpha_j \left[C_m - C_{im,j}\right], \quad j = 1, 2, \cdots, n, \tag{2}$$

where C_m and $C_{im,j}$ $[ML^{-3}]$ represent the aqueous concentrations in the well-mixed mobile zone and the j-th well-mixed immobile zone, respectively; β_j [dimensionless] is the capacity coefficient usually defined as the ratio of porosities of the j-th immobile and the mobile phases; v $[LT^{-1}]$ is the velocity vector; D $[L^2T^{-1}]$ is the dispersion coefficient tensor; n [dimensionless] is the number of distinct immobile phases; and α_j $[T^{-1}]$ is the first-order mass transfer rate (also called the rate coefficient) associated with the j-th immobile zone. When $n = 1$, Equation (1) reduces to the single-rate mass transfer model.

The summation term in the left-hand side of (1) can be expressed as a convolution, leading to the time-nonlocal form [16,26]:

$$\frac{\partial C_m}{\partial t} + g(t) * \frac{\partial C_{im,j}}{\partial t} = -\nabla \cdot [v\, C_m - D\, \nabla C_m], \tag{3}$$

where the symbol "$*$" denotes convolution, and $g(t)$ $[T^{-1}]$ is a memory function defined by the weighted sum of the exponential decay from individual immobile zones [16]:

$$g(t) = \int_0^{\infty} \alpha\, b(\alpha)\, \exp(-\alpha t)\, d\alpha, \tag{4}$$

where $b(\alpha)$ $[T]$ is a density function of first-order rate coefficients.

In terms of similarities with the other nonlocal methods discussed below, the MRMT model captures the time nonlocality caused by the diffusion-limited transport of solutes in immobile zones. A practical advantage of this approach is that the memory function has explicit hydrogeological meaning and thus it may be calculated, fitted, or even predicted [27]. Finally, the parameters v, D, and $g(t)$ can be spatially variable, so the MRMT method can capture the local variation of solute transport velocity caused by heterogeneity.

2.2. Tempered Time Fractional Advection–Dispersion Equation Model

The tt-fADE is one analytic technique that accounts for the time nonlocality of the medium, and simultaneously accounts for convergence of a stochastic solute particle motion process (i.e., a CTRW) to a limit distribution. In particular, if the distribution of trapping times between the movement of solute particles has an infinite mean, then the overall transport equation has one fractional-order derivative representing "dispersion" in time, leading to the standard time fractional advection–dispersion equation (t-fADE) model. Assuming a power-law memory function to describe random waiting times in the immobile zones [28]:

$$g(t) = \frac{t^{-\gamma}}{\Gamma(1-\gamma)}, \tag{5}$$

where $\Gamma(\cdot)$ is the Gamma function, and the exponent $0 < \gamma \le 1$ (when γ approaches to 1, $\Gamma(1-\gamma)$ approaches to infinity, making no memory effects $g(t) = 0$, and the model would behave like the traditional ADE model with a retardation coefficient $1 + \beta$). Then by definition,

$$\frac{\partial C_m(x,t)}{\partial t} * g(t) = \frac{\partial C_m(x,t)}{\partial t} * \frac{t^{-\gamma}}{\Gamma(1-\gamma)} = \frac{\partial^{\gamma} C_m(x,t)}{\partial t^{\gamma}}, \tag{6}$$

is a Caputo fractional derivative of order γ. Inserting (6) into the MRMT model (1) and assuming that the solute is initially placed in the mobile zone only, one obtains the following standard t-fADE describing the mobile and immobile solute transport:

$$\frac{\partial C_m}{\partial t} + \beta \frac{\partial^{\gamma} C_m}{\partial t^{\gamma}} = -\nabla \cdot [v\, C_m - D\, \nabla C_m] - \beta\, C_m(x, t=0)\, \frac{t^{-\gamma}}{\Gamma(1-\gamma)}, \tag{7}$$

$$\frac{\partial C_{im}}{\partial t} + \beta \frac{\partial^{\gamma} C_{im}}{\partial t^{\gamma}} = -\nabla \cdot [v\, C_{im} - D\, \nabla C_{im}] + C_m(x, t=0)\, \frac{t^{-\gamma}}{\Gamma(1-\gamma)}, \tag{8}$$

where C_{im} denotes the overall chemical concentration in all immobile domains.

Meerschaert et al. [23] generalized the t-fADE (7) and (8) by introducing an exponentially truncated power-law function, which is an incomplete Gamma function, as the memory function:

$$g(t) = \int_t^{\infty} e^{-\lambda s} \frac{\gamma s^{-\gamma-1}}{\Gamma(1-\gamma)}\, ds, \tag{9}$$

where $\lambda > 0$ $[T^{-1}]$ is the truncation parameter in time. This modification leads to the tt-fADE:

$$\frac{\partial C_m}{\partial t} + \beta e^{-\lambda t} \frac{\partial^{\gamma}}{\partial t^{\gamma}} \left[e^{\lambda t} C_m\right] - \beta \lambda^{\gamma} C_m = -\nabla \cdot [v\, C_m - D\, \nabla C_m] - \beta C_m^0 \int_t^{\infty} e^{-\lambda \tau} \frac{\tau^{-\gamma-1}}{\Gamma(1-\gamma)}\, d\tau \tag{10}$$

$$\frac{\partial C_{im}}{\partial t} + \beta e^{-\lambda t} \frac{\partial^{\gamma}}{\partial t^{\gamma}} \left[e^{\lambda t} C_{im}\right] - \beta \lambda^{\gamma} C_{im} = -\nabla \cdot [v\, C_{im} - D\, \nabla C_{im}] + C_m^0 \int_t^{\infty} e^{-\lambda \tau} \frac{\tau^{-\gamma-1}}{\Gamma(1-\gamma)}\, d\tau \tag{11}$$

where $C_m^0 = C_m\ (x, t = 0)$ denotes the initial source located only in the mobile phase. At a time

$$t << 1/\lambda, \tag{12}$$

the tail of the mobile-phase breakthrough curve (BTC) declines as a power law function:

$$C_m(x,t) \propto t^{-1-\gamma}, \tag{13}$$

while at a much later time $t >> 1/\lambda$, the slope of the mobile-phase BTC approaches negative infinity (i.e., the late-time BTC tail declines exponentially). Therefore, the value of λ controls the transition of the BTC late-time tail from a power-law function to exponential function [29].

The use of the tt-fADE (10) and (11) to model mobile/immobile anomalous solute transport is motivated by four factors: (1) the equation governs the limits of known stochastic processes; (2) it describes a combination of first-order mass transfer models and reduces to known mobile/immobile equations in the integer order case; (3) the equation has tractable solutions that model the significant features of solute plume evolution in time and space; and (4) the equation is parsimonious, with no more parameters than the standard MRMT model (1) with multiple pairs of rate and capacity coefficients. To summarize, the tt-fADE has one major limitation compared to the MRMT model (1) and the CTRW model discussed below: the memory function embedded in the tt-fADE is a specific form (9), while the memory function used in the MRMT and the CTRW model can have different forms. Potential effects of this difference on the models' abilities to estimate real-world physical behavior is explored below, in Section 4, and discussed in Section 5.

2.3. Continuous Time Random Walk Framework

Derivation of the CTRW model in hydrologic sciences [21] starts from the generalized master equation with the kernel Φ defined as (in Laplace space ($t \to s$)) [30]:

$$\widetilde{\Phi}(x,s) = \frac{s\,\widetilde{\Psi}(x,s)}{1-\widetilde{\phi}(s)}, \tag{14}$$

where the tilde "~" denotes the Laplace transform, $\widetilde{\Psi}(x,s)$ is the Laplace transform of the joint density of jump length and duration, and $\widetilde{\phi}(s)$ is the Laplace transform of the transition time or duration density $\phi(t) = \int\limits_{-\infty}^{+\infty} \Psi(x,t)dx$ used in the master equation [31]

$$\hat{\widetilde{p}}\,(k,s) = \frac{1-\widetilde{\phi}(s)}{s}\,\frac{1}{1-\widetilde{\phi}(s)\,\hat{f}(k)}, \tag{15}$$

where the right-hand side term is valid for independent jump size and transition time, and $\widetilde{f}(k)$ denotes the jump size density.

Berkowitz et al. [21] defined a memory function M to replace $\widetilde{\phi}(s)$ in (15)

$$\widetilde{M}\,(s) = \frac{t_1\,s\,\widetilde{\phi}(s)}{1-\widetilde{\phi}(s)}, \tag{16}$$

where t_1 [T] denotes a "*typical median transition time*" for particles. Inserting (16) into (15), and taking the Laplace and Fourier inverse transform, the following well-known CTRW framework was obtained [21]:

$$\frac{\partial\,p(x,t)}{\partial\,t} = \int_0^t M(t-\tau)\left[-v_\psi\,\frac{\partial}{\partial x} + D_\psi\,\frac{\partial^2}{\partial x^2}\right] p(x,\tau)\,d\tau, \tag{17}$$

where v_ψ denotes the average velocity, and D_ψ is the dispersion coefficient. When deriving the CTRW (17) from the master Equation (15), the jump size density $\widetilde{f}(k)$ in (15) needs to be expanded as

$$\hat{f}\,(k) \approx 1 - \mu\,i\,k + \frac{\alpha^2}{2}(i\,k)^2, \tag{18}$$

resulting in the spatially-averaged velocity v_ψ and dispersion coefficient D_ψ in (17):

$$v_\psi = \frac{\mu}{t_1}, \tag{19}$$

$$D_\psi = \frac{\sigma^2}{2t_1}, \tag{20}$$

Therefore, here the "*typical median transition time*" t_1 is the mean waiting time:

$$t_1 = \int_0^{+\infty} t\,\phi(t)\,dt, \tag{21}$$

The CTRW memory function M defined by (16) has various forms to capture various breakthrough curves (BTCs). One popular form is the exponentially truncated power-law, defined by the transition time ϕ (see Equation (16) in [32]):

$$\phi(t) = B\,\frac{\exp(-t/t_2)}{(1+t/t_1)^{1+\xi}}, \tag{22}$$

where the factor $B = \{t_1 \tau_2^{-\xi} \exp\left(\tau_2^{-1}\right)\Gamma * (-\xi, \tau_2^{-1})\}^{-1}$ (with $\tau_2 = t_2/t_1$.) keeps the integral of $\phi(t)$ to be 1. For simplicity and direct comparison between models, we constrain the exponent ξ to be $0 < \xi < 1$ in this study. The Laplace transform of (22) is (see Equation (17) in [32])

$$\widetilde{\phi}(s) = \frac{\Gamma\left(-\xi,\ \tau_2^{-1} + t_1 s\right)}{\Gamma\left(-\xi,\ \tau_2^{-1}\right)}\,(1 + \tau_2\, s\, t_1)^{\xi}\, \exp(t_1 s), \tag{23}$$

A simple manipulation shows the relationship between the MRMT memory function $g(t)$ and the CTRW memory function $M(t)$ in Laplace space:

$$\widetilde{M}(s) = \frac{1}{\theta_M + \theta_I \widetilde{g}(s)}, \tag{24}$$

where θ_M [dimensionless] and θ_I [dimensionless] are the porosity in the mobile and (total) immobile domains, respectively. Inserting (16) and (23) into (24), we obtain:

$$\begin{aligned} \widetilde{g}(s) &= \frac{1-\widetilde{\phi}(s)}{\theta_1\, t_1\, s\, \widetilde{\phi}(s)} - \frac{\theta_M}{\theta_I} \\ &= \frac{1}{\theta_1\, t_1\, s}\, \frac{\Gamma(-\xi, \tau_2^{-1})}{\Gamma(-\xi, \tau_2^{-1} + t_1 s)\,(1+\tau_2 s t_1{}^{\xi}\, \exp(t_1 s)} - \frac{1}{\theta_I s t_1} - \frac{\theta_M}{\theta_I} \end{aligned}, \tag{25}$$

There is no known analytical solution for g in real time t, except for the following asymptote at late time $t >> t_1 \tau_2$:

$$g(t) \propto \exp(-\frac{t}{t_1\, \tau_2}) = \exp(-\frac{t}{t_2}), \tag{26}$$

Based on (26), we obtain the late-time growth rate for p

$$p(t_{late}) \propto -\frac{\partial\, g(t)}{\partial\, t} \propto \exp(-\frac{t}{t_2}), \tag{27}$$

Therefore, for time $t >> t_2$, the non-Fickian transport transitions to Fickian diffusion. The cutoff time scale t_2 in (22), as explained by [32], "*corresponds to the largest heterogeneity length scale*". In another numerical study by Willmann et al. [33], t_2 was also called "*the late cutoff time*". The above analysis shows that t_2 is functionally equivalent to the inverse of the truncation parameter λ used in the tt-fADE model (10) and (11).

For the intermediate time $t_1 << t << t_2$, one can obtain the memory function g by solving (25) numerically (e.g., using the numerical inverse Laplace transform). This was done by [32], who found that the transition probability scales as a power-law function

$$p(t) \propto t^{-\xi-1}, \text{ where } t_1 << t << t_2 \tag{28}$$

Comparing (28) and (13), we find that the power-law exponent ξ in the CTRW framework is functionally equivalent to the scale index γ in the tt-fADE model (10) and (11).

Hence, the parameters in the CTRW framework (17) (e.g., ξ and t_2) are related to the parameters in the tt-fADE model (10) and (11) (γ and λ), except for t_1 in (17), which may be estimated by the mean diffusive time. Parameters predicted by one model (such as the tt-fADE) may also help to improve the estimated parameters of the other (i.e., the CTRW framework).

3. Applications: Capturing Non-Fickian Transport in Multidimensional Porous Media

Here we apply the above time nonlocal transport models to capture non-Fickian transport observed in multidimensional, heterogeneous porous media. Well-controlled laboratory experiments of sand column transport were conducted, to provide data to evaluate the three nonlocal transport models reviewed above. We used a cylindrical organic glass (polymethyl methacrylate) tube filled with non-uniform "lightweight expanded clay aggregate" (Leca) beads to monitor solute transport in saturated porous media. Leca beads were selected since they contain high intra-granular porosity (and potential for sorption on the large surface area of clay particles), which can lead to retention for solute transport as often occurs in the field. The length of the Leca column was 100 cm with an internal diameter of 4 cm. The experimental apparatus was composed of inflow and outflow water tanks, a porous medium column, tubing, and a detection device (Figure 1). The diameter of the Leca beads varied from 1.0 to 2.0 mm. A pulse of Brilliant Blue FCF (BBF), which is a conservative organic compound typically used as colorant for foods, with a volume of 5 mL was injected into the column (from the bottom of the glass tube set vertically) at a concentration of 0.1 g/L, representing an instantaneous point source. For the relatively short time-scales of these experiments, we believe the mechanical dispersion of BBF is predominant, and effects of molecular diffusion are negligible. An ultraviolet visible light spectrophotometer was used to measure the absorbance of solute, and the absorbance was then converted to concentration. Continuous sampling provided tracer BTCs used to check the applicability of the nonlocal transport models.

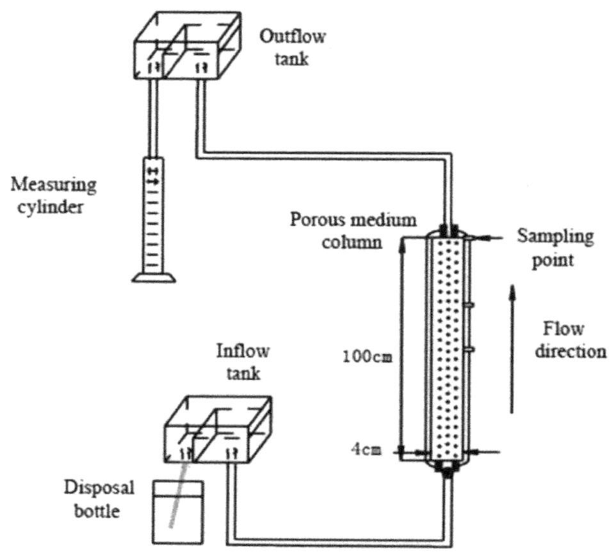

Figure 1. Experimental setup.

Six flow rates (increasing from 0.4, 0.6, 0.8, 1.0, 1.2, to 1.4 mL/s; see Table 1) were carried out in the experiment. For each flow rate, three experimental runs were conducted where the BBF BTCs were collected at three sections of the sand column, varying from 50, 70, to 100 cm. During each run with the similar flow rate, the hydraulic gradient between the inlet and outlet was kept constant in time, and therefore the flow velocity for the three experimental runs should be close to each other.

Table 1. Measured and fitted parameters (using the advection–dispersion equation (ADE) model with equilibrium sorption) for Brilliant Blue FCF (BBF) transport through the "lightweight expanded clay aggregate" (Leca) beads at different flow rates and travel distances and field tests. In the legend, Q represents the flow rate; L denotes the travel distance (i.e., the length of the column); v^* ($=v/R$) is the average flow velocity divided by the retardation coefficient; D^* ($=D/R$) is the dispersion coefficient divided by the retardation coefficient; θ is the porosity; and *RMSE* stands for root mean square error between observed values and predicted values.

Experiment	Q (mL/s)	L (cm)	v^* (mm/s)	D^* (mm²/s)	Θ	*RMSE*
	Measured	Measured	Fitted	Fitted	Measured	Calculated
Lab	0.4	50	1.20	5.0	0.39	2.75
		70	1.12	7.0	0.39	1.54
		100	1.07	5.3	0.39	2.30
	0.6	50	1.45	6.8	0.39	2.16
		70	1.55	15	0.39	0.78
		100	1.63	9.2	0.39	1.99
	0.8	50	2.23	15	0.39	1.57
		70	2.20	22	0.39	0.97
		100	2.07	14	0.39	1.55
	1	50	2.67	20	0.39	1.85
		70	2.85	24	0.39	1.43
		100	2.90	16	0 39	2.92
	1.2	50	3.25	23	0 39	2.80
		70	3.22	28	0.39	1.53
		100	3.04	19	0.39	2.40
	1.4	50	3.48	40	0.39	1.09
		70	3.85	50	0.39	0.90
		100	3.70	38	0.39	1.33
Field	83.33	200	0.003	0.694	0.3	n/a
		400	0.004	0.984	0.3	n/a
		600	0.004	0.926	0.3	n/a

The measured BTCs all exhibit late-time tailing behavior (see Figure 2 and Figures S1–S5 in the Supplementary File), which is one of the major characteristics of non-Fickian transport.

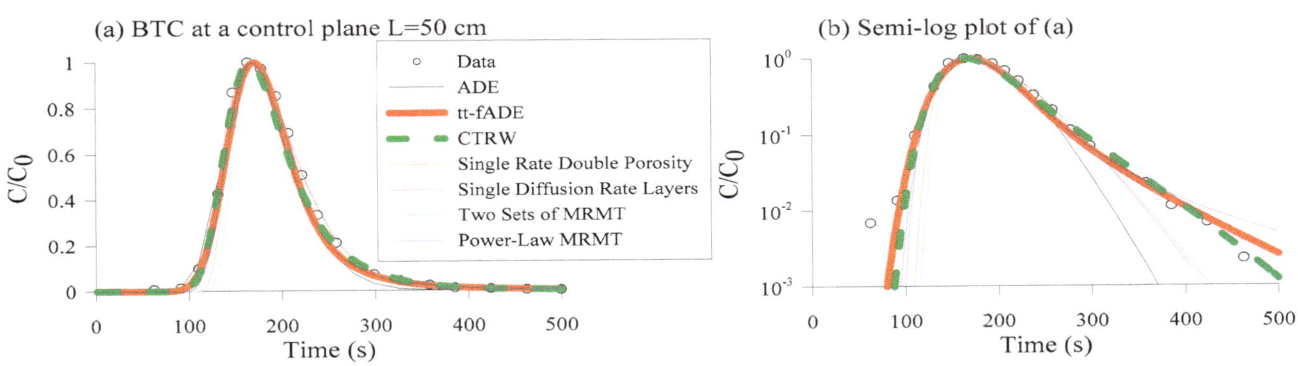

Figure 2. *Cont.*

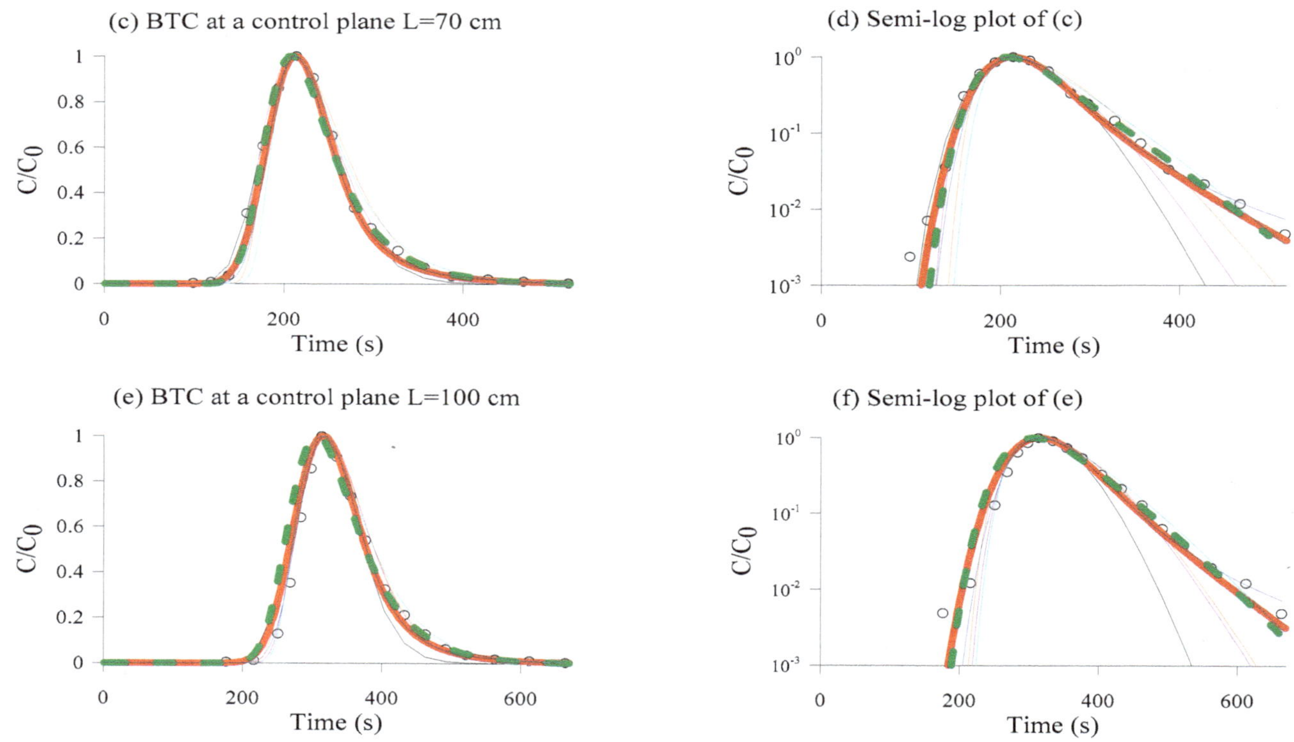

Figure 2. Laboratory tracer test: Comparison between the measured (symbols) and the modeled (lines) breakthrough curves (BTCs) using the advection–dispersion equation (ADE), the tempered time fractional advection–dispersion equation (tt-fADE) (red thick lines), the continuous time random walk (CTRW) (green dashed lines), and the multi-rate mass transfer (MRMT) model with the water flow rate $Q = 1.4$ mL/s.

4. Model Fit and Comparison

Comparisons of the measured and best-fit BTCs using the above three time-nonlocal transport models and the classical ADE are shown in Figure 2 and Figures S6–S10 in the Supplementary File. In this section, we briefly introduce the model fitting process, and then compare the model results.

4.1. The ADE Model with Equilibrium Sorption

For comparison, we first use the classic ADE with equilibrium sorption to quantify the measured BTCs. The governing equation for one-dimensional chemical transport in groundwater with advection, dispersion, and retardation is [34]:

$$R\frac{\partial C}{\partial t} = D\frac{\partial^2 C}{\partial x^2} - v\frac{\partial C}{\partial x}, \tag{29}$$

which has the following solution with an instantaneous point source at the origin:

$$C(x,t) = \frac{M}{2A\sqrt{\pi D^* t}} \exp\left[-\frac{(x - v^* t)^2}{4D^* t}\right], \tag{30}$$

where M $[M]$ represents the initial mass injected into the column; A $[L^2]$ is the cross-section area; and R [dimensionless] is the retardation coefficient. From a mathematical perspective, R acts as a rescaling factor in time. Hence, we reduce the three parameters v, D, and R in the ADE model to two parameters

v^* ($=v/R$) and D^* ($=D/R$). The fitted values for parameters v^* and D^* are listed in Table 1. These two parameters were calibrated manually based on visual inspection.

In our Leca-column transport experiments, the relatively large flow velocity led to a large Peclet number ($Pe >> 1$) and the dispersion coefficient D relates to v via $D = \alpha\, v$, where α [L] denotes the dispersivity. The relationship between dispersivity α and the travel distance in the laboratory experiments is shown in Figure 3. For a fixed flow rate, α varies with the travel distance without any fixed trend, perhaps due to varying dispersion over the relatively short travel distance.

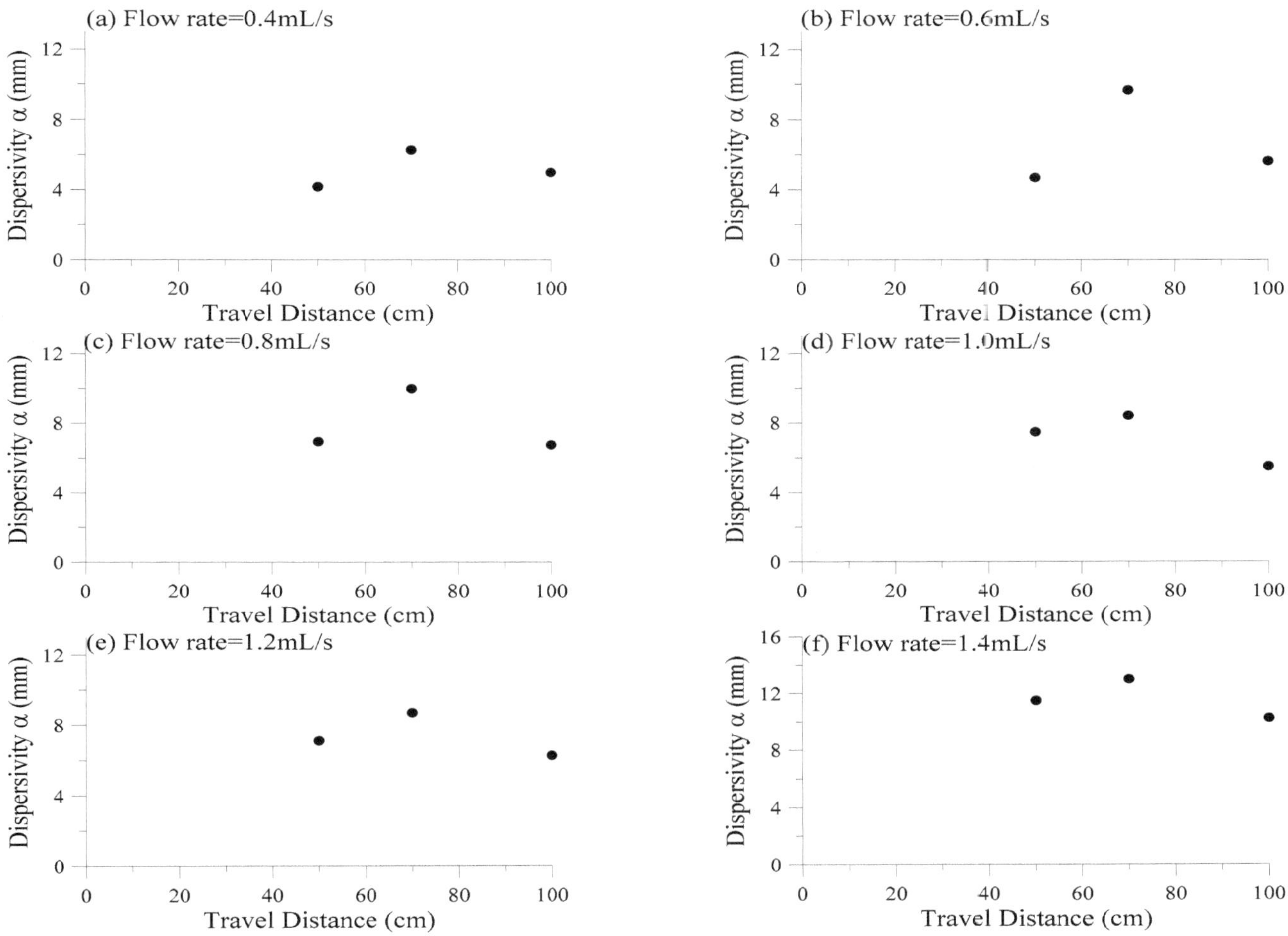

Figure 3. The best-fit dispersivity in the ADE model (29) versus the travel distance.

The relationship between dispersivity and flow rate in the laboratory experiments is shown in Figure 4. Although the best-fit dispersivity fluctuates with the flow rate, it generally increases with an increasing flow velocity. A larger flow velocity causes a wider spatial distribution of contaminant plume and increases the apparent dispersivity in the ADE model. This phenomenon is consistent with known non-Fickian solute transport behavior and indicates that the ADE model is performing as expected [33].

In the BTCs measured at three travel distances with six different flow rates in the laboratory experiments and field tracer tests, the ADE model (29) captures the rapid increase of early-time BTCs, but overestimates the decline of the late-time BTC tails although the delay of transport due to retardation was considered in the ADE (29). This implies that the delayed transport observed in our laboratory experiments and field tests is more complex than equilibrium adsorption described in (29), where the sorbed or immobile concentration is a simple linear function of the dissolved concentration.

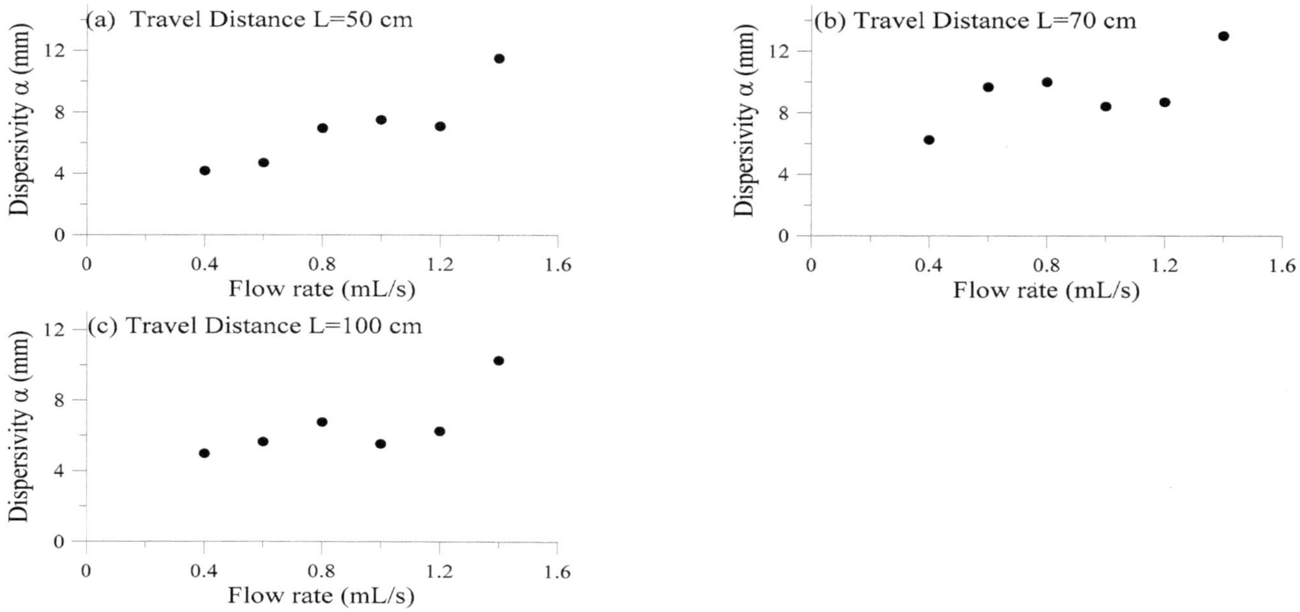

Figure 4. The best-fit dispersivity in the ADE model versus the water flow rate.

4.2. The MRMT Model

The MRMT model (1) has various specific forms, which might be useful for applications and therefore require further evaluation. For example, the immobile domain can be characterized as multiple layers with a distribution of diffusion rate coefficients or first-order mass-transfer rates, or be simplified as a layer with a single diffusion rate which can be attractive in applications due to its simplicity in manipulation. There is, however, no solid physical justification for the selection of any of these forms, given simply the limited information for a sand column like the one used in our laboratory. For systematic analyses of all potential mass transfer models, we selected the following four MRMT subsets with different mass-transfer formulations:

(1) MRMT model 1 (single mass-transfer rate): the immobile zones can be simplified by a single, homogeneous first-order mass-transfer rate (which is also the single-rate double-porosity model);
(2) MRMT model 2 (single diffusion rate): the immobile zones have a single diffusion rate for all layers;
(3) MRMT model 3 (two mass-transfer rates): the immobile zones have two sets of rate coefficients;
(4) MRMT model 4 (multiple mass-transfer rates): the immobile zones have a power-law distribution of (first-order mass-transfer) rate coefficients.

We use the STAMMT-L version 3.0 code [35] to solve the above four MRMT models. Best-fit results, which were calibrated manually, are shown in Figure 2 and Figures S6–S10 and discussed below.

4.2.1. MRMT Model 1 with a Single Mass-Transfer Rate

The general mobile-immobile (MIM) model with diffusion in the immobile zone can be written as [20]:

$$\frac{\partial C_{m,t}}{\partial t} + T(x,t) = -v\nabla C_{m,t} + D\nabla^2 C_{m,t}, \tag{31}$$

$$\frac{\partial C_{im,t}}{\partial t} = \frac{D_a}{r^{n-1}}\nabla\left[r^{n-1}\nabla C_{im,t}\right], \tag{32}$$

where $T(x,t)$ $[ML^{-3}T^{-1}]$ is a transient term accounting for rate-limited mass transfer between the mobile and immobile domains, and Da is the apparent diffusion coefficient. Here n denotes the dimensionality of the problem, and $n = 1, 2, 3$ denotes diffusion into layers, cylinders, and spheres, respectively. For a single rate first-order mass-transfer approximation, $T(x,t)$ can be expressed as [35]:

$$T(x,t) = \beta \frac{\partial C_{im,t}}{\partial t}, \tag{33}$$

$$\frac{\partial C_{im,t}}{\partial t} = \alpha \left[C_{m,t} - C_{im,t}\right], \tag{34}$$

which is the single rate version of the MRMT model (1).

The best-fit parameters using the single-rate MIM model (31)–(34) are listed in Table 2. The fitting exercise showed the sensitivity of model parameters to the travel distance and water flow rate in the laboratory experiments. First, the dispersivity α_L does not change with the travel distance (Table 2), since the plume expansion with time is captured by the mass transfer term in the model. This is different from the standard ADE model, where the dispersivity must increase with the travel distance to capture scale-dependent dispersion.

Table 2. The fitted parameters for the single mass-transfer rate mobile-immobile model (i.e., Equations (31)–(34)) at different flow rates and travel distances. In the legend, L denotes the travel distance (i.e., the length of the column); α_L is the dispersivity; v is the flow velocity; β_{tot} is the total capacity coefficient; and α stands for the mass transfer rate.

Experiment	Q (mL/s)	L (cm)	α_L (mm)	v (mm/s)	β_{tot}	α (s^{-1})
	Measured	Measured	Fitted	Fitted	Fitted	Fitted
Lab	0.4	50	1.5	1.4	0.50	0.05
		70	1.5	1.35	0.50	0.04
		100	1.5	1.35	0.50	0.04
	0.6	50	1.5	1.68	0.50	0.05
		70	1.5	1.71	0.40	0.04
		100	1.5	1.68	0.31	0.03
	0.8	50	1.5	2.38	0.45	0.06
		70	1.5	2.32	0.40	0.04
		100	1.5	2.15	0.33	0.05
	1	50	1.5	2.75	0.45	0.06
		70	1.5	2.93	0.43	0.06
		100	1.5	2.93	0.35	0.07
	1.2	50	1.5	3.15	0.30	0.06
		70	1.5	3.15	0.30	0.06
		100	1.5	3.13	0.28	0.06
	1.4	50	1.5	3.52	0.35	0.06
		70	1.5	3.95	0.35	0.06
		100	1.5	3.98	0.35	0.06
Field	83.33	200	210	0.004	0.50	0.14
		400	230	0.004	0.50	0.10
		600	230	0.004	0.50	0.03

Second, the average velocity used in the model is slightly larger than the measured BTC peak velocity and the ADE velocity (note that the ADE velocity is also larger than the real BTC peak velocity), probably due to the assumption that there might be an immobile domain interacting with the mobile domain. The parameters v and D in the MRMT and ADE models have different meanings and hence may not have the same values. Parameters v and D in the MRMT model refer to the mobile domain [36], and therefore the velocity in the MRMT model is generally larger than the ADE velocity, while the opposite is expected for the dispersion coefficient D. This holds true for all the MRMT formulations.

Third, the capacity coefficient β either remains constant or decreases very slightly with the travel distance, implying that the medium heterogeneity might not significantly change with the medium's length.

Fourth, the rate coefficient α increases slightly and the capacity coefficient β decreases slightly with an increasing flow rate at each control plane. This subtle change is likely due to the assumption that the faster water flows correspond to less volumetric proportion of immobile domains. A faster flow may decompose immobile domains and enhance the mass exchange between mobile and immobile domains, resulting in a larger mass exchange rate α.

4.2.2. MRMT Model 2 with a Single Diffusion Rate

The layered diffusion model is a specific case of the MRMT model [20] with the following rate and capacity coefficients:

$$\alpha_j = \frac{(2j-1)^2\,\pi^2}{4}\frac{D_a}{a^2},\quad j = 1, 2, \cdots, N_{im}, \tag{35}$$

$$\beta_j = \frac{8}{(2j-1)^2\,\pi^2}\beta,\quad j = 1, 2, \cdots, N_{im}, \tag{36}$$

The best-fit parameters for model (35) and (36) are listed in Table 3. There is no apparent correlation between the mass transfer rate and the travel distance or water flow velocity. The same conclusion is found for the capacity coefficient.

Table 3. The fitted parameters for the single diffusion rate multi-rate mass transfer (MRMT) model at different flow rates and travel distances. In the legend, α_d is the diffusion rate coefficient ($\alpha_d = D_a/a^2$, where Da is the apparent diffusion coefficient and a is the layer half-thickness).

Experiment	Q (mL/s)	L (cm)	α_L (mm)	v (mm/s)	β_{tot}	α_d (s^{-1})
	Measured	Measured	Fitted	Fitted	Fitted	Fitted
Lab	0.4	50	1.5	1.55	0.60	0.02
		70	1.5	1.48	0.60	0.02
		100	1.5	1.43	0.60	0.02
	0.6	50	1.5	1.73	0.50	0.03
		70	1.5	2.22	0.80	0.02
		100	1.5	2.12	0.61	0.02
	0.8	50	1.5	2.69	0.61	0.03
		70	1.5	2.85	0.70	0.02
		100	1.5	2.55	0.55	0.03
	1	50	1.5	3.00	0.55	0.03
		70	1.5	3.30	0.55	0.02
		100	1.5	3.25	0.48	0.05
	1.2	50	1.5	3.73	0.48	0.04
		70	1.5	3.69	0.48	0.03
		100	1.5	3.25	0.31	0.03
	1.4	50	1.5	4.25	0.60	0.03
		70	1.5	4.90	0.60	0.03
		100	1.5	4.10	0.40	0.02
Field	83.33	200	210	0.004	0.50	0.05
		400	210	0.004	0.50	0.04
		600	210	0.005	0.50	0.03

4.2.3. MRMT Model 3 with Two Mass-Exchange Rates

MRMT model 3 contains two pairs of coefficients: two rate coefficients (α_1 and α_2) and two capacity coefficients (β_1 and β_2) corresponding to the first and the second immobile domains, respectively.

The best-fit parameters are listed in Table 4. The dispersivity α_L remains constant, since solute plume expansion (due likely to solute retention) is mainly captured by mass exchange between mobile and the two immobile zones. Or in other words, the two pairs of parameters, α_j ($j = 1, 2$) and β_j ($j = 1, 2$), control the mass exchange. Their values fluctuate (without predictable trends) with the travel distance and flow rate in the laboratory experiments, although α_2 decreases with increasing α_1 for the same flow rate. There is no efficient way to directly measure the rate coefficient or capacity coefficient, which creates a challenge for relating MRMT model parameters to observable physical phenomena, and for providing information to MRMT models based on ancillary observations in heterogeneous media.

Table 4. The fitted parameters for the two-set MRMT model at different flow rates and travel distances. In the legend, α_1 and β_1 represent the mass transfer rate and capacity coefficient of the first immobile domain, respectively; and α_2 and β_2 are the mass transfer rate and capacity coefficient of the second immobile domain, respectively.

Experiment	Q (mL/s)	L (cm)	α_L (mm)	v (mm/s)	α_1 (s^{-1})	β_1	α_2 (s^{-1})	β_2
	Measured	Measured	Fitted	Fitted	Fitted	Fitted	Fitted	Fitted
Lab	0.4	50	1.5	1.12	0.08	0.08	0.015	0.15
		70	1.5	1.10	0.08	0.09	0.015	0.15
		100	1.5	1.03	0.10	0.09	0.008	0.10
	0.6	50	1.5	1.31	0.08	0.07	0.020	0.13
		70	1.5	1.54	0.15	0.08	0.020	0.25
		100	1.5	1.55	0.15	0.08	0.015	0.15
	0.8	50	1.5	1.96	0.05	0.15	0.030	0.08
		70	1.5	2.08	0.05	0.20	0.015	0.10
		100	1.5	1.88	0.07	0.10	0.018	0.10
	1	50	1.5	2.30	0.05	0.12	0.030	0.15
		70	1.5	2.50	0.07	0.13	0.025	0.13
		100	1.5	2.57	0.20	0.11	0.020	0.11
	1.2	50	1.5	2.85	0.07	0.12	0.020	0.13
		70	1.5	2.87	0.06	0.11	0.027	0.13
		100	1.5	2.87	0.07	0.10	0.025	0.01
	1.4	50	1.5	3.33	0.08	0.17	0.030	0.17
		70	1.5	3.80	0.09	0.17	0.030	0.17
		100	1.5	3.75	0.08	0.15	0.030	0.16
Field	83.33	200	210	0.004	0.4	0.2	0.1	0.2
		400	210	0.004	0.4	0.2	0.1	0.2
		600	210	0.004	0.4	0.1	0.1	0.1

4.2.4. MRMT Model 4 with Power-Law Distributed Rate Coefficients

The density of rate coefficient for MRMT model 4 can be defined as [16]:

$$b(\alpha) = b(\alpha_{min}, \alpha_{max}, k) = \begin{cases} \beta_{tot} \frac{(k-2)\alpha^{k-3}}{\alpha_{max}^{k-2} - \alpha_{min}^{k-2}} & k \neq 2 \\ \beta_{tot} \frac{1}{\ln(\alpha_{max}/\alpha_{min})\alpha} & k = 2 \end{cases}, \text{ for } \alpha_{min} \leq \alpha \leq \alpha_{max} \quad (37)$$

where α_{min} $[T^{-1}]$ denotes the minimum rate coefficient, α_{max} $[T^{-1}]$ is the upper boundary of the rate coefficient, and k is the exponent.

In the fitting parameters shown in Table 5, the dispersivity α_L remains stable for the same reason mentioned above for the other MRMT models. Flow velocity used in this model can be approximated by the peak velocity for the measured BTC. The total capacity coefficient (β_{tot}) does not significantly change with the travel distance. The exponent k controls the slope of the late-time BTC in a log-log plot, and hence a larger k denotes faster decline of the late-time BTC. The best-fit k slightly increases with an increasing flow rate, due likely to the relatively faster decline of the late-time solute concentration under a larger water flow rate.

Table 5. The best-fit parameters for the power-law MRMT model 4 at different flow rates and travel distances. In the legend, k stands for the slope of the late-time tail, and α_{min} and α_{max} stand for the lower and upper boundary of the mass transfer rates, respectively.

Experiment	Q (mL/s)	L (cm)	α_L (mm)	v (mm/s)	β_{tot}	k	α_{min} (s^{-1})	α_{max} (s^{-1})
	Measured	Measured	Fitted	Fitted	Fitted	Fitted	Fitted	Fitted
Lab	0.4	50	1.6	1.40	0.50	2.250	0.010	0.6
		70	1.6	1.35	0.50	2.250	0.010	0.6
		100	1.6	1.35	0.50	2.250	0.010	0.6
	0.6	50	1.6	1.69	0.50	2.292	0.011	0.6
		70	1.6	1.85	0.60	2.292	0.012	0.6
		100	1.6	2.10	0.60	2.292	0.012	0.6
	0.8	50	1.6	2.83	0.70	2.290	0.013	0.6
		70	1.6	2.95	0.80	2.290	0.012	0.6
		100	1.6	3.05	0.80	2.290	0.013	0.6
	1	50	1.6	3.59	0.80	2.300	0.013	0.6
		70	1.6	4.05	0.85	2.300	0.013	0.6
		100	1.6	4.16	0.80	2.300	0.014	0.6
	1.2	50	1.6	4.63	0.80	2.335	0.014	0.6
		70	1.6	4.63	0.80	2.335	0.014	0.6
		100	1.6	4.44	0.75	2.335	0.014	0.6
	1.4	50	1.6	4.95	0.85	2.358	0.014	0.6
		70	1.6	5.48	0.80	2.358	0.015	0.6
		100	1.6	5.50	0.80	2.358	0.015	0.6
Field	83.33	200	210	0.004	0.5	2.25	0.01	0.6
		400	210	0.004	0.5	2.25	0.01	0.6
		600	210	0.005	0.5	2.25	0.01	0.6

It is also noteworthy that, on one hand, the minimum mass transfer rate (α_{min}) in (37) controls the maximum waiting time for solute particles, which is functionally equivalent to the inverse of the cutoff time scale t_2 in the CTRW framework and the truncation parameter λ in the tt-fADE model. In general, α_{min} increases with an increasing flow rate (Table 5). Faster flow may accelerate the mass exchange between mobile and immobile domains, generating shorter mean residence times for solute particles in the immobile domain and leading to a greater mass transfer rate. On the other hand, the maximum mass transfer rate (α_{max}) in (37) defines the shortest waiting time (for solute particles between two displacements), whose impact on the late-time transport dynamics can be overwhelmed by the other smaller rates. Numerical results also show that α_{max} apparently does not change with the travel distance and flow rate. Hence, α_{max} can be kept constant for all cases (Table 5).

As shown in Figure 2 and Figures S6–S10, the MRMT model 4 captures the BTC late-time tail much better than the other mass-transfer formations with fewer rate coefficients. However, compared with the tt-fADE model, the simulated tail of the MRMT model 4 tends to be slightly heavier at the end of the modeling time. In other words, it slightly underestimates the mass transfer rate at the late time. Note that the memory function in the tt-fADE is not exactly the same as that in the MRMT model 4. The tt-fADE has exponentially-truncated power-law rate coefficients, while the MRMT model 4 simply deletes any rate coefficient larger than α_{max} and smaller than α_{min}. This subtle difference in memory functions between the tt-fADE and MRMT model 4 might be the reason for the differences observed here. In addition, the MRMT model 4 (with six model parameters) requires one more parameter than the tt-fADE model.

As shown in Figure 2 and Figures S6–S10, both the MRMT model 1 and model 2 can capture most characteristics of the observed BTCs, except for the late-time tailing. Therefore, we need more than one mass-transfer rate to fit the solute transport in the one-dimensional heterogeneous sand column and field tracer tests. Increasing the number of immobile domains improves the model's performance at late times, but more immobile domains can complicate the model application. To be specific, adding one more immobile domain adds two unknown parameters (a pair of capacity coefficient and mass exchange rate coefficient for that immobile domain).

4.3. The tt-fADE Model (10) and (11)

The failure of the standard ADE model and the single rate mass transfer model in capturing the late-time BTC tailing motivates the application of the other time nonlocal transport models such as the tt-fADE model (10) and (11). A numerical solver of (10) and (11) can be found in [37]. The best-fit results using this model are shown in Figure 2 and Figures S1–S5, where both the peak and the late time tailing can be captured simultaneously.

The best-fit parameters are shown in Table 6, where the best-fit velocity v is slightly larger than the measured peak velocity v_{peak}, due to the fractional-order capacity coefficient β in the left-hand side of the tt-fADE Equations (10) and (11) when $\gamma \to 1$:

$$v = (1 + \beta)\, v_{peak}, \tag{38}$$

Table 6. The best-fit parameters for the tempered time fractional advection–dispersion equation (tt-fADE) model (10) and (11) at different flow rates and travel distances. In the legend, α is the dispersivity; γ is the time/scale index; β is the fractional-order capacity coefficient; and λ is the truncation parameter.

Experiment	Q (mL/s)	L (cm)	v (mm/s)	D (mm^2/s)	α (mm)	γ	β (s$^{\gamma-1}$)	λ (s^{-1})	*RMSE*
	Measured	Measured	Fitted	Fitted	$\alpha = D/v$	Fitted	Fitted	Fitted	Calculated
Lab	0.4	50	1.10	0.2	0.18	0.250	0.011	0.010	0.379
		70	1.05	0.2	0.19	0.250	0.011	0.010	0.171
		100	1.00	0.2	0.20	0.250	0.011	0.010	0.343
	0.6	50	1.30	0.2	0.15	0.292	0.013	0.011	0.461
		70	1.40	0.4	0.29	0.292	0.013	0.012	0.296
		100	1.49	0.4	0.27	0.292	0.013	0.012	0.489
	0.8	50	1.91	0.5	0.26	0.290	0.018	0.013	0.577
		70	1.95	0.6	0.31	0.290	0.018	0.012	0.398
		100	1.84	0.6	0.33	0.290	0.018	0.013	0.443
	1	50	2.25	1.8	0.80	0.300	0.014	0.013	0.529
		70	2.40	1.8	0.75	0.300	0.014	0.013	0.218
		100	2.40	1.8	0.75	0.300	0.014	0.014	0.326
	1.2	50	2.86	1.0	0.35	0.335	0.019	0.014	0.405
		70	2.84	1.0	0.35	0.335	0.019	0.014	0.190
		100	2.81	1.0	0.36	0.335	0.019	0.014	0.260
	1.4	50	3.18	3.0	0.94	0.358	0.026	0.014	0.636
		70	3.55	3.0	0.85	0.358	0.026	0.015	0.380
		100	3.45	3.0	0.87	0.358	0.026	0.015	0.329
Field	83.33	200	0.004	0.69	198	0.358	0.17	0.01	n/a
		400	0.005	0.69	148	0.358	0.17	0.01	n/a
		600	0.005	0.69	148	0.358	0.17	0.01	n/a

Figures 5 and 6 show the fitted parameters versus the flow rate and travel distance in the laboratory experiments. The best-fit dispersivity α does not change significantly with the travel distance (Figure 5), which is different from the scale-dependent dispersion observed by other studies [38].

This discrepancy might be due to the following reasons: (1) the travel distance is too short to observe any stable trend of dispersion (tracer injection at the inlet may cause a boundary effect on transport and random deviation from the local mean velocity), and (2) the system is advection dominated (as a typical laboratory sand-column transport experiment) and therefore the estimates of D might be too uncertain to clearly show a subtle trend. In addition, the best-fit dispersivity increases with an increasing flow rate. The faster the water flows, the wider the plume becomes, requiring a larger dispersivity. This is consistent with the fitting results of the other models mentioned above.

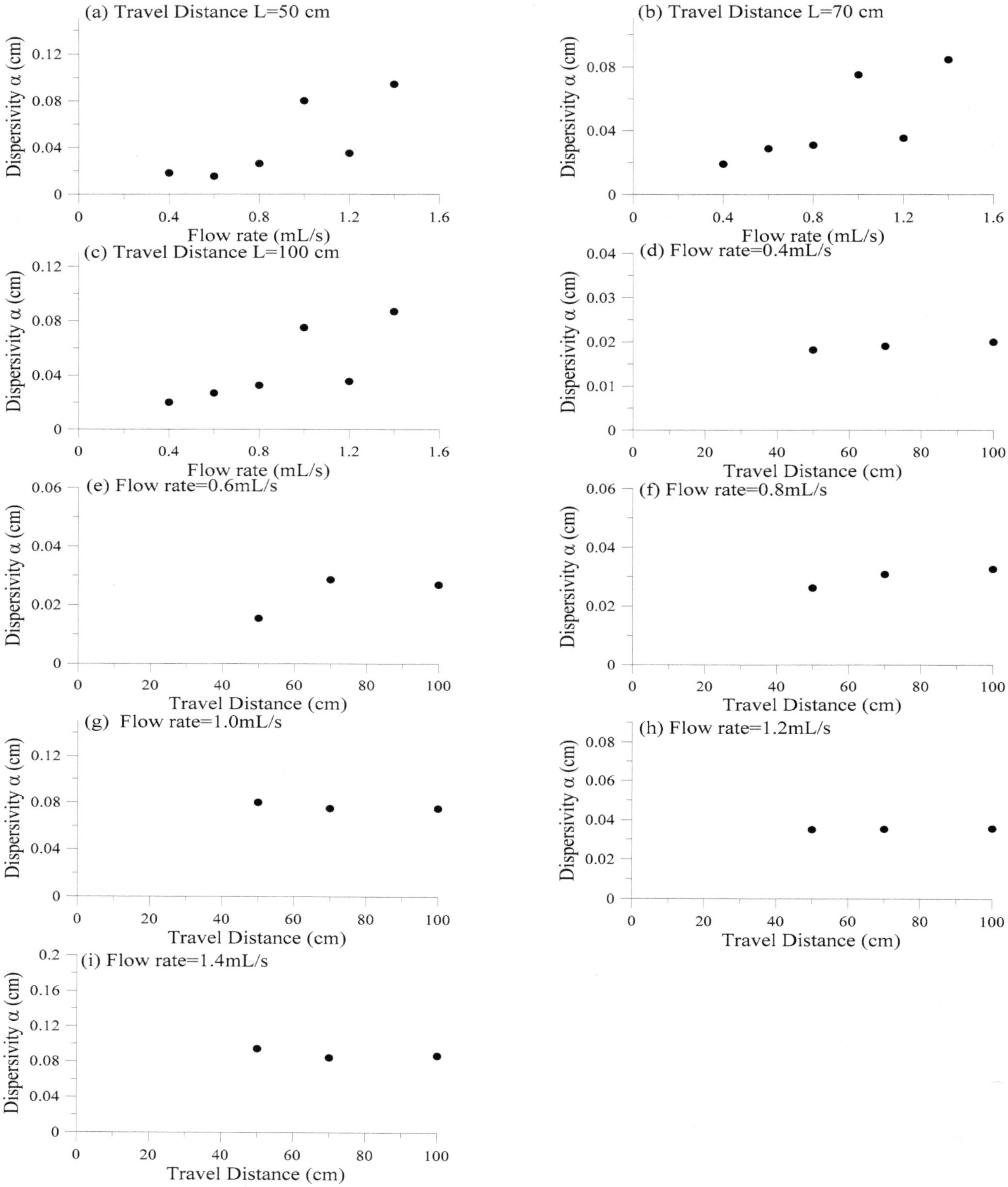

Figure 5. Dispersivity in the tt-fADE model changes with the travel distance and flow rate.

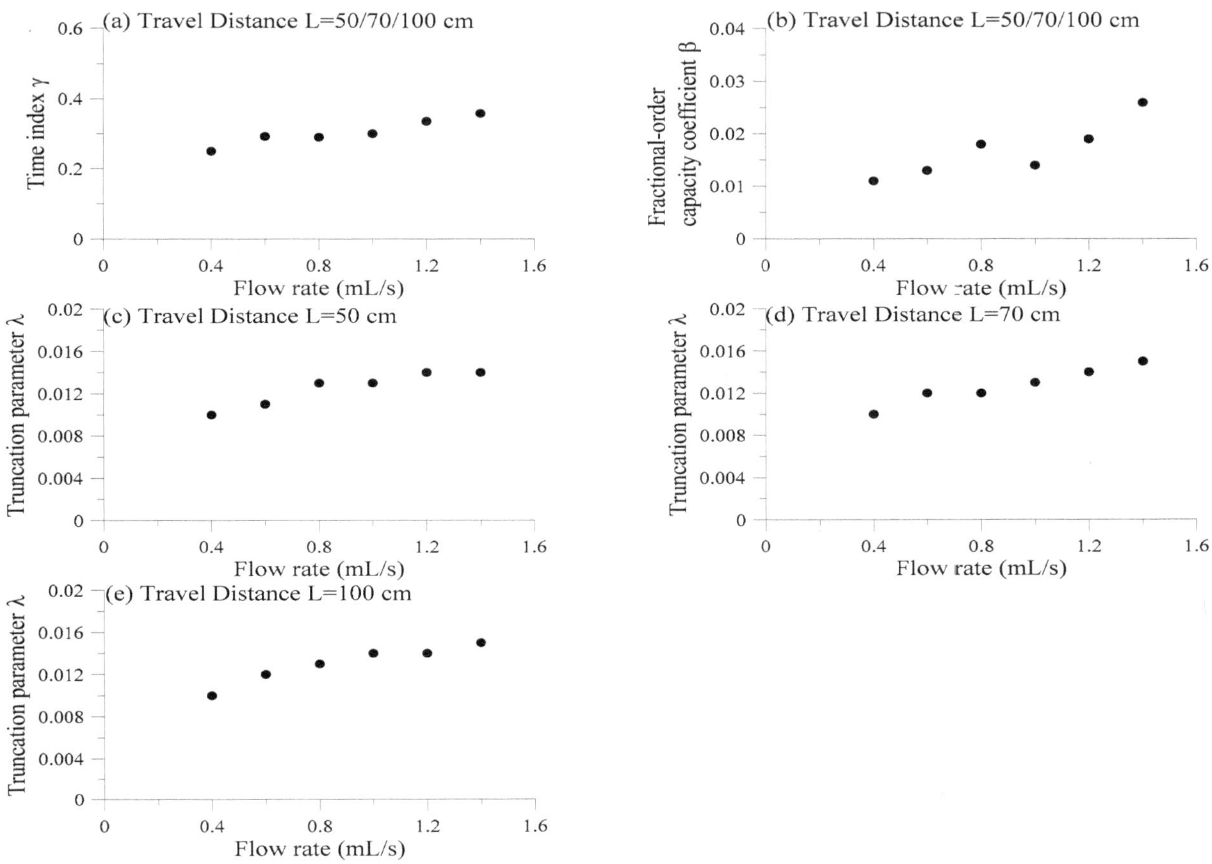

Figure 6. The tt-fADE model parameters change with the travel distance and flow rate.

The other model parameters, including the time index γ (representing residence time in the immobile domain), the capacity coefficient β (the ratio of immobile to mobile volume), and the truncation parameter λ (controlling the transition from power-law to exponential tailing), vary with the flow rate (Figure 6). First, the time index γ increases with an increasing flow rate, since a larger γ represents a shorter mean residence time for solute particles in the immobile domain (here we assume that the mean residence time in the immobile domain decreases with an increasing flow velocity). Second, the capacity coefficient β also increases with the flow rate, showing that the variation of the time index might overshadow the variation of the capacity coefficient. Third, the truncation parameter λ increases with the flow rate, implying that the non-Fickian transport converges to its Fickian asymptote more quickly if the flow rate is larger. This trend might be due to the decreased immobile portion with a higher flow rate in the short sand column. In addition, the truncation parameter λ is relatively small, in order to capture the relatively heavy late-time tails.

The above three parameters (γ, β, and λ), however, generally remain constant at different travel distances if the flow rate remains unchanged (Table 6). The packing procedure for the sand column was designed to yield a macroscopically homogeneous texture, and, indeed, the results imply that the statistics of medium properties (such as the sand size distribution, specific surface area, effective porosity, and/or internal structure) are spatially uniform. For example, a uniform fractional capacity coefficient β suggests that the ratio of mobile to immobile volume is constant in space. This phenomenon holds true for the field tracer tests. The spatial uniformity of these parameters simplifies the fitting procedure. For each flow rate with different sample distances, the time index and the capacity coefficient can be calibrated a single time for one sampling distance, and then kept constant for the other sampling distances.

Therefore, when using the tt-fADE model (10) and (11), we only need to fit three parameters: velocity, dispersion coefficient, and truncation parameter. Fitting is further simplified by using the observed BTC peak velocity as a lower bound when predicting the velocity in the tt-fADE model.

4.4. The CTRW Model

The fitted results using the CTRW model (17) are shown in Figure 2 and Figures S1–S5. We used the CTRW MATLAB Toolbox Version 3.1 developed by [39]. The CTRW framework with the truncated power-law transition time $\phi(t)$ (22) was used, as suggested by [21], since it can efficiently capture the transition from non-Fickian to Fickian transport [32].

This CTRW model contains five unknown parameters, ξ, t_1, t_2, v_ψ, and D_ψ. The time t_1 expressed by formula (21) represents the approximated mean transition time. The power law behavior in the BTC begins from t_1 and ends at t_2. Knowledge of parameters gained in the fitting exercise of the tt-fADE model in Section 4.3 improves the predictability of the CTRW model. In particular, we set the power-law exponent ξ in the CTRW model equal to the time index γ in the tt-fADE model (10) and (11), and approximate the cutoff time scale t_2 in the CTRW framework using the inverse of the truncation parameter λ in the tt-fADE (see Section 2.3). Here a smaller ξ represents more disorder of the host system. The remaining three parameters, including the velocity v_ψ, the dispersion coefficient D_ψ, and the mean waiting time t_1 in (17), can be fitted using the observed BTCs. The fitted parameters using the CTRW model are listed in Table 7.

Table 7. The best-fit parameters for the continuous time random walk (CTRW) model. In the legend, v_ψ is the CTRW transport velocity, which can be different from the average pore velocity v; D_ψ is the dispersion coefficient with the subscript ψ indicating CTRW interpretation; ξ is the power-law exponent; t_1 is the mean transition time; and t_2 is the truncation time scale. ξ is converted from γ in the tt-fADE model (10) and (11), and t_2 is converted from λ in the tt-fADE model.

Experiment	Q (mL/s)	L (cm)	v_ψ (mm/s)	D_ψ (mm^2/s)	ξ	$\log_{10}t_1$ (s)	$\log_{10}t_2$ (s)	*RMSE*
	Measured	Measured	Fitted	Fitted	Fixed	Fitted	Fixed	Calculated
Lab	0.4	50	0.465	0.38	0.250	1.70	2.00	0.239
		70	0.637	0.69	0.250	1.70	2.00	0.447
		100	0.900	14.00	0.250	1.70	2.00	0.862
	0.6	50	0.505	0.50	0.292	1.70	1.94	0.932
		70	0.735	1.47	0.292	1.70	1.93	0.859
		100	1.130	2.00	0.292	1.70	1.93	1.201
	0.8	50	0.575	0.75	0.290	1.80	1.90	1.317
		70	0.840	3.43	0.290	1.80	1.92	0.491
		100	1.150	4.00	0.290	1.80	1.90	0.976
	1	50	0.660	1.00	0.300	1.80	1.90	2.039
		70	0.994	2.94	0.300	1.81	1.89	0.780
		100	1.410	4.00	0.300	1.81	1.86	0.265
	1.2	50	0.760	1.75	0.335	1.81	1.84	0.972
		70	1.050	3.43	0.335	1.81	1.84	0.457
		100	1.500	4.00	0.335	1.81	1.84	0.913
	1.4	50	0.820	2.75	0.358	1.80	1.84	1.237
		70	1.288	5.88	0.358	1.80	1.83	0.713
		100	1.800	12.00	0.358	1.80	1.82	0.439
Field	83.33	200	0.30	0.0300	0.358	1	2	n/a
		400	0.18	0.0090	0.358	1	2	n/a
		600	0.12	0.0025	0.358	1	2	n/a

5. Discussion

5.1. Comparison of Transport Models

The transport models discussed in Section 4 can be classified into two main groups. The first group includes the ADE with equilibrium adsorption, the single mass-transfer rate MIM model, and the single diffusion rate MRMT model, which capture the observed early-time BTC and its peak, but underestimate the persistent late-time tail of each BTC. The second group includes the MRMT model with two or a power-law distributed rate coefficients, the tt-fADE, and the CTRW model with a truncated power-law memory function, which can capture the overall trend and positive skewness of the BTCs. The two rate coefficients in the MRMT model, however, might not be adequate to capture heavy late-time tailing for solute transport observed in the field. There is no efficient way to predict the capacity coefficient and the mass transfer rate in the MRMT model, especially when there are multiple sets of mass transfer rates due to various retention capabilities in natural soils with spatial variable hydraulic properties. In contrast, the tt-fADE and the CTRW model parameters are more directly relatable to physical features [29].

MRMT model: As articulated by [24], the MRMT model relates closely to the CTRW framework in capturing solute retention times. Particularly, the mean transition time t_1 in the CTRW framework is related to the inverse of the mean rate coefficient in the MRMT model with power-law distributed rate coefficients. Our fitting exercise in Section 4.4 shows that the BTC is not sensitive to α_{max}. The mean transition time t_1 in the CTRW framework and the maximum boundary α_{max} used in the MRMT model might not be needed when capturing the late-time tailing of solute transport (note that the cutoff time t_2 in the CTRW model or the minimum rate coefficient α_{min} in the MRMT model play a more important role than t_1 or α_{max} in affecting the late-time BTC), and therefore they may be removed from the fitting parameters to simplify the model applications. Note that the tt-fADE model does not need the lower-bound of retention times. In addition, the MRMT model with power-law distributed rate coefficients tends to slightly overestimate the late-time tailing in BTCs (Figure 2 and Figures S6–S10), implying that the actual mass transfer rates may decline faster than a power-law function at late times.

CTRW model: The CTRW framework has a complex relationship to the tt-fADE model. On one hand, as discussed in Section 2.3 and checked in Section 4.4, the power-law exponent ξ in the CTRW framework is functionally equivalent to the scale index γ in the tt-fADE, and the cutoff time scale t_2 in CTRW is equivalent to the inverse of the truncation parameter λ in the tt-fADE. On the other hand, the velocity and dispersion coefficient in the CTRW framework significantly differ from those in the tt-fADE model. Similarly, they are not directly related to the solution of the traditional ADE. Applications in Section 4.4 show that the average CTRW transport velocity v_ψ (1.303 mm/s) is ~59% less than the average real peak velocity (3.18 mm/s) of the observed BTC and ~62% less than the average best-fit velocity (3.39 mm/s) in the tt-fADE model at the flow rate Q = 1.4 mL/s. For the field tracer tests, the CTRW transport velocity v_ψ (0.3 mm/s) is two orders of magnitude larger than the real peak velocity (0.003 mm/s) of the observed BTC, which is similar to the best-fit velocity (0.004 mm/s) in the tt-fADE model. The velocity used in the tt-fADE model can be calculated by using Equation (38), instead of fitting. This procedure further eliminates the number of parameters in the tt-fADE model and makes the fitting more convenient. In addition, the dispersion coefficient in the CTRW framework D_ψ cannot be kept constant under the same flow rate for different travel distances like the dispersion coefficient in the tt-fADE model for both laboratory experiments and field tracer tests. Therefore, the spatially averaged velocity defined by Formula (19) for the CTRW framework may differ from the actual pore-scale velocity. In the CTRW model, different from the traditional ADE, the advective, dispersive, and diffusive transport mechanisms are combined in the random walk formalism. The advective component and the dispersive component are calculated by spatial moments of the same joint probability density function (PDF) for particle transitions and hence cannot be disconnected [21]. In particular, according to Equation (19), velocity in the CTRW model can be estimated by determining the characteristic time and mean distance, which is the first moment of the PDF of transition displacement. It is, however, difficult to predict the effective velocity v_ψ without detailed knowledge of the porous medium. It is also noteworthy that the generalized master Equation (21) does not separate the effects of the spatially varying velocity field on solute particle displacement into an advective part and a dispersive part. The concept of CTRW therefore does not build an explicit relationship between real velocity and model velocity, and the same is true for the dispersion coefficient. In addition, the power-law exponent ξ can also affect the overall magnitude of solute plume expansion, an effect that intermingles with the dispersion coefficient D_{ψ} and the effective velocity v_{ψ}. The above analysis is consistent with the result in [40] that many parameters, particularly v_{ψ} and ξ, in the CTRW model are correlated with each other. Their fitting exercises showed that different parameter combinations can lead to the same mean Eulerian velocity predicted by the model, implying that the estimated model parameters were not unique.

tt-fADE model: In the tt-fADE model, the best-fit velocity v can be larger than the plume peak's velocity v_{peak} because the effective velocity used in the tt-fADE is adjusted by the elapsed time for solutes spent in retention, as represented by the fractional-order capacity coefficient β on the right-hand side of Equation (38). In this study, laboratory experiments with six flow rates and field tracer tests show that the best-fit velocity in the tt-fADE model is close to the measured BTC's peak velocity, since the capacity coefficient shown in Equation (38) is relatively small. Because the transport velocity used in the tt-fADE model (10) and (11) is not significantly different from the BTC's peak velocity, the tt-fADE model uses an independent parameter, the time index γ, to control the power-law distribution of the late-time BTC. This parameterization of the tailing is relatively simple as compared to the CTRW model, with three parameters, v_{ψ}, D_{ψ}, and ξ that contribute to time-nonlocal non-Fickian dispersion. We calculate the root mean square error (RMSE) for the laboratory experiments. Comparison of the RMSE (see Tables 1, 6 and 7) of the two models and the ADE model demonstrates that they both perform better than the standard ADE, especially in describing tailing behavior in a heterogeneous medium. When the Formula (38) is used, the tt-fADE model is more convenient to apply in practice than the CTRW framework, since the former contains fewer parameters.

Why the fitted four-parameter tt-fADE model may provide the best performance? This might be due to two reasons. First, according to the physical derivation of the tt-fADE in Section 2.2, the tt-fADE separates motion in the mobile zone (using the basic transport parameters v and D to define advection and Gaussian diffusive displacement) and retention in the immobile domains. When the tracer particle moves in the mobile domain, its average speed is v, with a finite time required to finish the jump. This physical separation might be reasonable in hydrogeological media. In the CTRW framework, the memory function defines the random waiting time between two subsequent jumps, while the motion can finish instantaneously (in other words, no physical time is required for the tracer particle to move). This physical discrepancy may also cause the discrepancy of the two basic transport parameters v and D between the two models, in addition to the spatial average parameters required by the CTRW framework. Second, the tt-fADE does not specify the lower-bound of the waiting time (using an additional parameter such as the lower-limit t_1 in the CTRW framework), since the slow advection can also affect the late-time BTC tail. Mass exchange can apparently affect the late-time BTC tail only when the diffusive time scale is much longer than the advective time scale, as pointed out by [16]. This may explain why t_1 might not be needed in the CTRW framework. Further real-world tests are needed to check the above hypotheses.

5.2. Parameter Sensitivity

One example of parameter sensitivity is tested here for the tt-fADE (10) and (11). Sensitivity of the BTCs to variations of the four main parameters (D, γ, β, λ) in the tt-fADE model is shown in Figure 7.

First, the dispersion coefficient D has a subtle impact on the overall shape of BTCs. Decreasing D from 0.005 cm^2/s to 0.0005 cm^2/s results in similar BTCs after normalization (i.e., re-scaling), implying that trapping due to the immobile zones may account at least partially for the spatial expansion of solute plumes. When D increases from 0.005 cm^2/s to 0.05 cm^2/s, the BTC becomes wider and its shape slightly changes (Figure 7a,b).

Second, the time index γ controls the power-law slope of the late-time BTC. For example, when γ increases from 0.29 to 0.50 (representing the decrease of probability for long retention times), the late-time BTC becomes steeper, approaching relatively fast to its Gaussian asymptote (Figure 7c,d). When γ decreases from 0.29 to 0.05, the BTC's late-time tail becomes heavier (i.e., with a gentler slope), although the overall BTC looks narrower (likely due to the normalization of BTCs). The simulated BTC with the time index γ equal to 0.29 gives the best fit.

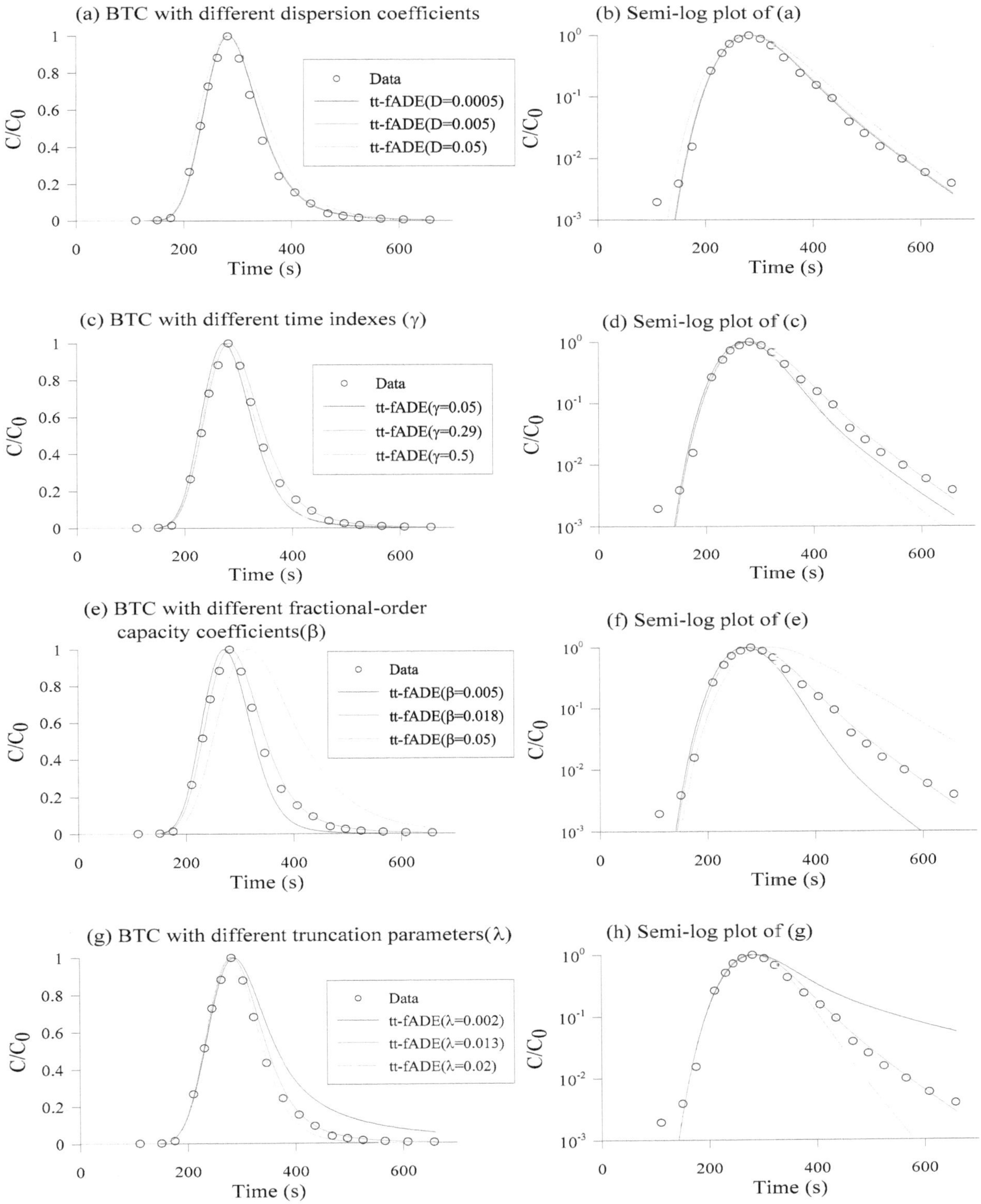

Figure 7. Sensitivity analysis for coefficients in the tt-fADE model.

Third, the fractional-order capacity coefficient β shifts the BTC and expands the BTC's late-time tail. When β decreases (representing a decrease of the immobile zone volume or the immobile solute mass at equilibrium), the BTC becomes narrower and shifts to the left (representing a larger effective velocity). A faster drop is apparent in the late-time BTC tail with a smaller β. An opposite change of the BTC can be seen for an increasing β (Figure 7e,f).

Fourth, the truncation parameter λ affects the speed for the late-time BTC to transfer from a heavy tailed, power-law slope to an exponential tail. A larger truncation parameter means an earlier transform from non-Fickian to Fickian transport (Figure 7g,h). For the largest λ tested, the resultant BTC is the closest to the solution of the ADE model, as expected because the tt-fADE reduces to the ADE model for $\lambda \to \infty$.

5.3. Application to Field Transport

A field trace transport test was conducted recently by Zheng [41], which provides field data to evaluate further the time nonlocal transport models and compare with the laboratory column experiments. The test site is in the Zhangjiawan Village, Zhangjiawan Town, southeast of the Tongzhou District, Beijing, China, with the longitude of 116.72° and latitude 39.848°. This experimental site is in the Chaobai River alluvial plain. The average annual precipitation in the vicinity is about 533 mm, and the evaporation is 1822 mm. It has a multi-layer aquifer structure. The aquifer is composed of gravelly coarse sand, coarse sand, fine sand, silty clay and clay layer in the test field and surrounding area. The hydraulic conductivity coefficient shows obvious heterogeneity, indicating that the aquifer develops a small-scale preferential flow channel network.

The subsurface network consisted of one injection well, one pumping well, and three monitoring wells with continuous multi-tubing (denoted as well 13, 23, and 33, respectively) (see Figure 8 for the study site).

All the five wells are along the same line of the general groundwater flow direction, and therefore a one-dimensional model may be used to approximate the overall transport. The injection well and the pumping well are separated by 8 m, and the three observation wells have a uniform interval of 2 m. Groundwater flows from the injection well to well 13, 23, 33, and to the pumping well.

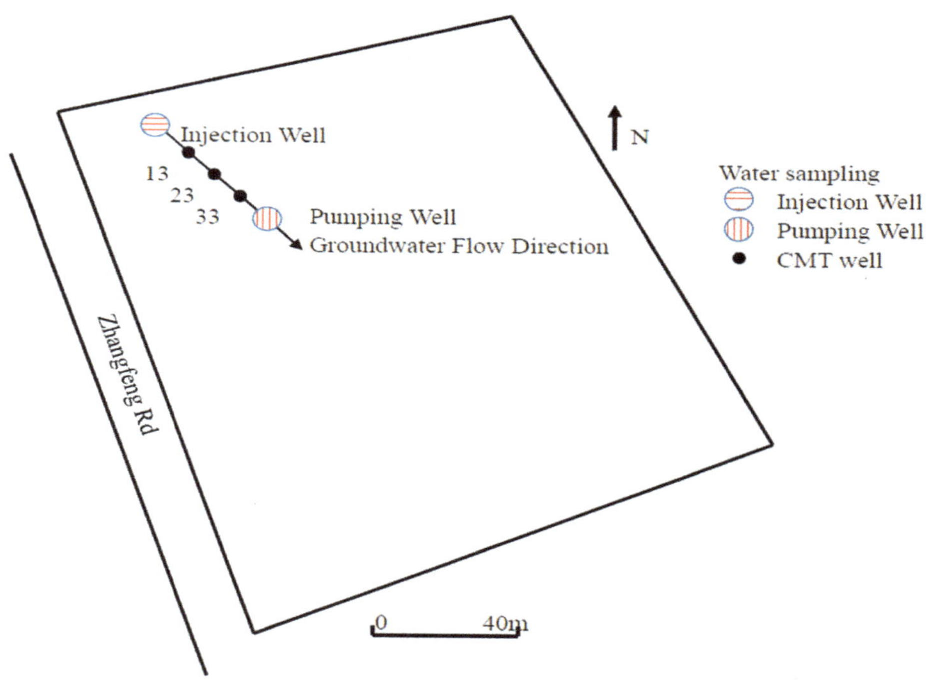

(**a**) Zhangjiawan test site location and distribution of wells

Figure 8. *Cont.*

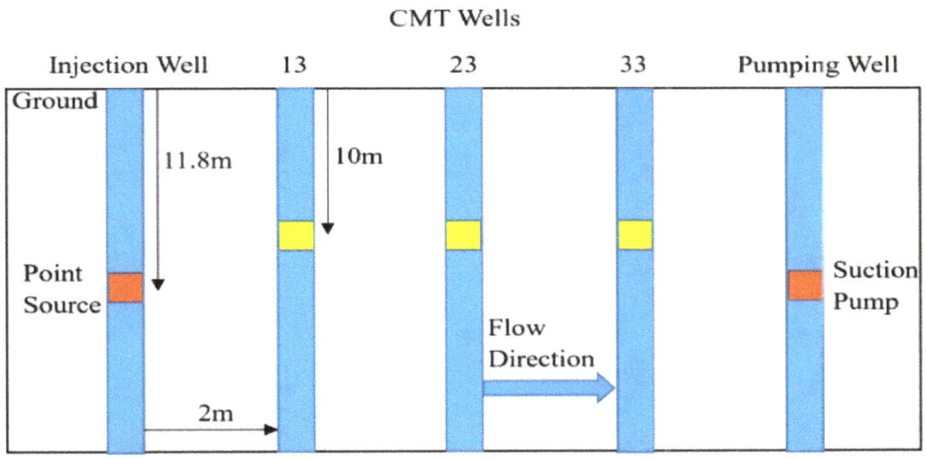

(**b**) Illustration for point source artificial hydraulic gradient tracing tests

Figure 8. Study site map showing injection well, pumping well, and continuous multichannel tubing (CMT) wells.

Sodium bromide, a commonly used conservative tracer, was injected into the injection well at a depth of 11.8 m, and groundwater samples were taken from the three observation wells (Well 13, 23, 33) located downstream at a depth of ~10 m. The concentration of the injected solution was 1288 mg/L. The injection rate was about 0.3 m^3/h, and the injection duration was 6 h. A peristaltic pump was used to collect the samples in a chronological order, and the sample concentration was measured using a MP523-06 bromide ion concentration meter.

The measured BTCs exhibit apparent late-time tailings (see Figure 9), similar to that observed in the laboratory column transport. Applications show that the time nonlocal models can capture part of the late-time BTC tail, but not the whole tail of BTC containing apparent noise. Due to the noise, it is also impossible to obtain a reliable RMSE. Although the apparent noise causes high uncertainty in model fitting, both the CTRW framework and the tt-fADE can capture a heavier late-time tail than the MRMT model with two sets of rate coefficients, and the measured BTCs do contain a high concentration at the very end of the sampling period (Figure 9).

In addition, all the measured BTCs show apparent early time tail, which cannot be captured by the tt-fADE or the CTRW framework with the time index between 0 and 1. It is, however, not a surprise, since the early arrivals are most likely due to fast motion of tracer particles along preferential flow paths, while the delayed arrivals are caused by solute retention due to mass exchange between the mobile and relatively immobile zones. At the field site, high-permeable sand constitutes the layer connecting the injection well and the three monitoring wells, likely forming the preferential channels. The time nonlocal transport models considered in this study were developed to capture solute retention, hence missing the early tail of the BTC. The spatiotemporal fADE may capture both the early and late time tails in the BTC [28,42], which will be explored in a future study.

It is also noteworthy that the flow velocity in real aquifers is several orders of magnitude smaller than that used in the laboratory experiments. Hence, the field transport is diffusion dominated, while the laboratory transport is advection dominated. This discrepancy might imply that the late-time BTC tail persists for groundwater flow with a broad range of Peclet numbers, which can be characterized by the time nonlocal models. However, for groundwater flow with a small Peclet number and potential preferential flow paths, the early time BTC tail may occur, which cannot be efficiently captured by the time nonlocal transport models such as the tt-fADE or the CTRW framework with an index less than 1.

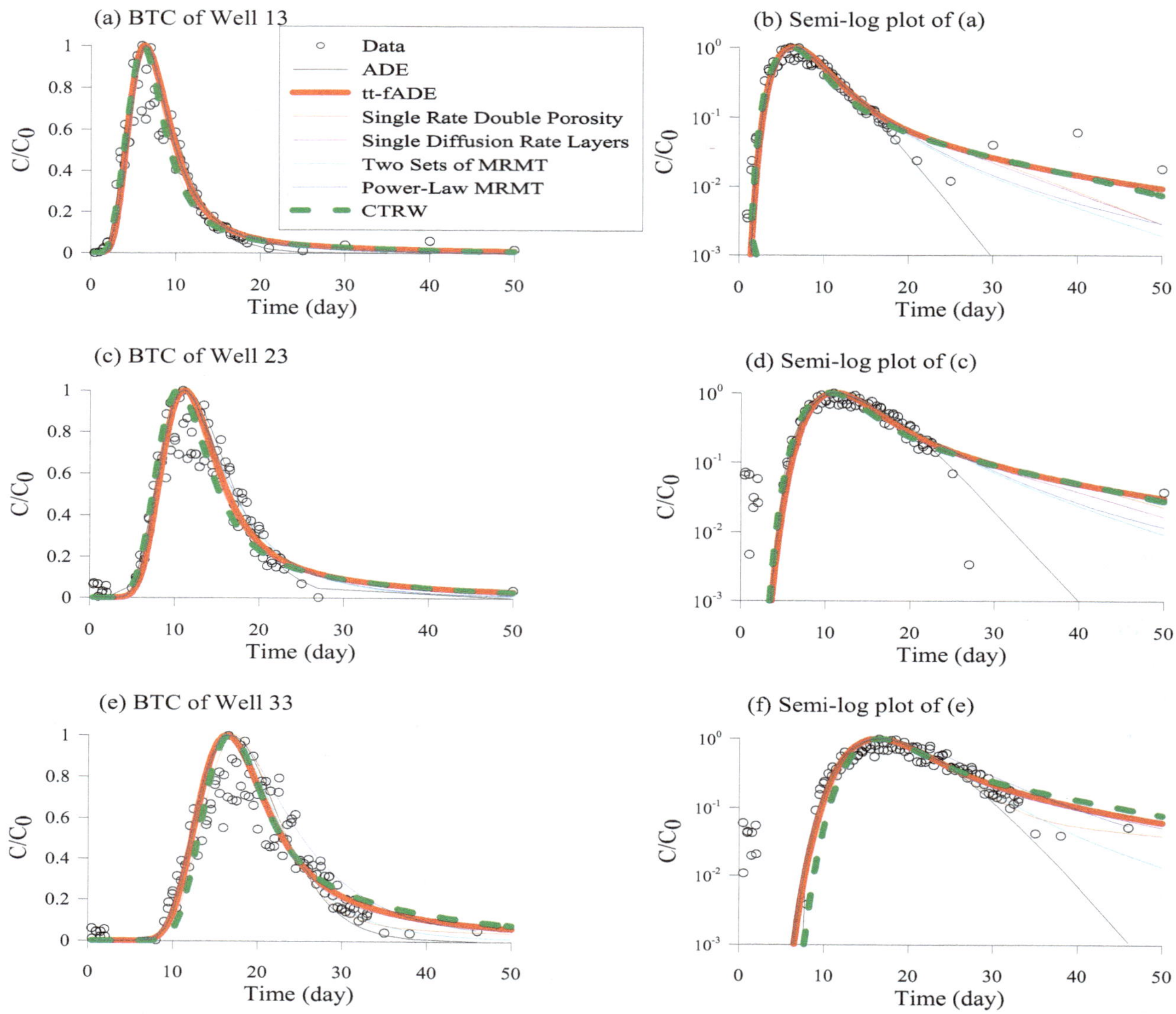

Figure 9. Field tracer test: comparison between the measured field data (symbols) and the modeled (lines) breakthrough curves using the ADE, the tt-fADE (red thick lines), the CTRW (green dashed lines), and the MRMT models.

6. Conclusions

This study compared three time-nonlocal transport models by combining theoretical analyses and applications for laboratory sand column transport experiments and field tracer tests. The models revisited by this study include the MRMT model with various specific forms, the tt-fADE model, and the CTRW framework. Four major conclusions were obtained for these models, which can be used by practitioners to select the appropriate model and improve practical applications, and can improve our understanding of the nature of non-Fickian transport in heterogeneous media.

First, the sand column packed in the laboratory and soil in the field may contain multiple immobile domains with different mass transfer capabilities. This assumption and the laboratory measurements challenge the applicability of the classical ADE model with equilibrium adsorption (i.e., instantaneous sorption/desorption) and the single-rate mobile–immobile model in capturing the tracer BTCs with a late-time tail, which declines at a rate slower than exponential. The MRMT models with multiple rates do capture the BTC's late time tail typical for non-Fickian transport, as revealed before, but the MRMT solutions with power-law distributed mass-exchange rates cannot capture the nuance of observed transition from power-law to exponential decline of the late-time concentration. The increase of the number of unpredictable parameters (both the rate coefficient and the capacity coefficient) may create a challenge for the applicability of the MRMT model.

Second, the tt-fADE model and the CTRW framework are similar in functionality, but differ in detailed parameters. (1) Both the CTRW framework and the tt-fADE model can capture a complex BTC with late time tailing. Both the CTRW framework and the tt-fADE model assume an exponentially-truncated power-law memory function, to capture the gradual transition from power-law to exponential decline of late-time concentration in the observed BTCs. (2) The tt-fADE parameters can be linked to the CTRW framework parameters. For example, the power-law exponent ξ in the CTRW framework is functionally equivalent to the scale index γ in the tt-fADE model (as revealed before), and the cutoff time scale t_2 in the CTRW framework is also equivalent to the inverse of the truncation parameter λ in the tt-fADE (not shown specifically before). Hence the predictability obtained by the tt-fADE model can also improve the predictability of the CTRW framework, and vice-versa. (3) Compared to the tt-fADE model, the CTRW framework defines one additional parameter t_1, which represents the mean diffusive time, corresponding to the mean of the inverse of rate coefficients in the MRMT model. Model applications, however, showed that t_1 in the CTRW framework is insensitive to model results, and may be neglected to alleviate model fitting burdens.

Third, in the tt-fADE model, the real BTC's peak velocity can be used to estimate the lower-end of the model velocity, increasing the predictability of the tt-fADE for real-world applications. Hence, the tt-fADE model with less parameters may conveniently and accurately estimate the BTC late-time tailing under the conditions of the column experiments and field tracer tests.

Fourth, for tracer transport in the field, early arrivals are likely due to preferential flow paths. Super-diffusive jumps along preferential flow paths cannot be efficiently captured by a typical time-nonlocal transport model focusing on solute retention with a time index less than one. Fast motion (which can exhibit direction-dependent scaling rates) and delayed transport (which is dimensionless), although co-existing in some field sites, are driven by different mechanisms, and hence we recommend different physical components to capture these processes. This motivates the application of the spatiotemporal fADE [28,43], which will be re-visited in a future study.

Supplementary Materials

Figure S1: Comparison between the measured (symbols) and modeled (lines) breakthrough curves (BTCs) using the ADE model (black line), the tt-fADE model (red line) and the CTRW model (green line) with the experimental water flow rate Q = 0.4 mL/s. Figure S2: Comparison between the measured (symbols) and modeled (lines) breakthrough curves (BTCs) using the ADE model (black line), the tt-fADE model (red line) and the CTRW model (green line) with the experimental water flow rate Q = 0.6 mL/s. Figure S3: Comparison between the measured (symbols) and modeled (lines) breakthrough curves (BTCs) using the ADE model (black line), the tt-fADE model (red line) and the CTRW model (green line) with the experimental water flow rate Q = 0.8 mL/s. Figure S4: Comparison between the measured (symbols) and modeled (lines) breakthrough curves (BTCs) using the ADE model (black line), the tt-fADE model (red line) and the CTRW model (green line) with the experimental water flow rate Q =1.0 mL/s. Figure S5: Comparison between the measured (symbols) and modeled (lines) breakthrough curves (BTCs) using the ADE model (black line), the tt-fADE model (red line) and the CTRW model (green line) with the experimental water flow rate Q = 1.2 mL/s. Figure S6: Comparison between the measured (symbols) and the modeled (lines) breakthrough curves (BTCs) using the ADE, the tt-fADE, the MRMT, and the CTRW models with the water flow rate Q = 0.4 mL/s. Figure S7: Comparison between the measured (symbols) and the modeled (lines) breakthrough curves (BTCs) using the ADE, the tt-fADE, the MRMT, and the CTRW models with the water flow rate Q = 0.6 mL/s. Figure S8: Comparison between the measured (symbols) and the modeled (lines) breakthrough curves (BTCs) using the ADE, the tt-fADE, the MRMT, and the CTRW models with the water flow rate Q = 0.8 mL/s. Figure S9: Comparison between the measured (symbols) and the modeled (lines) breakthrough curves (BTCs) using the ADE, the tt-fADE, the MRMT, and the CTRW models with the water flow rate Q = 1.0 mL/s. Figure S10: Comparison between the measured (symbols) and the modeled (lines) breakthrough curves (BTCs) using the ADE, the tt-fADE, the MRMT, and the CTRW models with the water flow rate Q = 1.2 mL/s.

Author Contributions: Conceptualization, B.L. and Y.Z.; Methodology, B.L. and Y.Z.; Software, B.L. and Y.Z.; Validation, C.Z., C.T.G. and C.O.; Formal Analysis, B.L.; Investigation, Y.Z.; Resources, H.-G.S.; Data Curation, J.Q.; Writing-Original Draft Preparation, B.L. and Y.Z.; Writing Review & Editing, Y.Z.; Visualization, B.L.; Supervision, Y.Z.; Project Administration, Y.Z.; Funding Acquisition, H.-G.S. and Y.Z.

References

1. Anderson, M.P. Movement of contaminants in groundwater: Groundwater transport—Advection and dispersion. *Groundw. Contam.* **1984**, *20*, 37–45.
2. Bakshevskaia, V.A.; Pozdniakov, S.P. Simulation of hydraulic heterogeneity and upscaling permeability and dispersivity in sandy-clay formations. *Math. Geosci.* **2016**, *48*, 45–64. [CrossRef]
3. Chang, C.-M.; Yeh, H.-D. Investigation of flow and solute transport at the field scale through heterogeneous deformable porous media. *J. Hydrol.* **2016**, *540*, 142–147. [CrossRef]
4. Cvetkovic, V.; Fiori, A.; Dagan, G. Tracer travel and residence time distributions in highly heterogeneous aquifers: Coupled effect of flow variability and mass transfer. *J. Hydrol.* **2016**, *543*, 101–108. [CrossRef]
5. Bromly, M.; Hinz, C. Non-fickian transport in homogeneous unsaturated repacked sand. *Water Resour. Res.* **2004**, *40*, W07402. [CrossRef]
6. Huber, F.; Enzmann, F.; Wenka, A.; Bouby, M.; Dentz, M.; Schafer, T. Natural micro-scale heterogeneity induced solute and nanoparticle retardation in fractured crystalline rock. *J. Contam. Hydrol.* **2012**, *133*, 40–52. [CrossRef] [PubMed]
7. Naftaly, A.; Dror, I.; Berkowitz, B. Measurement and modeling of engineered nanoparticle transport and aging dynamics in a reactive porous medium. *Water Resour. Res.* **2016**, *52*, 5473–5491. [CrossRef]
8. Nordin, C.F.; Troutman, B.M. Longitudinal dispersion in rivers: The persistence of skewness in observed data. *Water Resour. Res.* **1980**, *16*, 123–128. [CrossRef]
9. Zaramella, M.; Marion, A.; Lewandowski, J.; Nützmann, G. Assessment of transient storage exchange and advection–dispersion mechanisms from concentration signatures along breakthrough curves. *J. Hydrol.* **2016**, *538*, 794–801. [CrossRef]
10. Bianchi, M.; Zheng, C. A lithofacies approach for modeling non-fickian solute transport in a heterogeneous alluvial aquifer. *Water Resour. Res.* **2016**, *52*, 552–565. [CrossRef]
11. Ziemys, A.; Kojic, M.; Milosevic, M.; Ferrari, M. Interfacial effects on nanoconfined diffusive mass transport regimes. *Phys. Rev. Lett.* **2012**, *108*, 236102. [CrossRef] [PubMed]
12. Karadimitriou, N.K.; Joekar-Niasar, V.; Babaei, M.; Shore, C.A. Critical role of the immobile zone in non-fickian two-phase transport: A new paradigm. *Environ. Sci. Technol.* **2016**, *50*, 4384–4392. [CrossRef] [PubMed]
13. Van Genuchten, M.T.; Wierenga, P.J. Mass transfer studies in sorbing porous media I. Analytical solutions. *Soil Sci. Soc. Am. J.* **1976**, *40*, 473–480. [CrossRef]
14. Sardin, M.; Schweich, D.; Leij, F.J.; van Genuchten, M.T. Modeling the nonequilibrium transport of linearly interacting solutes in porous media: A review. *Water Resour. Res.* **1991**, *27*, 2287–2307. [CrossRef]
15. Cvetkovic, V.; Dagan, G. Transport of kinetically sorbing solute by steady random velocity in heterogeneous porous formations. *J. Fluid Mech.* **1994**, *265*, 189–215. [CrossRef]
16. Haggerty, R.; McKenna, S.A.; Meigs, L.C. On the late-time behavior of tracer test breakthrough curves. *Water Resour. Res.* **2000**, *36*, 3467–3479. [CrossRef]
17. Valocchi, A.J. Validity of the local equilibrium assumption for modeling sorbing solute transport through homogeneous soils. *Water Resour. Res.* **1985**, *21*, 808–820. [CrossRef]
18. Jury, W.A.; Horton, R. *Soil Physics*; John Wiley: Hoboken, NJ, USA, 2004; 370p.
19. Coats, K.K.H.; Smith, B.B.D.; van Genuchten, M.T.; Wierenga, P.J. Dead-end pore volume and dispersion in porous media. *Soil Sci. Soc. Am. J.* **1964**, *4*, 73–84. [CrossRef]
20. Haggerty, R.; Gorelick, S.M. Multiple-rate mass transfer for modeling diffusion and surface reactions in media with pore-scale heterogeneity. *Water Resour. Res.* **1995**, *31*, 2383–2400. [CrossRef]
21. Berkowitz, B.; Cortis, A.; Dentz, M.; Scher, H. Modeling non-fickian transport in geological formations as a continuous time random walk. *Rev. Geophys.* **2006**, *44*. [CrossRef]
22. Meerschaert, M.M.; Scheffler, H.-P. *Limit Distributions for Sums of Independent Random Vectors: Heavy Tails in Theory and Practice*; Wiley: Hoboken, NJ, USA, 2001; Volume 321.
23. Meerschaert, M.M.; Zhang, Y.; Baeumer, B. Tempered anomalous diffusion in heterogeneous systems. *Geophys. Res. Lett.* **2008**, *35*, L17403. [CrossRef]
24. Dentz, M.; Berkowitz, B. Transport behavior of a passive solute in continuous time random walks and multirate mass transfer. *Water Resour. Res.* **2003**, *39*, 1111. [CrossRef]

25. Benson, D.A.; Meerschaert, M.M. A simple and efficient random walk solution of multi-rate mobile/immobile mass transport equations. *Adv. Water Resour.* **2009**, *32*, 532–539. [CrossRef]
26. Schumer, R.; Benson, D.; Meerschaert, M.; Baeumer, B. Fractal mobile/immobile solute transport. *Water Resour. Res.* **2003**, *39*. [CrossRef]
27. Zhang, Y.; Benson, D.A.; Baeumer, B. Predicting the tails of breakthrough curves in regional-scale alluvial systems. *Ground Water* **2007**, *45*, 473–484. [CrossRef] [PubMed]
28. Schumer, R. Multiscaling fractional advection–dispersion equations and their solutions. *Water Resour. Res.* **2003**, *39*, 1–11. [CrossRef]
29. Zhang, Y.; Green, C.T.; Baeumer, B. Linking aquifer spatial properties and non-fickian transport in mobile–immobile like alluvial settings. *J. Hydrol.* **2014**, *512*, 315–331. [CrossRef]
30. Klafter, J.; Silbey, R. Derivation of the continuous-time random-walk equation. *Phys. Rev. Lett.* **1980**, *44*, 55–58. [CrossRef]
31. Montroll, E.; Weiss, G. Random walks on lattices. II. *J. Math. Phys.* **1965**, *6*, 167. [CrossRef]
32. Dentz, M.; Cortis, A.; Scher, H.; Berkowitz, B. Time behavior of solute transport in heterogeneous media: Transition from anomalous to normal transport. *Adv. Water Resour.* **2004**, *27*, 155–173. [CrossRef]
33. Willmann, M.; Carrera, J.; Sánchez-Vila, X. Transport upscaling in heterogeneous aquifers: What physical parameters control memory functions? *Water Resour. Res.* **2008**, *44*. [CrossRef]
34. Van Genuchten, M.T.; Alves, W. *Analytical Solutions of the One-Dimensional Convective-Dispersive Solute Transport Equation*; Technical Bulletin; United States Department of Agriculture: Washington, DC, USA, 1982.
35. Haggerty, R. *STAMMT-L 3.0. A Solute Transport Code for Multirate Mass Transfer and Reaction along Flowlines*; Sandia National Laboratories: Albuquerque, NM, USA, 2009.
36. Fiori, A.; Becker, M.W. Power law breakthrough curve tailing in a fracture: The role of advection. *J. Hydrol.* **2015**, *525*, 706–710. [CrossRef]
37. Zhang, Y.; Meerschaert, M.M.; Baeumer, B.; LaBolle, E.M. Modeling mixed retention and early arrivals in multidimensional heterogeneous media using an explicit lagrangian scheme. *Water Resour. Res.* **2015**, *51*, 6311–6337. [CrossRef]
38. Pickens, J.F.; Grisak, G.E. Scale-dependent dispersion in a stratified granular aquifer. *Water Resour. Res.* **1981**, *17*, 1191–1211. [CrossRef]
39. Cortis, A.; Berkowitz, B. Computing "anomalous" contaminant transport in porous media: The ctrw matlab toolbox. *Ground Water* **2005**, *43*, 947–950. [CrossRef] [PubMed]
40. Fiori, A.; Zarlenga, A.; Gotovac, H.; Jankovic, I.; Volpi, E.; Cvetkovic, V.; Dagan, G. Advective transport in heterogeneous aquifers: Are proxy models predictive? *Water Resour. Res.* **2015**, *51*, 9577–9594. [CrossRef]
41. Zheng, C.; Hill, M.C.; Cao, G.; Ma, R. Mt3dms: Model use, calibration, and validation. *Trans. ASABE* **2012**, *55*, 1549–1559. [CrossRef]
42. Benson, D.A.; Wheatcraft, S.W.; Meerschaert, M.M. Application of a fractional advection–dispersion equation. *Water Resour. Res.* **2000**, *36*, 1403–1412. [CrossRef]
43. Benson, D.A.; Wheatcraft, S.W.; Meerschaert, M.M. The fractional-order governing equation of lévy motion. *Water Resour. Res.* **2000**, *36*, 1413–1423. [CrossRef]

13

Expanded Application of the Passive Flux Meter: In-Situ Measurements of 1,4-Dioxane, Sulfate, Cr(VI) and RDX

Alexander A. Haluska [1], Meghan S. Thiemann [1], Patrick J. Evans [2], Jaehyun Cho [1] and Michael D. Annable [1,*]

1 Department of Environmental Engineering Sciences, University of Florida, 217 A.P. Black Hall, P.O. Box 116450, Gainesville, FL 32611, USA; haluska@ufl.edu (A.A.H.); meghan.thiemann@gmail.com (M.S.T.); jaehyun.cho@essie.ufl.edu (J.C.)

2 CDM Smith, 14432 S.E. Eastgate Way, Suite 100, Bellevue, WA 98007, USA; evanspj@cdmsmith.com

* Correspondence: annable@ufl.edu

Abstract: Passive flux meters (PFMs) have become invaluable tools for site characterization and evaluation of remediation performance at groundwater contaminated sites. To date, PFMs technology has been demonstrated in the field to measure midrange hydrophobic contaminants (e.g., chlorinated ethenes, fuel hydrocarbons, perchlorate) and inorganic ions (e.g., uranium and nitrate). However, flux measurements of low partitioning contaminants (e.g., 1,4-dioxane, hexahydro-1,3,5-trinitro-s-triazine (RDX)) and reactive ions-species (e.g., sulfate (SO_4^{2-}), Chromium(VI) (Cr(VI)) are still challenging because of their low retardation during transport and quick transformation under highly reducing conditions, respectively. This study is the first application of PFMs for in-situ mass flux measurements of 1,4-dioxane, RDX, Cr(VI) and SO_4^{2-} reduction rates. Laboratory experiments were performed to model kinetic uptake rates and extraction efficiency for sorbent selections. Silver impregnated granular activated carbon (GAC) was selected for the capture of 1,4-dioxane and RDX, whereas Purolite 300A (Bala Cynwyd, PA, USA) was selected for Cr(VI) and SO_4^{2-}. PFM field demonstrations measured 1,4-dioxane fluxes ranging from 13.3 to 55.9 mg/m^2/day, an RDX flux of 4.9 mg/m^2/day, Cr(VI) fluxes ranging from 2.3 to 2.8 mg/m^2/day and SO_4^{2-} consumption rates ranging from 20 to 100 mg/L/day. This data suggests other low-partitioning contaminates and reactive ion-species could be monitored using the PFM.

Keywords: TCE; RDX; sulfate; PFM; mass flux; remediation

1. Introduction

The freely dissolved concentrations of contaminants and metals in sediment pore water are critical measurements for assessing their fate and transport in groundwater systems. However, accurate measurement of metals and organic contaminants using traditional groundwater sampling methods (e.g., low flow point sampling) is challenging due to the differences in solubilities and solid phase partitioning behaviors. Concentrations of compounds in groundwater withdrawn from wells are controlled partly by the transfer of mass to flowing water from other phases: (1) Mass absorbed to aquifer solids; and/or (2) mass trapped in immobile pockets. Transfers of masses can result in underestimating extracted groundwater concentrations because this traditional method cannot account for vertical or horizontal spatial variability in the distribution, nor can it account for how concentrations may change over time due to seasonal variability (e.g., increased rainfall, tidal changes, etc.) [1]. However, passive sampling approaches have emerged as a promising method since they allow both temporally and spatially averaged concentrations.

Diffusion-based samplers inhibit advective transport processes through porous casing material and are usually deployed for a limited exposure time due to variations in environmental conditions (e.g., water level, groundwater flow rates) [2]. Examples of diffusion-based samplers currently used include ceramic dosimeters, bag samplers, dialysis membrane samplers, polyethene samplers, peepers and polymer-based samplers [3–9]. However, such samplers have difficulty reaching equilibrium between the pore water and sorbent material due to mass transfer limitations through the depletion layer and the absence of active mixing [10–12]. Permeation-based passive samplers rely on groundwater flow to control advective transport processes as it passes through the sampler, while Passive Flux Meters (PFMs) have proven to be the only sampler that is able to effectively measure mass fluxes near source groundwater [2]. Additionally, PFMs and integral pump tests were reported to produce similar estimates of mass flux, demonstrating the accuracy of the device [13,14].

The standard PFM is a self-contained permeable sorbent-filled cartridge installed in wells and boreholes that provides simultaneous in-situ time-average measurement of mass flux (J_c) and water flux (q_o) with depth under ambient hydraulic gradients in saturated porous media [15,16]. The sorbent material (e.g., granular activated carbon (GAC), anionic exchange resin, etc.) is impregnated with water-soluble tracers that elute at proportional groundwater flow rates. The sorbent material also serves to intercept and retain any aqueous phase compounds present in groundwater flowing through a well [17,18]. This design has been adapted and modified for measurements of J_c and q_o at fractured-rock sites, the hyporheic zone, streams, rivers and lakes [19–22]. To date, PFMs have been shown to measure chlorinated ethenes (e.g., Perchloroethene (PCE), Trichloroethene (TCE), *cis*-dichloroethene (*cis*-DCE), vinyl chloride (VC)), brominated ethenes (e.g., tetrabromoethane (TBA), tribromoethene (TBE)], methyl *tert*-butyl ether (MTBE)), perchlorate, arsenic, chromate, chromium(VI) (Cr(VI)), uranium, phosphate and nitrogen [16,23–30]. However, this technology has not yet been proven to measure low partitioning compounds (e.g., Hexahydro-1,3,5-trinitro-1,3,5-triazine (RDX), 1,4-dioxane) and reactive ion species in field sites with highly reducing conditions (e.g., sulfate (SO_4^{2-}), Cr(VI)).

1,4-dioxane is a ubiquitous contaminant commonly found at current and former military installations and industrial sites. 1,4-dioxane is used as a stabilizer in chlorinated solvents and has a relatively high solubility in water ($C_s = 4.31 \times 10^5$ mg/L) and low retardation value [31–33]. Thus, the release of 1,4-dioxane behaves similarly to the release of a non-reactive tracer from the DNAPL (Dense Non-Aqueous Phase Liquid) source zone, resulting in transport of 1,4-dioxane ahead of chlorinated ethenes. This can produce an extensive 1,4-dioxane plume disconnected from the source zone. Additionally, the low retardation values signify the challenge of capturing it using sorbents typically used in PFMs. The partitioning between the GAC typically used in PFMs and 1,4-dioxane is estimated to have a retardation value (R) of about 10–20, which is based on a reported log K_{ow} of -0.27 [34]. A field study was conducted at McClellan Air Force Base comparing rigid polypropylene (RPP) diffusion-based passive sampler concentrations of 1,4-dioxane to low-flow purge grab sample concentrations [35]. RPP passive samplers detected 1,4-dioxane but appeared to underpredict aqueous phase concentrations in comparison to the low flow method [35]. To the best of the authors' knowledge, permeation-based passive samplers for either RDX or 1,4-dioxane have not been reported in the literature.

RDX is a high energy explosive found at numerous military installations and artillery ranges [36]. The US EPA (United States Environmental Protection Agency) has classified RDX as a possible contaminant. RDX has a moderate solubility ($C_S = 37.5$ mg/L at pH 6.2 °C and 20 °C) and a low affinity for carbon ($K_{oc} = 6.26$ to 42 L/kg), suggesting that it is fairly mobile and difficult to capture from the aqueous phase [37,38]. However, several methods have been developed to detect RDX in groundwater systems. Jar aquifer tests, assessing RDX uptake into a polyethene diffusion-based passive sampler using different soils (e.g., sandy loam soil, silty loam soil and sand), have shown that organic carbon content impacts the uptake behavior but may not accurately reflect the amount of RDX present in the soil [39]. Polar organic chemical integrative samplers (POCIS), diffusion-based samplers, have recently been developed and deployed at underwater munitions sites and have detected concentrations of RDX ranging from 4.3 to 11.0 μg/L [40]. GAC has also shown to be a useful material for the removal of RDX from aqueous phase solutions [41–43].

The biological formation of iron-sulfide minerals has been shown to dramatically enhance rates of reductive dechlorination of chlorinated ethenes [44–51]. Factors that are important in the promotion of these biogeochemical transformations include SO_4^{2-} concentration, hydraulic residence time, electron donor availability and the presence of iron oxides [52,53]. The optimal combination of these factors promotes a high volumetric SO_4^{2-} consumption rate and a high rate of reactive iron sulfide generation [53]. However, few monitoring tools exist to quantify these in-situ transformation rates. Currently, biogeochemical reaction rates are monitored by measuring the groundwater concentration in microcosms, soil core analysis, or calorimetrically [33,54,55]. While groundwater concentrations can provide an approximation of the rate, it often underestimates the rate [56]. Microcosms are often used to estimate biodegrading rates but are not representative of the environment since they are often manipulated in the lab [33,57,58]. No device currently exists to directly and spatially determine in-situ biogeochemical transformation rates.

Chromium, a widespread toxic metal contaminant, has been used for industrial processes since 1816 [59]. Cr(VI) is relatively soluble, fairly toxic and highly unstable in reducing environments more so than trivalent chromium [Cr(III)] [60]. Beside monitoring aqueous phase concentrations, several passive samplers have been developed to monitor Cr(VI). These include passive capillary samplers, suction cup/zero tension lysimeters and NALGENE® polyethylene diffusion-based passive samplers [61,62]. The complex chemistry of Cr(VI) in groundwater makes interpretation of field-based transport data difficult to understand, implying diffusion-based samplers may have high uncertainties associated with them [63]. Lysimeters and capillary passive samplers are more qualitative than quantitative, as they sample an undefined volume, do not provide flux data and macroporous flow can cause mass transfer issues with the devices [62,64]. Depending on the environmental conditions, the NALGENE® passive sampler may only provide semi-quantitative results, as the equilibrium between the sampler and aqueous phase may never be reached [10]. Box-aquifer tests were previously conducted, and average measurement errors for Cr(VI) mass flux of 12% were measured, suggesting that this permeation-based passive sampler can overcome the limitation of the diffusion-based samplers, as previously described [24]. However, an in-situ PFM field measurement of Cr(VI) has not yet been reported in the literature.

PFMs have been used extensively for the last decade for high-resolution monitoring of both water and contaminant fluxes for long-term monitoring assessment and contaminant source remediation efforts. This high-resolution flux data can be used at field sites to target remedial actions where contaminant fluxes are greatest, potentially resulting in significant cost savings. Expanded application of the PFM technology allows remediation practitioners better tools to characterize sites and to make more informed decisions on the implementation of in-situ remedial strategies.

The PFM technology was expanded and demonstrated in both laboratory studies and in field trials to measure fluxes of RDX, Cr(VI), 1,4-dioxane and in-situ biogeochemistry transformation rates of SO_4^{2-}. Anionic exchange resin and GAC have been demonstrated to be effective at removing Cr(VI) and RDX, respectively, from water [24,41]. Modified PFMs of alternating layers of different sorbents were deployed at a DOD (Department of Defense) contaminated radionuclide site for post-remedial flux monitoring of Cr(VI) and RDX. The standard and modified versions of the PFM were assessed in laboratory batch and aquifer tests for capture of 1,4-dioxane. The standard version performed well in laboratory studies and was deployed at a chloroethene contaminated site undergoing remediation (confidential sites) to measure 1,4-dioxane mass fluxes. The first application, to the best of our knowledge, of a PFM using anionic exchange resin for measuring both SO_4^{2-} reduction rates and water fluxes is presented here. Laboratory batch equilibrium studies and column studies were used to assess the sorbent materials ability to uptake SO_4^{2-}. The PFM was then deployed at Altus Air

Force Base (AFB) to monitor biowalls remediation performance for removal of chlorinated solvents by assessing volumetric SO_4^{2-} consumption rates.

2. Material and Methods

2.1. Passive Flux Meter Technology

The experimental basis, theoretical basis and field performance data validating the PFM have been previously described [14–17]. Briefly, the PFM is a self-contained, permeable sorbent filled cartridge installed in wells and boreholes that provide simultaneous J_c and q_o under ambient hydraulic gradients. The activated carbon serves to intercept and retain dissolved hydrophobic organic contaminants present in groundwater flow through the well. The sorbent material is also impregnated with known amounts of four water-soluble tracers. These tracers are leached from the sorbent at rates proportional to the fluid flux.

After a specified period of exposure to groundwater flow, the PFM is removed from the well screen. Next, the sorbent is extracted to quantify the mass of all organic contaminants intercepted by the PFM and the remaining masses of all resident tracers. The contaminant masses are used to calculate time-averaged contaminant mass fluxes (C_f) or J_c, while residual resident tracer masses are used to calculate time-averaged water flux. Depth variations of fluxes can be measured by vertically segmenting the exposed sorbent at specified depth intervals. Retardation factors for soluble tracer impregnated onto GAC have been described elsewhere [16].

The q_o ($L^3/L^2/T$) and J_c ($M/L^2/T$) are given by Equations (1) and (2) [15–17]:

$$q_o = \frac{1.67(1 - M_R)r\theta R}{t} Jc \tag{1}$$

$$J_c = \frac{1.67m_c}{\pi rbt} \tag{2}$$

where r is the radius of the PFM cylinder, θ is the water content in the PFM, R is the retardation factor of the resident water-soluble tracers on the sorbent, M_R is the relative mass of resident tracer remaining on the PFM sorbent at a particular well depth, t is the sampling time, mc is the mass of the contaminate sorbed onto the tracer, and b is the length of the sorbent matrix segment.

The PFM flux data can then be used to estimate C_f (M/L^3) over the well screen interval (e.g., PFM sorbent section) given by Equation (3) [65]:

$$C_f = \frac{\int J_{ddz}}{\int q_{ddz}} \tag{3}$$

where C_f is independent of groundwater fluxes, meaning it is not subject to effects of flow convergence towards or divergence around the PFM. C_f represents a time-averaged contaminant concentration estimate over the deployment period, whereas traditional groundwater sampling techniques yield an instantaneous contaminant concentration that is estimated only at the time the groundwater samples were collected [13,65].

2.2. Sorbent Materials

Numerous sorbents have been created and utilized for the removal of both organic and inorganic compounds from wastewater and groundwater. Silver impregnated coconut-based 12 × 40 mesh GAC (Royal Oak Enterprises, Dunnellon, FL, USA), Amberlite XAD16 (Sigma Aldrich, St. Louis, MO, USA), Amberlite XAD4 (Sigma Aldrich, St. Louis, MO, USA) and Superlite DAX8 (Sigma Aldrich, St. Louis, MO, USA) were investigated as possible sorbent materials for the capture of 1,4-dioxane. Coconut-based GAC was considered because GACs from pecan and walnut shells have shown higher levels of adsorption than other GACs [66,67]. In previous laboratories studies, Amberlite

XAD16, Amberlite XAD4 and Superlite DAX8 were proven to be effective at removing aromatic carbon from water and wastewater effluent [66,68,69]. Batch-equilibrium studies were conducted on each sorbent material to determine the adsorption capacity. Those showing promise for the capture of 1,4-dioxane were subjected to column studies and box aquifer tests for further analysis.

GAC has been shown to effectively absorb RDX in column studies, pump and treat systems, and fluidized bed reactors despite its relatively high solubility and low affinity for organic carbon [40,70,71]. However, GAC has a finite adsorption capacity. GAC was only examined for extraction efficiency, since previous literature studies have already proven its effectiveness at adsorption/desorption of RDX.

Both GAC (Royal Oak Enterprises, Dunnellon, FL, USA) and Purolite A300 (BroTech, Philadelphia, PA, USA) were used as sorbent materials in the PFM to remove Cr(VI) from aqueous phase solutions. Cr(VI) occurs in the environment as either chromate (CrO_4^{2-}) or dichromate ($Cr_2O_7^{2-}$), which are relatively mobile in soils [72]. GAC has been shown to effectively absorb Cr(VI) in batch-equilibrium studies and ex-situ filtration systems [73–79]. Purolite A300 has never been tested, to the best of our knowledge, to remove Cr(VI) from aqueous solutions. However, Purolite A300 has been used to effectively remove nitrate (NO_3^-) from aqueous phase solutions, which has a weaker charge than Cr(VI) species [80,81]. Thus, it is likely that Cr(VI) species may be even more strongly sorbed to Purolite A300. Screening extraction experiments (data not shown) showed recovery of Cr(VI) was possible from both GAC and Purolite A300, which has been described in Extraction and Analytical Procedures section of this paper.

Purolite A300 (BroTech, Philadelphia, PA, USA) and 500A (BroTech, Philadelphia, PA, USA) were assessed for the removal of SO_4^{2-} from aqueous phase solutions. Purolite A300 was shown in both batch studies and column studies to effectively remove SO_4^{2-} from aqueous phase solutions, suggesting that it is a good candidate for sorbent selections [82]. Fix-bed experiments assessing the ability of Purolite A500 to remove SO_4^{2-} from wastewater effluent showed that it had a loading capacity of 25 to 30 mg SO_4^{2-}/L, making this an excellent candidate for possible use as a sorbent material in the PFM [83]. Both Purolite A300 and A500 underwent batch equilibrium studies to determine adsorption capacity and column loading tests to estimate retardation factors and dynamic sorption capacities. Note, Table S1 in the supplemental section summarizes the resins tested.

2.3. Extraction and Analytical Procedures

The deployment and retrieval of the PFMs and analyses of the silver-impregnated granular activated carbon (SI-GAC) has been previously described [15]. As part of the analysis, a pre-weighed amount of PFM-deployed GAC sample was placed in 10 mL of extraction solution and rotated for 24 h on a Glass-Col Rotator (Glas-Col, Terre Haute, IN, USA). 1,4-dioxane, chlorinated ethenes, ethene and five "resident" tracers [methanol; ethanol; isopropyl alcohol; *tert*-butyl alcohol; and 2,4-dimethyl-3-pentanol] were extracted with dichloromethane and then analyzed using a gas chromatograph (GC) equipped with a flame ionization detector and an autosampler (Perkin-Elmer, Waltham, MA, USA). Cr(VI) was extracted from sorbent material using 1 M sodium chloride (NaCl) solution and analyzed by UF/IFAS Analytical Service Laboratory (Gainesville, FL) using HPLC/ICP-MS. This extraction technique was found to give Cr(VI) recoveries of 87 $\pm$ 5% (data not shown). Cr(VI) and RDX were extracted from sorbent material using acetonitrile and analyzed by a U.S. Department of Energy Laboratory using a high-performance liquid chromatography—ultraviolet detector. This extraction technique was found to give RDX recoveries of 14 $\pm$ 2% (data not shown). SO_4^{2-} was extracted from sorbent material using 1M NaCl solution and was analyzed using ion chromatograph [28].

2.4. Batch Adsorption Isotherms and Recovery

Batch adsorption isotherms were developed for SO_4^{2-} and 1,4-dioxane using a method described elsewhere [15,24,28,30]. 2 g of Purolite A500 and Purolite A300 were added to 30 mL SO_4^{2-} solutions ranging from 0 to 8200 mg/L. 0.1 to 10 g of GAC and 0.1 to 5 g of nonionic exchange resin (e.g., Purolite

XAD16, Puroloite XAD4, Superlite DAX8) were added to a 1,4-dioxane solution of 1030 mg/L. The mixtures were rotated for a minimum of 24 h and then filtered through a 0.45-μm glass filter. Sorbed contaminants were recovered from the drained resin using extraction solvents and rotated for an additional 24 h. Filtered and sorbent recovery solutions were analyzed using the methods as described in the Extraction and Analytical Procedures section. Note, batch studies were not performed on Cr(VI) and RDX since previously mentioned literature studies showed GAC was a good sorbent for the capture of these materials.

2.5. Column Studies

Column studies were only conducted for SO_4^{2-} and 1,4-dioxane using Purolite A300 and GAC, respectively [15,24,28,30]. Small glass columns (2.5 cm inner diameter and 5 cm long, Kontes Co., Vineland, NJ, USA) were wet-packed with GAC and Purolite A300. GAC packed columns were then flushed with a contaminant solution consisting of 65 mg/L 1,4-dioxane, 68 mg/L methylene chloride (CH_2Cl_2), 57 mg/L *cis*-DCE and 36 mg/L methanol (CH_3OH) at a steady solution flow of 1.7 mL/min. CH_2Cl_2, CH_3OH and *cis*-DCE served as controls for performance assessment and to see how multiple contaminants may impact breakthrough times since 1,4-dioxane is rarely the only contaminant found at chlorinated compound contaminated sites [67]. The Purolite A300 packed column was flushed with 3000 mg/L SO_4^{2-} solution at a steady solution flow of 0.5 mL/min. Effluent samples were analyzed for contaminants and tracers—as described in the Extraction and Analytical Procedures.

2.6. Bench-Scale Aquifer Testing of 1,4-Dioxane PFM

Bench-scale aquifer model experiments were conducted as previously described [28]. A rectangular stainless-steel box (38 cm × 30.7 cm × 12 cm) was packed with commercial grade sand (Sunniland Corporation, Longwood, FL, USA). One-cylinder PVC well screen (16 cm long, 3.2 cm inner diameter, 4.2 cm outer diameter) was placed upright at the center of the chamber [28]. Teflon tape was placed over the top part of the well screen to minimize loss of 1,4-dioxane and to create a screen well of 10 cm. The water table was set to a height of 10 cm [28]. Coarse gravel was packed at the injection and extraction ends to facilitate uniform flow through the model [28]. A 1-inch layer of bentonite clay was used to cover the sand to minimize contaminant loss from volatilization. A 4-L aspirator bottle (Kimax Co. Vineland, NJ, USA) was used to contain a 1,4-dioxane solution and maintain a constant head [28]. The flushing experiments were conducted with an injection solution of 1.8 mg/L 1,4-dioxane.

The standard PFM was constructed using sewn fabric socks (10 cm long and 3.2 cm inner diameter) filled with GAC. Before packing, the GAC was pre-equilibrated with an aqueous solution of alcohol tracers consisting of methanol, ethanol, isopropyl alcohol (IPA), *tert*-butyl alcohol (TBA) and 2,4-dimethyl-3-pentanol (2,4-DMP) [15,16,28]. A GAC-filled fabric sock was placed in the PVC packing tube [28]. Mechanical vibration was then used to compact the resin into place. During packing the GAC was subsampled to measure the initial concentrations of resident tracers [28]. The PFM was then slipped into the model aquifer well screen [28]. Upon termination of the experiment the model standard PFM was retrieved, homogenized and then subsampled for 1,4-dioxane and tracer loss analysis, as has been previously described [28]. Six PFMs were constructed and deployed in the model aquifer under flow rates ranging from 2.2 to 18 mL/min.

2.7. Field Studies

2.7.1. RDX and Cr(VI) Measurements

A U.S. Department of Energy nuclear weapons facility (confidential site) has been a storage site for both conventional and nuclear weapons for the last 65 years. Pre-1980's, industrial waste disposal activates included on-site disposal of high explosive industrial wastewater into unlined ditches. Contaminated groundwater leached into the aquifer below. The aquifer has been contaminated with RDX, octahydro-1,3,5,7-tetranitro-1,3,5,7-tetrazocine (HMX), 2,4,6-trinitrotoluene (TNT), 2,4-dinitrotoluene

(2,4-DNT), 1,2-dichloroethane (1,2-DCA), TCE, PCE and Cr(VI). The groundwater is approximately 90 m below the surface, 30 m above an uncontaminated drinking water source and the saturated thickness of the contaminated aquifer is less than 4.5 m. The stated objective of the deployment at this site was to demonstrate the PFM technology and identify where fluxes of contaminants were greatest so future remediation efforts could be targeted to prevent migration of both RDX and Cr(VI) off-site. Of these, considerable attention was focused on the high explosive RDX and Cr(VI) because they were the most widespread and prevalent at the site.

Eight PFMs were installed in wells within, upgradient and downgradient of the treatment zone in both in-situ performance monitoring (ISPM) and in-situ bioremediation (ISB) wells. Each deployed PFM was constructed with alternating segments of GAC and Purolite A300. PFMs 1-5 were installed within the treatment zone, PFM 6 was installed upgradient of the treatment zone and both PFM 7 and 8 were installed downgradient of the treatment zone.

2.7.2. 1,4-Dioxane Measurements

PFMs were deployed at a chlorinated solvent contaminated site being remediated by a U.S. based environmental remediation firm (confidential) that will be referred to as Site 1. Site 1 is a former chemical plant that had been in operation since 1923 but was subsequently decommissioned. Contamination was determined to be from areas where drums, sludge, boiler ash and other wastes were landfilled or buried. At Site 1, PFMs were deployed in February 2013 in 13 wells containing benzene, chloroform flux and carbon tetrachloride.

2.7.3. Sulfate Measurements

PFMs were deployed at biowalls OU-1 and SS017 in Altus Air Force Base, Oklahoma in order to determine volumetric SO_4^{2-} consumption rates as part of a performance remediation study. Detailed site maps are presented elsewhere [53]. The volumetric SO_4^{2-} consumption rate was calculated from SO_4^{2-} flux data using the following equation:

$$R_{SO_4} = \left(F_{SO_4^{2-}}^{upgradient} - F_{SO_4^{2-}}^{biowall}\right) / W_{biowall} \quad (4)$$

where R_{SO_4}. is the volumetric SO_4^{2-} consumption rate, $F_{SO_4^{2-}}^{upgradient}$ is the SO_4^{2-} flux in a well upgradient of the biowall, $F_{SO_4^{2-}}^{biowall}$ is the SO_4^{2-} flux in the biowall and $W_{biowall}$ is the width of the biowall parallel to groundwater flow.

SS17 site is located in the southwest portion of Altus AFB and contains a VOC plume comprised mostly of TCE and reductive dechlorination products. Groundwater is characterized by high SO_4^{2-} concentrations ranging from 1000 to 2000 mg/L. The biowall was installed in 2005 and materials used for construction included 42% mulch, 32% sand, 15% gravel and 11% cotton gin waste. The section of the biowall where the PFMs were deployed were spiked with magnetite at the time of construction to increase the iron content from 1600 mg/kg to 120,000 mg/kg to promote biogeochemical reduction of chlorinated ethenes present at the site. Emulsified vegetable oil (EVO) injections occurred in 2008 to promote SO_4^{2-} reduction so that iron-sulfur clusters could be formed. PFMs were deployed upgradient of the biowall (Well BB04U) and in the wall (Well BB04W). PFMs were deployed in November 2011 and May 2012.

The Operable Unit 1 (OU1) site in the vicinity of a landfill, containing a TCE plume that has SO_4^{2-} concentrations similar to SS17. The biowall was installed in 2002 and constructed with 50% tree mulch, 10% composted cotton gin waste and 40% river sand. The biowall was amended in 2011 with EVO, lactate, ferrous lactate and hematite. PFMs were deployed in upgradient wells (EPAU108, EPAUMP01, EPAU105) and in wall wells (EPA107, EPA106 and EPA104). PFMs were deployed in May 2011 and October 2011.

3. Results

3.1. RDX and Cr(VI)

PFMs were deployed in September 2016 right after cessation of active bio-stimulation occurred. RDX was only detected in PFM 4 (Figure 1) with an average mass flux of 4.9 $mg/m^2/day$, which equates to a C_f of 174 μg/L. Groundwater samples taken just before deployment of PFM 4 were in good agreement since an RDX concentration of 180 μg/L was measured, which is a 3.4% difference from the PFM C_f value. RDX was not detected in any of the other PFMs deployed, suggesting that the PFM 4 is the only hotspot and may be undergoing degradation prior to leaving the site, since it was not detected in any downgradient samples. Cr(VI) was detected in PFMs 4, 6 and 7, but not in PFMs 1, 2, 3, 5 and 8. The absence of Cr(VI) suggests that it was transformed to Cr(III), an insoluble form of chromium. Average Cr(VI) mass fluxes from PFM 4, 6 and 7, based on recovery of Cr(VI) from Purolite A300, were determined to be 2.3, 2.3 and 2.8 $mg/m^2/day$ (Figure 1). Average Cr(VI) mass fluxes from PFMs 4 and 7, based on recovery of Cr(VI) from GAC, were determined to be 1.0 and 0.77 $mg/m^2/day$. Note, no Cr(VI) was detected from GAC samples from PFM 6. Groundwater samples, obtained prior to PFM deployment, were only available for PFM 4 and were determined to be 82 μg/L. PFM C_f values based on Purolite A300 and GAC were 83 and 38 μg/L, respectively, which are 1.2% and 74% differences from groundwater sample data. Thus, Purolite A300 appeared to perform better than GAC at the capture of Cr(VI) in aqueous phase environmental samples. Additionally, this data suggests that Cr(VI) fluxes are relatively stable both within and downgradient of the treatment zone. Thus, more target injections in and around these wells may be needed to stimulate the bacteria necessary to reduce Cr(VI) to an insoluble form.

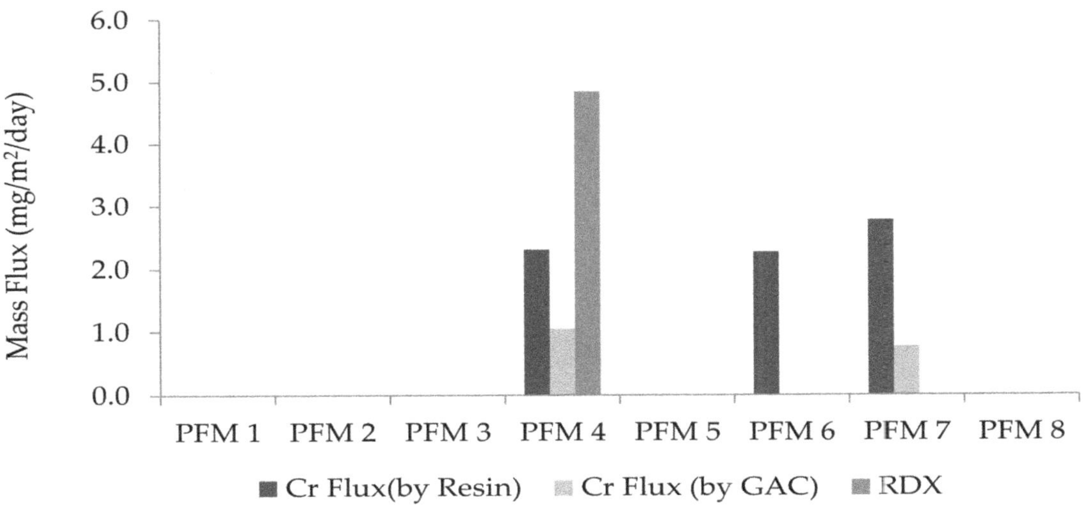

Figure 1. Average mass flux for each well measured at DOD Nuclear Facility.

3.2. 1,4-Dioxane

Adsorption capacity and recovery efficiency of GAC, XAD16, XAD2 and DAX8 for 1,4-dioxane were evaluated. Adsorbed 1,4-dioxane mass is shown as a function of the equilibrium aqueous 1,4-dioxane concentration (Figure 2). Adsorption capacity of each sorbent is assessed by fitting data with a Freundlich isotherm model [28,75]:

$$C_s = (K_f C_e)^{1/n} \tag{4}$$

where C_s (M/L^3) is the equilibrium value of SO_4^{2-} mass adsorbed per unit mass of sorbent, C_e (M/L3) is the equilibrium 1,4-dioxane concentration in the aqueous phase, K_f is the empirical fitting parameters where $n > 1$. The Freundlich isotherm model fit for GAC was determined to be $K_f = 3.15$ and $n = 2.11$, with a Nash-Sutcliffe model efficiency (E) coefficient of 0.99. The Freundlich isotherm model fit for XAD16 was determined to be $K_f = 0.50$ and $n = 2.37$ with $E = 0.87$. The Freundlich isotherm model fit for XAD4 was determined to be $K_f = 0.23$ and $n = 1.7$ with $E = 0.99$. No fit could be determined for DAX8 likely due to a low sorption capacity of the resin for 1,4-dioxane. Further studies using lower concentrations of dioxane are needed to see if an isotherm can be developed for DAX8. The observed sorption capacity for GAC for 1,4-dioxane (73 mg/g at 838 mg/L C_{eq}) was much higher than the other resins used in these batch tests (8.8 mg/g at 862 mg/L C_{eq} for XAD16; 12.3 mg/g at 822 mg/L C_{eq} for XAD4).

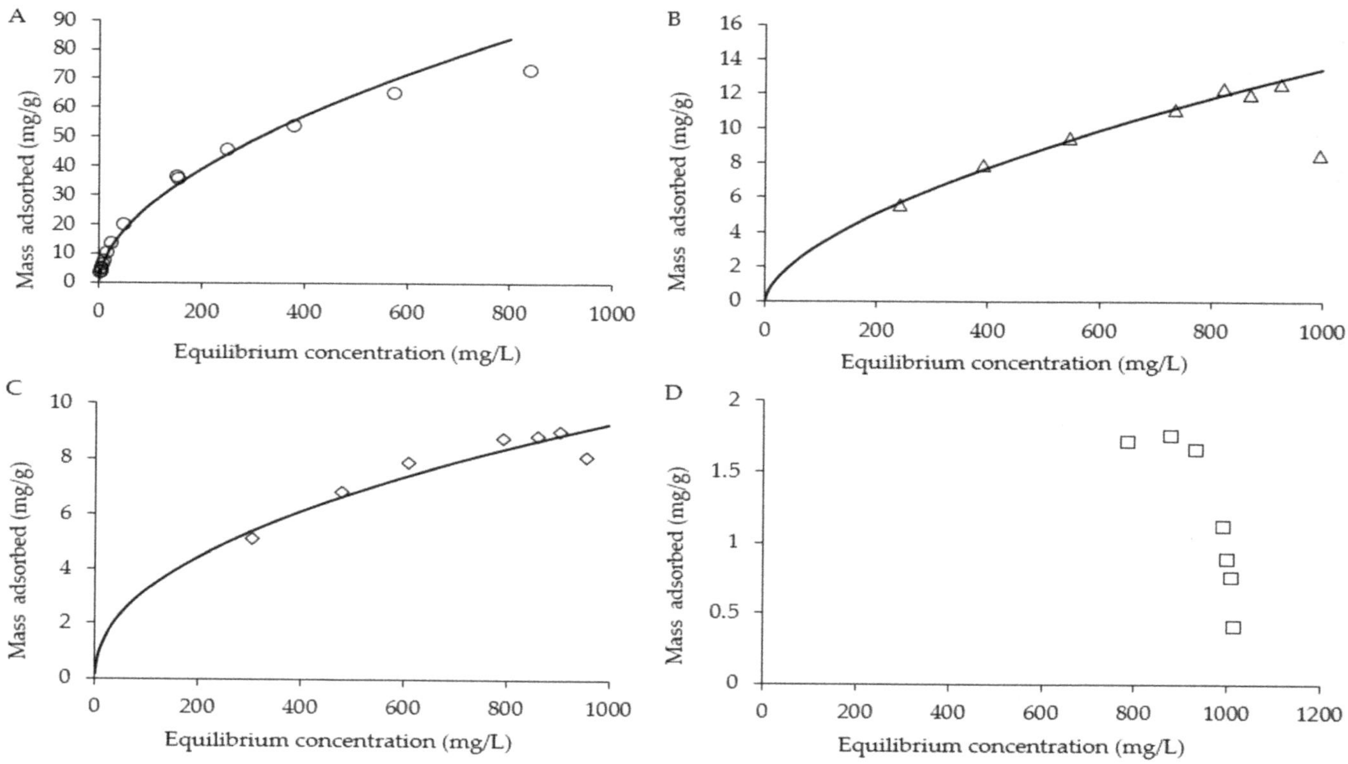

Figure 2. Adsorption isotherm of 1,4-dioxane on (**A**) GAC, (**B**) XAD4, (**C**) XAD16 and (**D**) DAX8.

The 1,4-dioxane flux measurement using the PFM is based on the cumulative 1,4-dioxane mass retained on the sorbent over the test period. However, other contaminants could potentially interfere with the sorption of 1,4-dioxane to the sorbent material, potentially leading to significant errors in flux estimations [15,24,28,30]. Thus, a multispecies column test was conducted using methanol, 1,4-dioxane, $C_2H_2Cl_2$, cis-DCE and commonly found contaminants at 1,4-dioxane contaminated sites. The column test was conducted at a steady water flow using GAC as a sorbent material (Figure 3). R values were determined using the following equation [15]:

$$R = \frac{m_c}{\pi r^2 b A_{RC} \theta C_f} \tag{5}$$

where m_c is the mass retained by the PFM (M), r is the radius of the column (L), b is the thickness of the column (L), ARC is the dimensionless fraction of sorptive matrix containing contaminants, C_f is the flux-averaged concentration of the contaminant of the aqueous phase (M/L^3) and θ is the dimensionless volumetric water content of the sorptive matrix. θ for wet-packed GAC was determined to be 0.55. The bulk volume of the wet-pack GAC in the column, equaling $\pi r^2 b A_{RC}$, 25.3 cm^3. The amount of contaminant per gram of GAC passed through the column at initial breakthrough,

50% breakthrough and 100% breakthrough and R values for each breakthrough are summarized in Table 1. The R values determined from 50% breakthrough for 1,4-dioxane was 57 time higher than that estimated for methanol, two and seven times lower than $C_2H_2Cl_2$ and cis-DCE, respectively, suggesting GAC is sufficient to retain the 1,4-dioxane over a typical PFM deployment and multiple contaminants in the aqueous phase did not significantly affect the capture of 1,4-dioxane during a typical deployment period. R values determined at initial and 100% breakthroughs showed similar trends. Additionally, the R values calculated on 50% breakthroughs were used to estimate water fluxes from the PFM in the bench-scale aquifer model test discussed below. Note, 50% breakthrough is often the used as a proxy to determine when PFM should be removed when deployed at contaminated sites [15].

Table 1. Methanol, 1,4-dioxane, methylene chloride, *cis*-1,2-dichloroethene properties determined from 1D column test with granular activated carbon (GAC).

Breakthrough	Methanol		1,4-Dioxane		Methylene Chloride		*cis*-1,2-Dichloroethene	
	Mass Retained by GAC	R Factor	Mass Retained by GAC	R Factor	Mass Retained by GAC	R Factor	Mass Retained by GAC	R Factor
Initial (mg/g)	0.04	1	6.4	62	16.6	147	71.3	713
50% (mg/g)	0.12	3	13.3	192	28.2	455	98.1	1700
100% (mg/g)	0.4	5	19.8	235	43.6	386	122.5	1167

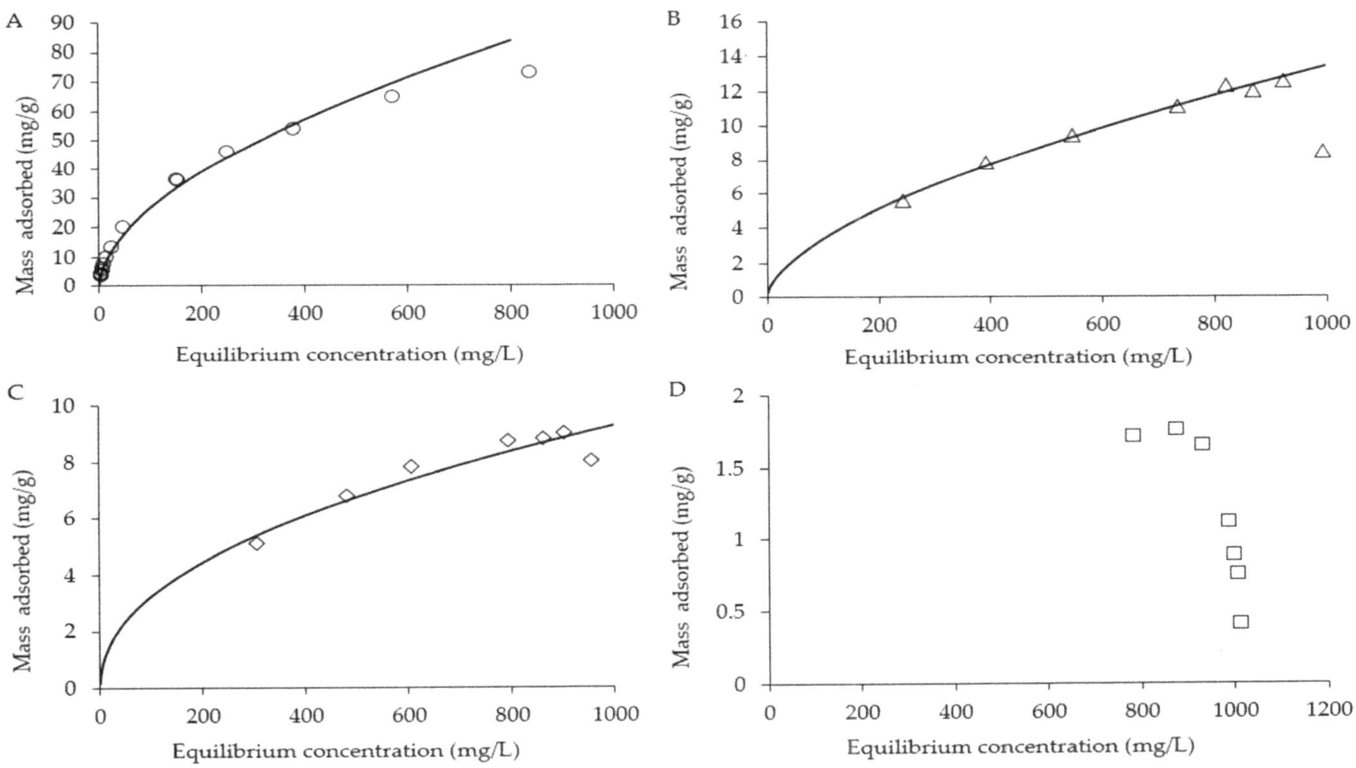

Figure 3. Breakthrough curves for (**A**) methanol, (**B**) 1,4-dioxane, (**C**) methylene chloride and (**D**) *cis*-dichloroethene for the 1-D multispecies column test.

A series of bench-scale aquifer model studies were conducted to evaluate the PFM for measuring simultaneous time-averaged water and 1,4-dioxane fluxes in porous media as previously described [15,28]. Flux-measurements were performed with a PFM deployed in a single screen well during each test. After deployment, PFMs were sampled and data from the resulting samples were averaged and compared with applied water and solute flux as has previously been done [15,28]. Figure 4A provides a comparison of the measured actual versus measured cumulative water flux based on the ethanol tracer. A high correlation (R^2 =0.99) between measured and actual fluxes was obtained.

Measured water fluxes averaged 6% higher than actual fluxes. Figure 4B compares the measured and true cumulative 1,4-dioxane fluxes. A high correlation (R^2 = 0.96) was obtained between measured and true cumulative contaminant fluxes. Measured fluxes averaged 8% greater than actual values.

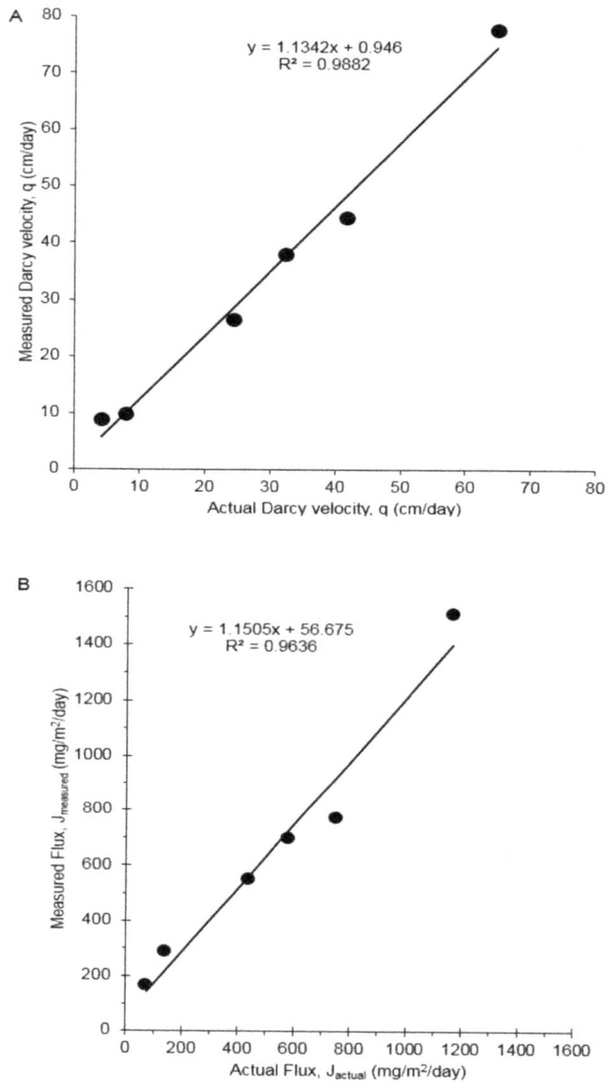

Figure 4. (**A**) Actual versus measured Darcy velocity with a standard PFM; (**B**) Actual vs measured 1,4-dioxane flux in box aquifer tests on the standard PFM.

PFMs were deployed at Site 1 in February 2013. 1,4-dioxane was detected in nine of the 13 wells with mass fluxes ranging from 1.2 to 295 mg/m^2/day (Figure 5). Figure 5A shows the vertical distribution of 1,4-dioxane mass flux and volumetric water flux. The vertical trends in 1,4-dioxane flux profile tend to agree with the water flux profile, in that as the water flux increases, 1,4-dioxane flux increases. The resulting flux-averaged concentration for PFM 7 was 2195 μg/L. The flux-averaged TCE and *cis*-DCE concentrations, estimated from the PFM results, were compared with groundwater data obtained in 2013–2014 (Table 2) and estimates ranged from 2 to 133% differences with all C_f values being much greater than traditional groundwater data. Similar comparisons have been made, and C_f estimates are likely to be more representative of in-situ contaminant concentrations than traditional groundwater samples [65]. Chloroform, carbon tetrachloride and benzene were also detected at this site (Figure 5B).

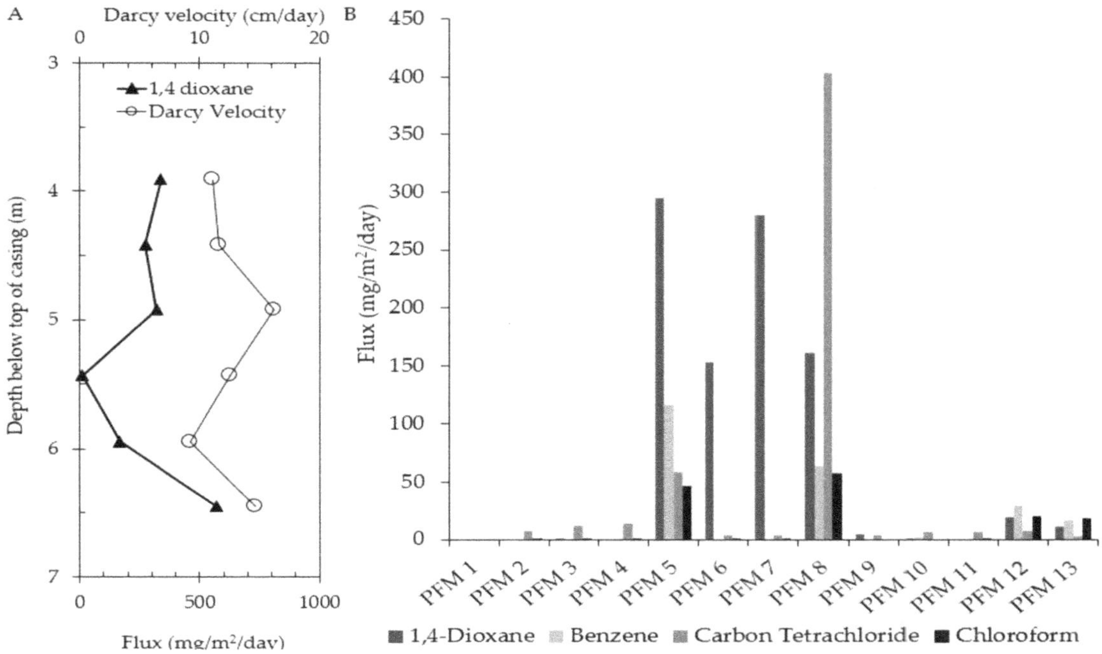

Figure 5. Mass flux measurements at Site 1 (confidential site). (**A**) Water and 1,4-dioxane flux profiles for PFM-7 and (**B**) Average mass flux of contaminations for each well at Site 2.

Table 2. Comparison of PFM Results (August 2015) with traditional groundwater data (2013–2014).

Well	PFM Flux-Averaged Concentration (μg/L)	Measured Aqueous Phase Concentration (μg/L)	Percent Difference (%)
PFM 1	<5	2.3	-
PFM 2	<5	330	-
PFM 3	18	8.6	68
PFM 4	<5	1.9	-
PFM 5	5932	1600	115
PFM 6	1663	1700	2
PFM 7	2195	990	76
PFM 8	1675	480	111
PFM 9	55	46	18
PFM 10	20	<1	-
PFM 11	<5	2.3	-
PFM 12	1384	280	133
PFM 13	602	310	64

BD = below detection.

3.3. *In-situ Biogenic SO_4^{2-} Reduction Rates*

Equilibrium batch studies were conducted using Puroline A500 and Purolite A300 (Figure 6). Adsorption capacity of each sorbent is assessed by fitting data with a Freundlich isotherm model as was previously described and done for 1,4-dioxane [28,75]. The Freundlich isotherm model fit for the Purolite A500 were K_f = 7.65, n =5.31 and the Nash-Sutcliffe model efficiency coefficient (E) = 0.98. The Freundlich isotherm model fit for the Purolite A500 acid form were K_f = 7.10, n =5.1 and E = 0.97. Based on this data, Purolite A300 was used in column studies.

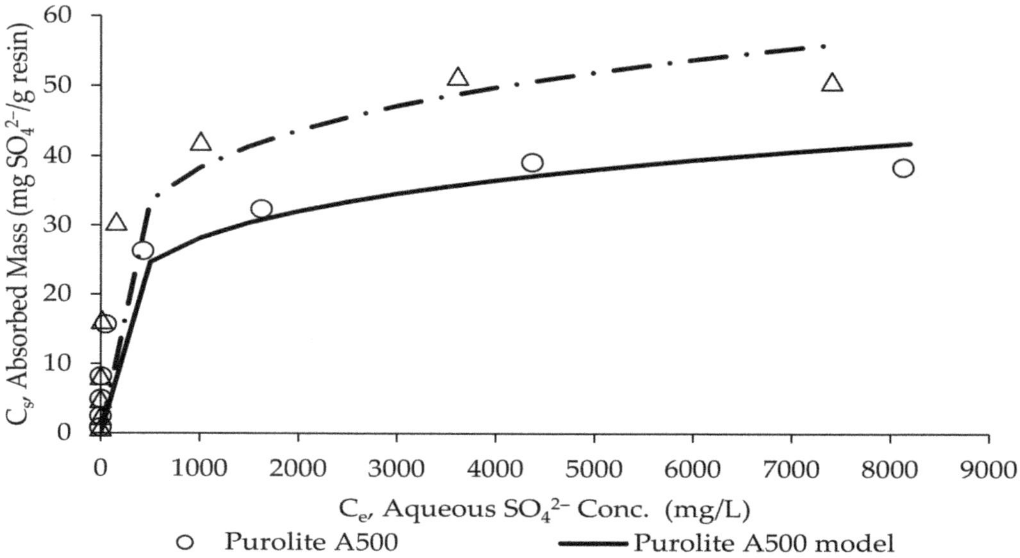

Figure 6. Adsorption isotherm of SO_4^- on Purolite A500 and Purolite A300 resins with a Freundlich model fit.

To assess the performance of Purolite A300 under dynamic conditions, a 1D column test was conducted. The column elution experiment generated concentration data as a function of column pore volumes (PV) of eluted solution (Figure 7). The R values at initial breakthrough, 50% breakthrough, and 100% breakthrough were determined (Table 3). These retardation values were estimated to range from 15 to 37, suggesting Purolite A300 is sufficient to entrap sulfate during the typical deployment of a PFM.

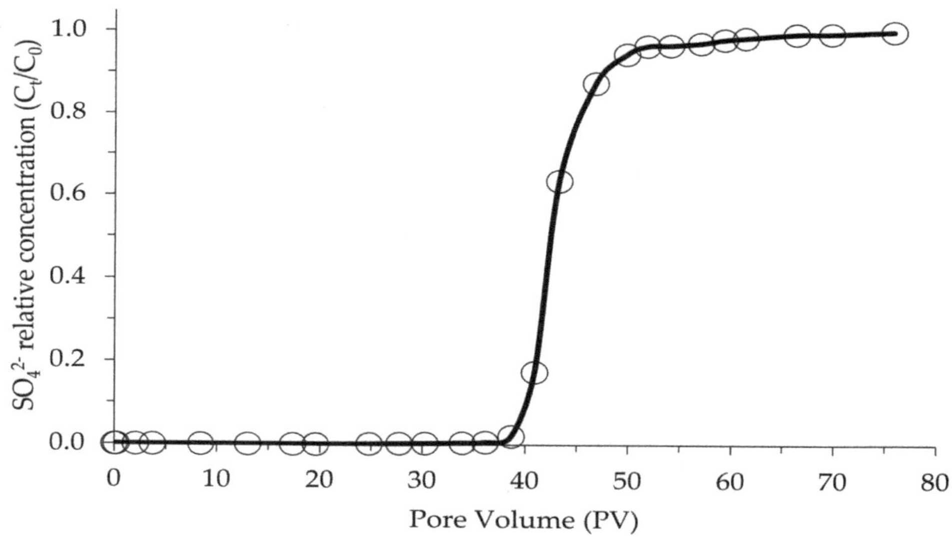

Figure 7. Breakthrough curves for SO_4^{2-} for Purolite A300 from 1D column test.

PFMs were constructed and deployed at Altus AFB. They were packed with alternating individual segments of Ag-GAC impregnated with water-soluble tracers and Purolite A300. The reason for the alternating layers was to test the individual capability of Purolite A300 to capture SO_4^{2-} under field conditions. GAC data was used to estimate groundwater flux. The PFMs were deployed at Altus AFB for about 2–3 weeks between 2011 and 2012 to monitor biowall performance at two chloroethene contaminated sites. The individual PFM profiles were too numerous to display, but Figure 8 shows the vertical distribution of SO_4^{2-} mass flux and volumetric water flux. Greater water flux results were observed with higher SO_4^{2-} fluxes. This data suggests that contaminant fluxes varied during sampling, suggesting that seasonal effects (e.g., rainfall) likely caused increases and decreases of SO_4^{2-} flux at the site. However, this data show that SO_4^{2-} mass fluxes can be measured in-situ. This data were used to calculate volumetric SO_4^{2-} reduction rates as a performance measure of the biowall as an ongoing remediation project at Altus Air Force Base.

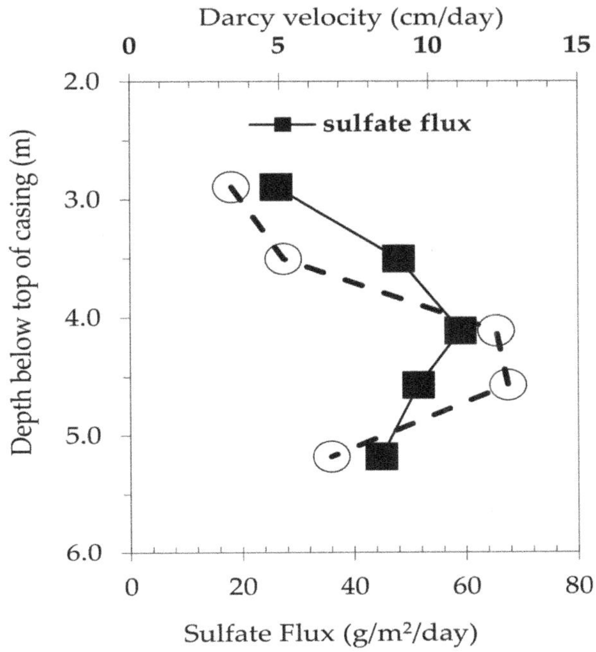

Figure 8. Mass flux measurements at Altus Air Force Base Oklahoma for well EPAUMP01.

Figure 9 presents the relationships among VOC removal, sulfide generation and SO_4^{2-} consumption in the four biowall sections, as previously described. Sulfide concentrations in the SS17 biowall were less than 10 mg/L (Figure 9A). Sulfide concentrations in the OU1 biowall sections prior to amendment varied between 90 (organic-amended section) and 260 mg/L (organic/iron-amended section). Sulfide concentrations in the OU1 unamended biowall section remained elevated, ranging from 140 to 200 mg/L. However, sulfide concentrations decreased in the OU1 amended biowall sections. Sulfide concentrations in the OU1 organic-amended and organic/iron-amended biowall sections decreased to minimum values of 13 and 1.6 mg/L, respectively (data not shown). Percent total molar VOC removal was high in the SS17 biowall section throughout the study, demonstrated no discernible trends and ranged from 93 to 100% (Figure 9B). VOC removal in the OU1 organic/iron-amended biowall section was also high with VOC increasing from 90−93% to 97−98%. VOC removals were initially low in the OU1 organic amended biowall but increased simultaneously with the decrease in sulfide concentrations. VOC removals were low in the OU1 unamended biowall section, which had consistently high sulfide concentrations. This data demonstrates a relationship between high sulfide concentration and an inhibition of VOC removal. This data also indicates that OU1 biowall amendment reduced sulfide concentrations, which facilitated VOC removal.

An unexpected observation is the high sulfide in the OU1 organic/iron-amended biowall section prior to amendment (260 mg/L) and the lack of significant inhibition of biotic reductive dechlorination (total VOC removal was 90 to 93%). A possible explanation is that biogeochemical transformation was occurring at a greater rate in this section relative to the other OU1 biowall sections prior to amendment. Nevertheless, amendment with organic/iron led to improved performance with respect to total VOC destruction, which increased from 97 to 98%.

The use of passive flux meters demonstrated that volumetric SO_4^{2-} consumption in the biowalls was lower in the OU1 unamended biowall section relative to the other biowall sections (Figure 9C). This lower consumption rate correlated with lower VOC removal (Figure 9D). Electron donor limitation may partially explain the lower VOC removal. However, sulfide generation was still high in the unamended biowall, suggesting that a lack of electron donor was not completely limiting biological activity. While volumetric SO_4^{2-} consumption was lower in the OU1 unamended biowall section, it was not sufficiently low to preclude the observed generation of high sulfide concentrations. Thus, a combination of high sulfide concentrations and lower volumetric SO_4^{2-} consumption prevented biogeochemical transformation, complete biotic reductive dechlorination and complete VOC removal in the OU1 unamended biowall section.

Table 3. Sulfate properties determined from 1D column test with Purolite A300.

Breakthrough	Sulfate	
	Breakthrough (mg/g)	**Retardation Factor**
Initial	50.5	27
50%	68	15
100%	70.4	37

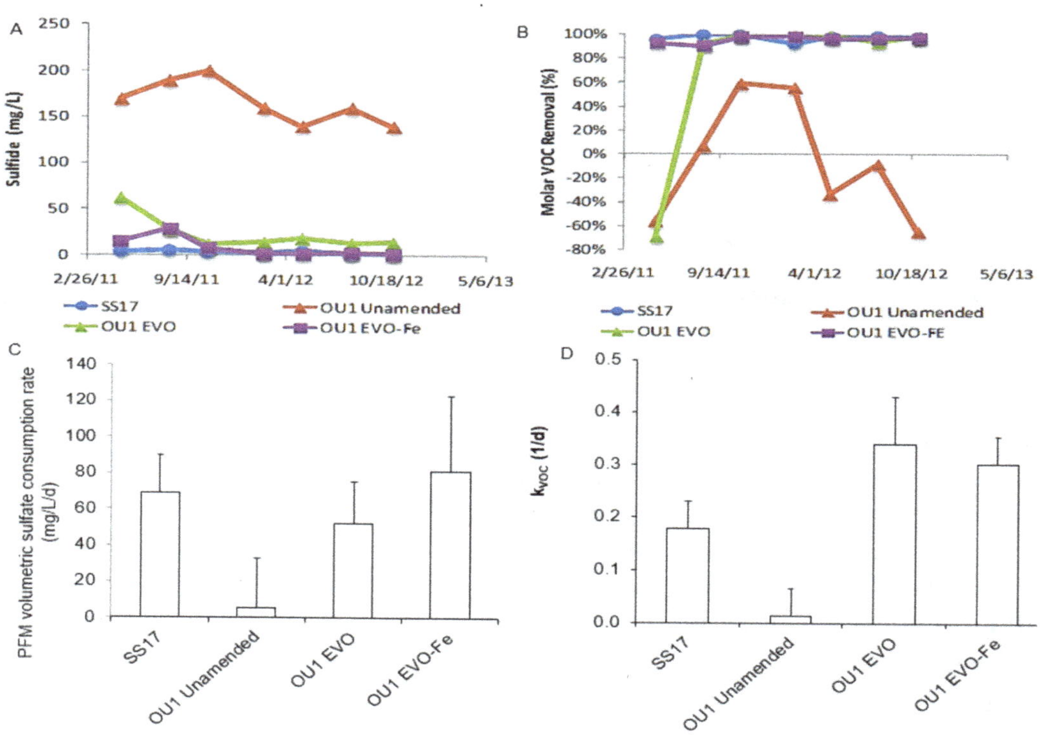

Figure 9. Comparison of VOC removal, sulfide production and SO_4^{2-} consumption. (**A**) Biowall sulfide concentrations; (**B**) Percent VOC removal trends; (**C**) Average volumetric SO_4^{2-} consumption rate for T2 through T7; and (**D**) Average percent VOC removal for T2 through T7. Positive error bars represent one standard deviation.

4. Discussion

This work, based on the results obtained from Cr(VI) and RDX standard PFMs deployed in the field, confirms the laboratory results for measuring Cr(VI) flux [24]. The field deployment of this PFM was able to sorb both RDX and Cr(VI) onto GAC, suggesting that competitive sorption does not undermine the ability of the PFM to capture and retain low partitioning target contaminants. Because

RDX is an explosive and should not be allowed to accumulate to large quantities, caution must be taken during measurement. This work is preliminary for RDX, as more work is needed to understand the sorption mechanism under different pH and geochemical conditions. Additionally, flux-averaged concentrations should be compared with aqueous phase concentrations to validate the measurement of the PFM. This demonstrates that explosive and inorganic compounds can be detected using the standard PFM.

The laboratory experiments and the field demonstrations showed that the PFM can effectively measure both water and 1,4-dioxane contaminant mass fluxes. This data suggests that other low partitioning compounds such as perfluorooctanesulfonic acid or ibuprofen could be better tracked in groundwater systems. Since 1,4-dioxane is predominantly associated with dilute plumes that are often detached from the original source zone, measurement of the flux distribution within the aquifer could become an early detection system of further contamination and would allow for the better identification of source zones in upgradient wells where 1,4-dioxane has been detected. Additionally, being able to detect 1,4-dioxane downgradient will give remediation practitioners sufficient time to design and implement a remediation strategy to either mitigate or contain contaminants upgradient from spreading any further.

The biowalls at Altus AFB SS17 and OU1 provided a unique opportunity to evaluate monitoring tools for biogeochemical transformation. The SS17 biowall was demonstrated to be functioning with a biogeochemical transformation pattern based on a consistently high total molar VOC destruction without accumulation of reductive dechlorination and with a high SO_4^{2-} consumption rate, as determined by the passive flux meters. These results contrast with the low total molar VOC destruction observed in the unamended OU1 biowall section. Unlike the SS17 biowall, this biowall section had very low volumetric SO_4^{2-} consumption rates. High dissolved sulfide concentrations also probably inhibited complete biotic reductive dechlorination of TCE to ethene. Thus, high dissolved sulfide concentration was an important indicator of ineffective biotic and biogeochemical transformation processes of VOC destruction. Injection of organics into the organic-amended OU1 biowall section resulted in activation of passivated iron oxides as shown by decreased dissolved sulfide concentrations, even though volumetric SO_4^{2-} consumption rates were high.

The PFM was demonstrated in the field to measure both water fluxes and biogeochemical transformation rates in porous media. Passive flux meters have the potential to fill the gap in measuring biogeochemical transformation rates of other electron acceptors in groundwater remediation systems. Quantifying biogeochemical transformation rates in groundwater systems is an essential step to further evaluating remediation performance. In the future, biogeochemical transportation rates, as obtained by the PFM, have the possibility of being incorporated into a transport model to understand geochemical cycling in some systems. We anticipate further improvement in, and increased use of, passive flux meter approaches to advance conceptual models of biogeochemical cycling in groundwater systems undergoing active remediation. We demonstrated modifications which extended the PFM application from mass flux measurement to measurement of biogeochemical rates. Further applications will likely focus on the measuring of other biogeochemical transformation rates such as Fe-reduction. Ultimately, this approach will allow for determination of bulk in-situ degradation rates, potentially allowing practitioners to estimate the effectiveness of active remediation and possibly closure of the contaminated site could occur.

Author Contributions: M.D.A. was responsible for obtaining funding, designing experiments and aided in writing the manuscript. M.S.T. conducted 1,4-dioxane column and batch shake test on 1,4-dioxane. P.J.E. deployed PFMs for sulfate analysis and interpret sulfate-based field data. J.C. was responsible for building PFMs for deployment, conducting sulfate batch studies, conducting 1D sulfate column test and measuring contaminants. A.A.H. was responsible for analyzing/interpreting data and preparing the manuscript for publication.

Acknowledgments: Technical support was provided by CDM Smith and EnviroFLUX. The authors thank the anonymous reviewers for their feedback.

References

1. Rao, P.S.C.; Jawitz, J.W.; Enfield, C.G.; Falta, R.W.; Annable, M.D.; Wood, A.L. Technology Integration for Contaminated Site Remediation: Clean-Up Goals and Performance Criteria. In *Groundwater Quality: Natural and Enhanced Restoration of Groundwater Pollution*; International Association of Hydrological Sciences: Sheffield, UK, 2002; Volume 571–578, Available online: http://hydrologie.org/redbooks/a275/iahs_275_571.pdf (accessed on 16 September 2018).
2. Verreydt, G.; Bronders, J.; Van Keer, I.; Diels, L.; Vanderauwera, P. Passive samplers for monitoring vocs in groundwater and the prospects related to mass flux measurements. *Ground Water Monit. Remed.* **2010**, *30*, 114–126. [CrossRef]
3. Martin, H.; Piepenbrink, M.; Grathwohl, P. Ceramic dosimeters for time-integrated contaminant monitoring. *J Process Anal. Chem.* **2001**, *6*, 68–73.
4. Vroblesky, D.A.; Campbell, T.R. Equilibration times, compound selectivity, and stability of diffusion samplers for collection of ground-water VOC concentrations. *Adv. Environ. Res.* **2001**, *5*, 1–12. [CrossRef]
5. Harter, T.; Talozi, S. Evaluation of a simple, inexpensive dialysis sampler for small diameter monitoring wells. *Ground Water Monit. Remed.* **2004**, *24*, 97–105. [CrossRef]
6. Jalalizadeh, M.; Ghosh, U. In situ passive sampling of sediment porewater enhanced by periodic vibration. *Environ. Sci. Technol.* **2016**, *50*, 8741–8749. [CrossRef] [PubMed]
7. Webster, I.T.; Teasdale, P.R.; Grigg, N.J. Theoretical and experimental analysis of peeper equilibration dynamics. *Environ. Sci. Technol.* **1998**, *32*, 1727–1733. [CrossRef]
8. Vrana, B.; Mills, G.A.; Dominiak, E.; Greenwood, R. Calibration of the chemcatcher passive sampler for the monitoring of priority organic pollutants in water. *Environ. Pollut.* **2006**, *142*, 333–343. [CrossRef] [PubMed]
9. De Jonge, H.; Rothenberg, G. New device and method for flux-proportional sampling of mobile solutes in soil and groundwater. *Environ. Sci. Technol.* **2005**, *39*, 274–282. [CrossRef] [PubMed]
10. Jalalizadeh, M.; Ghosh, U. Analysis of measurement errors in passive sampling of porewater PCB concentrations under static and periodically vibrated conditions. *Environ. Sci. Technol.* **2017**, *51*, 7018–7027. [CrossRef] [PubMed]
11. Lohmann, R. Critical review of low-density polyethylene's partitioning and diffusion coefficients for trace organic contaminants and implications for its use as a passive sampler. *Environ. Sci. Technol.* **2015**, *49*, 3985. [CrossRef] [PubMed]
12. Apell, J.N.; Gschwend, P.M. Validating the use of performance reference compounds in passive samplers to assess porewater concentrations in sediment beds. *Environ. Sci. Technol.* **2014**, *48*, 10301–10307. [CrossRef] [PubMed]
13. Basu, N.B.; Suresh, P.; Rao, C.; Poyer, I.C.; Nandy, S.; Mallavarapu, M.; Naidu, R.; Davis, G.B.; Patterson, B.M.; Annable, M.D.; et al. Integration of traditional and innovative characterization techniques for flux-based assessment of dense non-aqueous phase liquid (DNAPL) sites. *J. Contam. Hydrol.* **2009**, *105*, 161–172. [CrossRef] [PubMed]
14. Brooks, M.C.; Wood, A.L.; Annable, M.D.; Hatfield, K.; Cho, J.; Holbert, C.; Rao, R.S.C.; Enfield, C.G.; Lynch, K.; Smith, R.E. Changes in contaminant mass discharge from dnapl source mass depletion: Evaluation at two field sites. *J. Contam. Hydrol.* **2008**, *102*, 140–153. [CrossRef] [PubMed]
15. Hatfield, K.; Annable, M.; Cho, J.H.; Rao, P.S.C.; Klammler, H. A direct passive method for measuring water and contaminant fluxes in porous media. *J. Contam. Hydrol.* **2004**, *75*, 155–181. [CrossRef] [PubMed]
16. Annable, M.D.; Hatfield, K.; Cho, J.; Klammler, H.; Parker, B.L.; Cherry, J.A.; Rao, P.S.C. Field-scale evaluation of the passive flux meter for simultaneous measurement of groundwater and contaminant fluxes. *Environ. Sci. Technol.* **2005**, *39*, 7194–7201. [CrossRef] [PubMed]
17. Klammler, H.; Hatfield, K.; Annable, M.D. Concepts for measuring horizontal groundwater flow directions using the passive flux meter. *Adv. Water Resour.* **2007**, *30*, 984–997. [CrossRef]
18. Klammler, H.; Hatfield, K.; da Luz, J.A.G.; Annable, M.D.; Newman, M.; Cho, J.; Peacock, A.; Stucker, V.; Ranville, J.; Cabaniss, S.A.; et al. Contaminant discharge and uncertainty estimates from passive flux meter measurements. *Water Resour. Res.* **2012**, *48*, 19. [CrossRef]
19. Padowski, J.C.; Rothfus, E.A.; Jawitz, J.W.; Klammler, H.; Hatfield, K.; Annable, M.D. Effect of passive surface water flux meter design on water and solute mass flux estimates. *J. Hydrol. Eng.* **2009**, *14*, 1334–1342. [CrossRef]
20. Layton, L.; Klammler, H.; Hatfield, K.; Cho, J.; Newman, M.A.; Annable, M.D. Development of a passive

sensor for measuring vertical cumulative water and solute mass fluxes in lake sediments and streambeds. *Adv. Water Resour.* **2017**, *105*, 1–12. [CrossRef]
21. Klammler, H.; Hatfield, K.; Newman, M.A.; Cho, J.; Annable, M.D.; Parker, B.L.; Cherry, J.A.; Perminova, I. A new device for characterizing fracture networks and measuring groundwater and contaminant fluxes in fractured rock aquifers. *Water Resour. Res.* **2016**, *52*, 5400–5420. [CrossRef]
22. Kunz, J.V.; Annable, M.D.; Cho, J.; von Tumpling, W.; Hatfield, K.; Rao, S.; Borchardt, D.; Rode, M. Quantifying nutrient fluxes with a new hyporheic passive flux meter (HPFM). *Biogeosciences* **2017**, *14*, 631–649. [CrossRef]
23. Lee, J.; Rao, P.S.C.; Poyer, I.C.; Toole, R.M.; Annable, M.D.; Hatfield, K. Oxyanion flux characterization using passive flux meters: Development and field testing of surfactant-modified granular activated carbon. *J. Contam. Hydrol.* **2007**, *92*, 208–229. [CrossRef] [PubMed]
24. Campbell, T.J.; Hatfield, K.; Klammler, H.; Annable, M.D.; Rao, P.S.C. Magnitude and directional measures of water and Cr(VI) fluxes by passive flux meter. *Environ. Sci. Technol.* **2006**, *40*, 6392–6397. [CrossRef] [PubMed]
25. Verreydt, G.; Annable, M.D.; Kaskassian, S.; Van Keer, I.; Bronders, J.; Diels, L.; Vanderauwera, P. Field demonstration and evaluation of the passive flux meter on a CAH groundwater plume. *Environ. Sci. Pollut. Res.* **2013**, *20*, 4621–4634. [CrossRef] [PubMed]
26. Johnston, C.D.; Davis, G.B.; Bastow, T.P.; Woodbury, R.J.; Rao, P.S.C.; Annable, M.D.; Rhodes, S. Mass discharge assessment at a brominated DNAPL site: Effects of known DNAPL source mass removal. *J. Contam. Hydrol.* **2014**, *164*, 100–113. [CrossRef] [PubMed]
27. Clark, C.J.; Hatfield, K.; Annable, M.D.; Gupta, P.; Chirenje, T. Estimation of arsenic contamination in groundwater by the passive flux meter. *Environ. Forensics* **2005**, *6*, 77–82. [CrossRef]
28. Cho, J.Y.; Annable, M.D.; Jawitz, J.W.; Hatfield, K. Passive flux meter measurement of water and nutrient flux in saturated porous media: Bench-scale laboratory tests. *J. Environ. Qual.* **2007**, *36*, 1266–1272. [CrossRef] [PubMed]
29. Kunz, J.V.; Annable, M.D.; Rao, S.; Rode, M.; Borchardt, D. Hyporheic passive flux meters reveal inverse vertical zonation and high seasonality of nitrogen processing in an anthropogenically modified stream (Holtemme, Germany). *Water Resour. Res.* **2017**, *53*, 10155–10172. [CrossRef]
30. Stucker, V.; Ranville, J.; Newman, M.; Peacock, A.; Cho, J.; Hatfield, K. Evaluation and application of anion exchange resins to measure groundwater uranium flux at a former uranium mill site. *Water Res.* **2011**, *45*, 4866–4876. [CrossRef] [PubMed]
31. Zenker, M.J.; Borden, R.C.; Barlaz, M.A. Occurrence and treatment of 1,4-dioxane in aqueous environments. *Environ. Eng. Sci.* **2003**, *20*, 423–432. [CrossRef]
32. Zhang, S.; Gedalanga, P.B.; Mahendra, S. Advances in bioremediation of 1,4-dioxane-contaminated waters. *J. Environ. Manag.* **2017**, *204*, 765–774. [CrossRef] [PubMed]
33. Alvarez, P.J.J.; Illman, W.A. *Bioremediation and Natural Attenuation*; John Wiley & Sons, Inc.: Hoboken, NJ, USA, 2006.
34. Howard, P. *Handbook of Environmental Fate and Exposure Data for Organic Chemicals*; Lewis Publishers, Inc.: Chelsea, MI, USA, 1990.
35. ITRC. *Protocol for Use of Five Passive Sampler to Sample for a Variety of Contaminants in Groundwater*; ITRC: Washington, DC, USA, 2007.
36. Paquet, L.; Monteil-Rivera, F.; Hatzinger, P.B.; Fuller, M.E.; Hawari, J. Analysis of the key intermediates of RDX (hexahydro-1,3,5-trinitro-1,3,5-triazine) in groundwater: Occurrence, stability and preservation. *J. Environ. Monit.* **2011**, *13*, 2304–2311. [CrossRef] [PubMed]
37. Lynch, J.C.; Myers, K.F.; Brannon, J.M.; Delfino, J.J. Effects of pH and temperature on the aqueous solubility and dissolution rate of 2,4,6-trinitrotoluene (TNT), hexahydro-1,3,5-trinitro-1,3,5-triazine (RDX), and octahydro-1,3,5,7-tetranitro-1,3,5,7-tetrazocine (HMX). *J. Chem. Eng. Data* **2001**, *46*, 1549–1555. [CrossRef]
38. Pascoe, G.A.; Kroeger, K.; Leisle, D.; Feldpausch, R.J. Munition constituents: Preliminary sediment screening criteria for the protection of marine benthic invertebrates. *Chemosphere* **2010**, *81*, 807–816. [CrossRef] [PubMed]
39. Zhang, B.H.; Smith, P.N.; Anderson, T.A. Evaluating the bioavailability of explosive metabolites, hexahydro-1-nitroso-3,5-dinitro-1,3,5-triazine (MNX) and hexahydro-1,3,5-trinitroso-1,3,5-triazine (TNX), in soils using passive sampling devices. *J. Chromatogr. A* **2006**, *1101*, 38–45. [CrossRef] [PubMed]
40. Rosen, G.; Wild, B.; George, R.D.; Belden, J.B.; Lotufo, G.R. Optimization and field demonstration of a passive sampling technology for monitoring conventional munition constituents in aquatic environments. *Mar. Technol. Soc. J.* **2016**, *50*, 23–32. [CrossRef]

41. Morley, M.C.; Henke, J.L.; Speitel, G.E. Adsorption of rdx and hmx in rapid small-scale column tests: Implications for full-scale adsorbers. *J. Environ. Eng.-ASCE* **2005**, *131*, 29–37. [CrossRef]
42. Heilmann, H.M.; Wiesmann, U.; Stenstrom, M.K. Kinetics of the alkaline hydrolysis of high explosives RDX and HMX in aqueous solution and adsorbed to activated carbon. *Environ. Sci. Technol.* **1996**, *30*, 1485–1492. [CrossRef]
43. Millerick, K.; Drew, S.R.; Finneran, K.T. Electron shuttle-mediated biotransformation of hexahydro-1,3,5-trinitro-1,3,5-triazine adsorbed to granular activated carbon. *Environ. Sci. Technol.* **2013**, *47*, 8743–8750. [CrossRef] [PubMed]
44. Brown, R.A.; Mueller, J.G.; Seech, A.G.; Henderson, J.K.; Wilson, J.T. Interactions between biological and abiotic pathways in the reduction of chlorinated solvents. *Remedation* **2009**, *20*, 9–20. [CrossRef]
45. Butler, E.C.; Hayes, K.F. Kinetics of the transformation of trichloroethylene and tetrachloroethylene by iron sulfide. *Environ. Sci. Technol.* **1999**, *33*, 2021–2027. [CrossRef]
46. Butler, E.C.; Hayes, K.F. Factors influencing rates and products in the transformation of trichloroethylene by iron sulfide and iron metal. *Environ. Sci. Technol.* **2001**, *35*, 3884–3891. [CrossRef] [PubMed]
47. Ferrey, M.L.; Wilkin, R.T.; Ford, R.G.; Wilson, J.T. Nonbiological removal of cis-dichloroethylene and 1,1-dichloroethylene in aquifer sediment containing magnetite. *Environ. Sci. Technol.* **2004**, *38*, 1746–1752. [CrossRef] [PubMed]
48. Jeong, H.Y.; Hayes, K.F. Reductive dechlorination of tetrachloroethylene and trichloroethylene by mackinawite (FeS) in the presence of metals: Reaction rates. *Environ. Sci. Technol.* **2007**, *41*, 6390–6396. [CrossRef] [PubMed]
49. Jeong, H.Y.; Anantharaman, K.; Han, Y.S.; Hayes, K.F. Abiotic reductive dechlorination of *cis*-dichloroethylene by Fe species formed during iron- or sulfate-reduction. *Environ. Sci. Technol.* **2011**, *45*, 5186–5194. [CrossRef] [PubMed]
50. Lee, W.; Batchelor, B. Abiotic reductive dechlorination of chlorinated ethylenes by iron-bearing soil minerals. 1. Pyrite and magnetite. *Environ. Sci. Technol.* **2002**, *36*, 5147–5154. [CrossRef] [PubMed]
51. Lee, W.; Batchelor, B. Abiotic, reductive dechlorination of chlorinated ethylenes by iron-bearing soil minerals. 2. Green rust. *Environ. Sci. Technol.* **2002**, *36*, 5348–5354. [CrossRef] [PubMed]
52. Evans, P.J.; Nguyen, D.; Chappell, R.W.; Whiting, K.; Gillette, J.; Bodour, A.; Wilson, J.T. Factors controlling in situ biogeochemical transformation of trichloroethene: Column study. *Ground Water Monit. Remed.* **2014**, *34*, 65–78. [CrossRef]
53. Whiting, K.; Evans, P.J.; Lebron, C.; Henry, B.; Wilson, J.T.; Becvar, E. Factors controlling in situ biogeochemical transformation of trichloroethene: Field survey. *Ground Water Monit. Remed.* **2014**, *34*, 79–94. [CrossRef]
54. Kennedy, L.; Everett, J.W.; Gonzales, J. Aqueous and mineral intrinsic bioremediation assessment: Natural attenuation. *J. Environ. Eng.-ASCE* **2004**, *130*, 942–950. [CrossRef]
55. Wilkin, R.T.; Bischoff, K.J. Coulometric determination of total sulfur and reduced inorganic sulfur fractions in environmental samples. *Talanta* **2006**, *70*, 766–773. [CrossRef] [PubMed]
56. Wiedemeier, T.H.; Swanson, M.A.; Moutoux, D.E.; Gordon, E.K.; Wilson, J.T.; Wilson, B.H.; Kampbell, D.H.; Haas, P.E.; Miller, R.N.; Hansen, J.E.; et al. *Technical Protocol for Evaluating Natural Attenuation of Chlorinated Solvents in Ground Water*; United States Environmental Protection Agency: Washington, DC, USA, 1998.
57. Nobre, R.C.M.; Nobre, M.M.M.; Campos, T.M.P.; Ogles, D. In-situ biodegradation potential of 1,2-DCA and VC at sites with different hydrogeological settings. *J. Hazard. Mater.* **2017**, *340*, 417–426. [CrossRef] [PubMed]
58. Madsen, E.L. Epistemology of environmental microbiology. *Environ. Sci. Technol.* **1998**, *32*, 429–439. [CrossRef]
59. Baruthio, F. Toxic effects of chromium and its compounds. *Biol. Trace Elem. Res.* **1992**, *32*, 145–153. [CrossRef] [PubMed]
60. Grevatt, P.C. *Toxicological Review of Trivalent Chromium*; Agency, USEP: Washington, DC, USA, 1998.
61. Driscoll, S.K.; McArdle, M.E.; Plumlee, M.H.; Proctor, D. Evaluation of hexavalent chromium in sediment pore water of the Hackensack river, New Jersey, USA. *Environ. Toxicol. Chem.* **2010**, *29*, 617–620. [CrossRef] [PubMed]
62. Hopp, L.; Peiffer, S.; Durner, W. Spatial variability of arsenic and chromium in the soil water at a former wood preserving site. *J. Contam. Hydrol.* **2006**, *85*, 159–178. [CrossRef] [PubMed]

63. James, B.R. The challenge of remediating chromium-contaminated soil. *Environ. Sci. Technol.* **1996**, *30*, A248–A251. [CrossRef] [PubMed]
64. Barbee, G.C.; Brown, K.W. Comparison between suction and free-drainage soil solution samplers. *Soil Sci.* **1986**, *141*, 149–154. [CrossRef]
65. Basu, N.B.; Rao, P.S.C.; Poyer, I.C.; Annable, M.D.; Hatfield, K. Flux-based assessment at a manufacturing site contaminated with trichloroethylene. *J. Contam. Hydrol.* **2006**, *86*, 105–127. [CrossRef] [PubMed]
66. Johns, M.M.; Marshall, W.E.; Toles, C.A. Agricultural by-products as granular activated carbons for adsorbing dissolved metals and organics. *J. Chem. Technol. Biotechnol.* **1998**, *71*, 131–140. [CrossRef]
67. Otto, M.; Nagaraja, S. Treatment technologies for 1,4-dioxane: Fundamentals and field applications. *Remedation* **2007**, *17*, 81–88. [CrossRef]
68. Navalon, S.; Alvaro, M.; Garcia, H. Analysis of organic compounds in an urban wastewater treatment plant effluent. *Environ. Technol.* **2011**, *32*, 295–306. [CrossRef] [PubMed]
69. Otero, M.; Zabkova, M.; Rodrigues, A.E. Comparative study of the adsorption of phenol and salicylic acid from aqueous solution onto nonionic polymeric resins. *Sep. Purif. Technol.* **2005**, *45*, 86–95. [CrossRef]
70. Maloney, S.W.; Adrian, N.R.; Hickey, R.F.; Heine, R.L. Anaerobic treatment of pinkwater in a fluidized bed reactor containing GAC. *J. Hazard. Mater.* **2002**, *92*, 77–88. [CrossRef]
71. Parette, R.; Cannon, F.S.; Weeks, K. Removing low ppb level perchlorate, RDX, and HMX from groundwater with cetyltrimethylammonium chloride (CTAC) pre-loaded activated carbon. *Water Res.* **2005**, *39*, 4683–4692. [CrossRef] [PubMed]
72. Kim, J.G.; Dixon, J.B. Oxidation and fate of chromium in soils. *Soil Sci. Plant Nutr.* **2002**, *48*, 483–490. [CrossRef]
73. Daoud, W.; Ebadi, T.; Fahimifar, A. Optimization of hexavalent chromium removal from aqueous solution using acid-modified granular activated carbon as adsorbent through response surface methodology. *Korean J. Chem. Eng.* **2015**, *32*, 1119–1128. [CrossRef]
74. Di Natale, F.; Lancia, A.; Molino, A.; Musmarra, D. Removal of chromium ions form aqueous solutions by adsorption on activated carbon and char. *J. Hazard. Mater.* **2007**, *145*, 381–390. [CrossRef] [PubMed]
75. Fetter, C.W. *Contaminant Hydrology*; Prentice Hall: Upper Saddle River, NJ, USA, 1999.
76. Han, I.; Schlautman, M.A.; Batchelor, B. Removal of hexavalent chromium from groundwater by granular activated carbon. *Water Environ. Res.* **2000**, *72*, 29–39. [CrossRef]
77. Piazzoli, A.; Antonelli, M. Feasibility assessment of chromium removal from groundwater for drinking purposes by sorption on granular activated carbon and strong base anion exchange. *Water Air Soil Pollut.* **2018**, *229*, 17. [CrossRef]
78. Satapathy, D.; Natarajan, G.S.; Patil, S.J. Adsorption characteristics of chromium(VI) on granular activated carbon. *J. Chin. Chem. Soc.* **2005**, *52*, 35–44. [CrossRef]
79. Singha, S.; Sarkar, U. Analysis of the dynamics of a packed column using semi-empirical models: Case studies with the removal of hexavalent chromium from effluent wastewater. *Korean J. Chem. Eng.* **2015**, *32*, 20–29. [CrossRef]
80. Song, H.O.; Yao, Z.J.; Shuang, C.D.; Li, A.M. Accelerated removal of nitrate from aqueous solution by utilizing polyacrylic anion exchange resin with magnetic separation performance. *J. Ind. Eng. Chem.* **2014**, *20*, 2888–2894. [CrossRef]
81. Song, H.O.; Yao, Z.J.; Wang, M.Q.; Wang, J.N.; Zhu, Z.L.; Li, A.M. Effect of dissolved organic matter on nitrate-nitrogen removal by anion exchange resin and kinetics studies. *J. Environ. Sci.* **2013**, *25*, 105–113. [CrossRef]
82. Primo, O.; Rivero, M.J.; Urtiaga, A.M.; Ortiz, I. Nitrate removal from electro-oxidized landfill leachate by ion exchange. *J. Hazard. Mater.* **2009**, *164*, 389–393. [CrossRef] [PubMed]
83. Guimaraes, D.; Leao, V.A. Fundamental aspects related to batch and fixed-bed sulfate sorption by the macroporous type 1 strong base ion exchange resin Purolite A500. *J. Environ. Manag.* **2014**, *145*, 106–112. [CrossRef] [PubMed]

14

Spatial Pattern of Bacterial Community Diversity Formed in Different Groundwater Field Corresponding to Electron Donors and Acceptors Distributions at a Petroleum-Contaminated Site

Zhuo Ning [1,2,3], Min Zhang [1,*], Ze He [1,3], Pingping Cai [1,4], Caijuan Guo [1] and Ping Wang [1,2,3]

1 Institute of Hydrogeology and Environmental Geology, Chinese Academy of Geological Sciences, Shijiazhuang 050061, China; ningzhuozhuo@163.com (Z.N.); heze25@163.com (Z.H.); cppyjy@163.com (P.C.); caizidongdong@163.com (C.G.); shuiwp@126.com (P.W.)

2 School of Chinese Academy of Geological Sciences, China University of Geosciences (Beijing), Beijing 100083, China

3 Key Laboratory of Groundwater Remediation of Hebei Province, Zhengding 050083, China

4 School of Resources and Environmental Engineering, HeFei University of Technology, Hefei 230009, China

* Correspondence: minzhang205@live.cn

Abstract: The benefits of an electron-transfer mechanism for petroleum biodegrading have been widely acknowledged, but few have studied the spatial pattern of microbial community diversity in groundwater fields, and few discuss the bacterial community's diversity in relation to electron donors-acceptors distribution, which is largely determined by groundwater flow. Eleven samples in different groundwater fields are collected at a petroleum-contaminated site, and the microbial communities are investigated using 16S rRNA gene sequences with multivariate statistics. These are mainly linked to the chemical composition analysis of electron donor indexes COD, BTEX and electron acceptor indexes DO, NO_3^-, Fe^{2+}, Mn^{2+}, and SO_4^{2-}, HCO_3^-. The spatial pattern of the bacterial community's diversity is characterized and the effect of the electron redox reaction on bacterial community formation in different groundwater field zones is elucidated. It is found that a considerable percentage (>65%) of the bacterial communities related to petroleum degrading suggest that petroleum biodegrading is occurring in groundwater. The communities are subject to the redox reaction in different groundwater field zones: The side plume zone and the upstream of the source zone are under aerobic redox or denitrification redox, and the corresponding bacteria are *Rhodoferax*, *Novosphingobium*, *Hydrogenophaga*, and *Comamonas*; the source zone and downstream of the source zone are under Fe^{3+}, Mn^{4+}, and SO_4^{2-} reduction redox, and the corresponding bacteria are Rhodoferax, Treponema, Desulfosporosinus, Hydrogenophaga, and Acidovorax. These results imply that groundwater flow plays a definitive role in the bacterial community's diversity spatial pattern formation by influencing the distribution of electron donor and acceptor.

Keywords: groundwater; bacterial community diversity; petroleum contamination; electron acceptor; electron donor; groundwater flow path

---

1. Introduction

Due to its toxicity to humans and other organisms, there is much concern about petroleum-contaminated groundwater [1,2]. Various methods, including pump-and-treat (PandT) and other in situ technologies have been used to remediate petroleum-contaminated groundwater [3]. As an environmentally sound and cost-effective technology in various applications, in situ bioremediation, especially natural attenuation, has gained considerable attention. During bioremediation, petroleum degrading microorganisms can mineralize petroleum components as their carbon source and electron donors, electron acceptors redox reaction occurs and the contaminants are removed [4]. Bacterial community diversity and electron-transfer research are essential for bioremediation [5,6].

Research on the diversity of bacterial communities has been carried out at petroleum-contaminated sites since the 1970s [7], and the amounts have grown rapidly as the 16S rRNA gene sequences mature and become universal; most were in soils and sediments [8–12]. A few researchers focusing on groundwater such as Anne Fahy and Zhaoxian Zheng, investigated the relationships between the bacterial community structures and the groundwater geochemistry respectively [13,14], and found that hydrocarbon metabolism would vary the diversity of the bacterial communities. Ai-xia Zhou researched the responses of microbial communities to seasonal fluctuations in groundwater level and found that groundwater-table fluctuations would affect the distribution, transport, and biodegradation of the contaminants [15]. Petroleum compounds can be transported from the source area in groundwater, with the result that the petroleum concentrations, redox conditions, biogeochemical processes, and bacterial communities would vary along the groundwater flow path [16]. This view can be supported by other related research, such as from Karolin Tischer, Etienne Yergeau and C.E. Main who have reported that contaminant concentrations have a significant influence on microbial communities in various environmental mediums [8,17,18]. During petroleum biodegradation, electron donors and acceptors dissolved in the groundwater are consumed. Generally speaking, electron acceptors are usually consumed in the following order: O_2, NO_3^-, Mn^{4+}, Fe^{3+}, SO_4^{2-}, and HCO_3^- (or CO_2), and these electron acceptors, other than HCO_3^- (or CO_2), are transported in groundwater from uncontaminated groundwater upstream under little vertical recharge conditions. The electron acceptors would be used according to their redox potential and the concentrations of electron acceptors often vary with the groundwater flow path in contaminated sites [19]. Research has shown that certain electron acceptors affect bacterial communities [20,21]. There is enough dissolved oxygen in the edge of contamination plume, and the bacteria are mainly aerobic, while in the source area the bacteria would mainly become methanogens and sulfate-reducing bacteria since the other electron acceptors were already exhausted [22]. There is not much doubt that redox zonation and microbial changes along the path of groundwater plumes are present.

However, for a petroleum-contaminated site bioremediation, microorganisms have cooperative and competitive relationships, not only along a groundwater path. To develop an effective remediation scheme, information about the abundance of petroleum degradation microorganisms from the overall bacterial community diversity is required. The spatial pattern of the diversity of the bacterial community in different groundwater fields of the entire contaminated groundwater area should be well described, and a particularly detailed response regarding the relationship between the bacterial community and electron-transfer will provide field case support for microbial functional gene identification.

Eleven groundwater samples from different places along and beside the groundwater flow path in a petroleum-contaminated aquifer were collected. Then high throughput sequencing of 16S rRNA genes was used to investigate the diversity of the bacterial communities in the samples. The relationships between the bacterial communities and the electron acceptors and donors in the different groundwater fields were assessed. The reasons for a different redox zonation are expected to be gained based on water geochemistry (electron acceptors and available hydrocarbons) and phylogenetic types of microorganisms in the groundwater.

2. Materials and Methods

2.1. Site Description and Sampling Procedure

The contaminated site was located in the northern part of the North China Plain, which was formerly a chemical plant that was contaminated when petroleum leaked from a storage tank more than forty years ago. While the pollution had been removed from the surface, the groundwater and subsurface soils were still seriously contaminated. It reported that the main contaminates were monocyclic aromatics and aliphatic hydrocarbon. More than 11 wells were drilled in the site to survey and monitor the groundwater contamination and then remediate it. The samples were collected before the remediation. The aquifer at the site was mainly composed of sandy gravel and sand, and there was no clay layer to prevent contamination of the vadose zone. Previous studies reported that the contaminated aquifer was unconfined and the depth to the groundwater table was approximately 25 m. The groundwater naturally flowed from northwest to southeast, regionally. Since the site was located in the urban area where there was usage of groundwater, the flow direction varied slightly with time. The flow was from west to east on the sampling days in the site. An area of 400 m^2 with 11 wells around the petroleum leak was established for the purposes of this study (Figure 1).

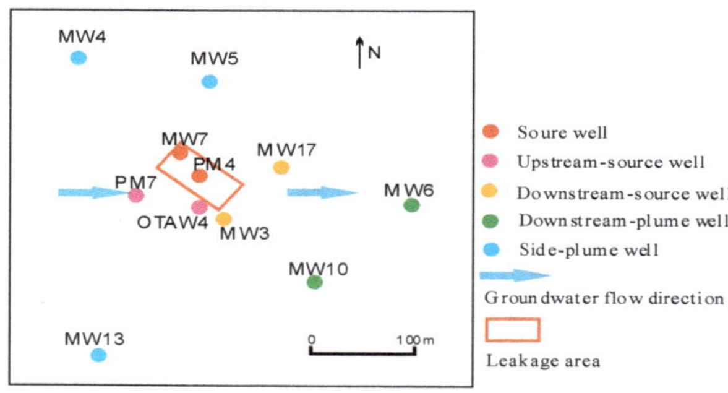

Figure 1. Diagram showing the petroleum-contaminated area, the monitoring wells, and the groundwater flow direction.

Information about the water temperature (t), pH, electrical conductivity (EC), dissolved oxygen (DO), and oxidation-reduction potential (ORP) in each well was recorded before collecting the groundwater samples. The groundwater samples were considered to be representative when the values of T, pH, EC, DO, and ORP in three successive samples were within ± 1 °C, ± 0.2, $\pm 3\%$, $\pm 10\%$ or ± 0.2 mg/L, and ± 20 mV, respectively. The groundwater samples were collected by sterile bailers. When sampling, groundwater 5 L was collected into a sterilized 5 L plastic bucket and stored on ice in an incubator. The water samples were transported to the laboratory, and the bacteria were collected into 5 PTFE filter membranes with a pore size of 0.22 μm by air pump filtration in one day. The filter membranes were stored at -80 °C until DNA extraction. Other portions of the samples were collected into 500 mL plastic bottles and 40 mL amber glass bottles for inorganic and organic analyses, respectively.

The sampling wells were divided into five groups by the groundwater flow direction and the location of the contamination source. Since the flow direction varied a little with time, the concentrations of contaminants were also considered during the grouping. The upstream-source group included samples PM7 and OTAW4, the source group included MW7 and PM4, the downstream-source group included MW3 and MW17, the downstream-plume group included samples MW6 and MW10, and the side-plume group included samples MW4, MW5, and MW13.

2.2. Chemical Analyses

Considering that monocyclic aromatics are related to their toxicity and relatively high solubility [23], and the chemical oxygen demand (COD) is always used to quantify the amount of organics in petroleum [24], the concentrations of monocyclic aromatics and COD were monitored as the contamination indexes during the study.

The concentrations of monocyclic aromatics, such as toluene, ethylbenzene, m-xylene, p-xylene, and o-xylene, were determined as outlined in US EPA Method 8260 [25]. Concentrations of other variables, namely chemical oxygen demand (COD), total dissolved solids (TDS), pH, Ca^{2+}, Mg^{2+}, Na^+,

K^+, NO_3^-, SO_4^{2-}, HCO_3^-, Cl^-, Fe^{2+}, Mn^{2+}, and CO_2, in the groundwater were measured following the Standard Methods [26].

2.3. DNA Extraction, PCR Amplification, Library Construction and Sequencing

DNA were extracted from each sample using an EZNA™ Soil DNA Kit (OMEGA bio-tek, Norcross, GA, USA) following the manufacturer's protocol. The V4–V5 region of bacterial 16S-rRNA genes was amplified using the universal primers 515F (GTGCCAGCMGCCGCGGTAA) and 926R (CCGTCAATTCMTTTRAGTTT) [27]. The PCR analysis was carried out in the following order: Initial denaturation at 98 °C for 2 min, 30 cycles of denaturation at 98 °C for 15 s, annealing at 55 °C for 30 s, extension at 72 °C for 30 s, and a final extension at 72 °C for 5 min. Libraries were sequenced by a sequencing platform (HiSeq 2500) at Personalbio-Shanghai, Shanghai, China.

2.4. Bioinformatics Analysis

Raw sequences were filtered and then high-quality reads were assigned to operational taxonomic units (OTUs) [28]. Then the OTUs were subsampled to the minimum reads. Various alpha-diversity indexes (observed species [Sobs], the Chao estimate, abundance-based coverage estimator [ACE], the Shannon and Simpson diversity indexes) were used to evaluate the species information.

To determine the influence of the petroleum on the bacteria, samples were divided into three groups according to the degree of COD contamination. The samples with COD concentrations less than 10 mg/L, between 10 mg/L and 100 mg/L, and greater than 100 mg/L were classified as having low contamination, medium contamination, and high contamination, respectively. The authors then used Venn diagrams to examine the bacterial communities in the groups by the contamination level and by the groundwater flow areas.

The authors carried out principal coordinate analysis (PCoA) of the microbial communities using unweighted unifrac with full trees at the genus level. The authors used redundancy analysis (RDA) to determine which environmental variables were associated with changes in the structures of the bacterial community. The environmental variables were divided into two groups. One group contained the contaminant compounds, organic index (COD), and the electron acceptors, such as DO, NO_3^-, SO_4^{2-} and HCO_3^-, or metabolic by-products, Mn^{2+} and Fe^{2+}, involved in microbial degradation. The other group contained the pH, TDS, and the major ions in groundwater, such as Ca^{2+}, Mg^{2+}, Na^+, K^+, NO_3^-, SO_4^{2-}, HCO_3^-, and Cl^-.

The bioinformatics analyses were carried out in the cloud platform of majorbio (http://www.i-sanger.com) according to the programs as the reference mentioned [29].

3. Results

3.1. The Distribution of Electron Acceptors-Donors and Other Chemical Parameters

The concentrations of electron acceptors and donors and other chemical parameters in the groundwater are shown in Table 1. The concentrations of the contaminants, i.e., electron donors and COD, show that the groundwater was seriously contaminated by petroleum and that the contamination varied in the different areas. The contamination was highest in the source area and decreased (in order) in the downstream-source, downstream-plume, upstream-source, and side-plume areas. Apart from the side-plume wells, the DO and NO_3^- concentrations were less than 2 mg/L and 12 mg/L, respectively. The concentrations of HCO_3^-, K^+, Na^+, Ca^{2+}, Mg^{2+}, Cl^-, TDS, and CO_2 were lower in the upstream-source wells.

Table 1. Hydrochemical parameters and the contaminant concentrations.

Location		Side-Plume			Upstream-Source		Source		Downstream-Source		Downstream-Plume	
Well		**MW5**	**MW4**	**MW13**	**PM7**	**OTAW4**	**MW7**	**PM4**	**MW3**	**MW17**	**MW6**	**MW10**
Electron donors (μg·L^{-1})	toluene	97	4	6	315	108	11,211	20,610	723	11,680	7689	287
	ethylbenzene	11.1	6.9	0.1	51.7	20	4316	583	832.9	2078	668	971.5
	m(p)-xylene	22.21	4.81	1.24	58.3	25	3001	3678	579.17	1636.5	1115	759.8
	o-xylene	22.22	2.13	0.66	56.67	8.3	3001	3680	320.2	1636.5	225	89.45
Electron acceptors (mg·L^{-1})	DO	2.38	2.19	1.95	0.87	0.76	1.29	0.84	1.3	0.85	1.97	1.67
	NO_3^-	16.77	68.96	4.6	1.78	<0.20	1.75	1.76	1.76	1.75	11.5	7.73
	SO_4^{2-}	107.2	277.6	39.67	38.15	69.22	83.79	66.7	21.54	16.68	163.3	98.7
	HCO_3^-	791.6	780.1	889.1	316.7	494.5	648.5	715.5	712.5	831.8	822.1	767.3
metabolic by-products (mg·L^{-1})	Fe^{2+}	0.049	0.018	0.372	0.127	0.061	0.489	0.806	0.587	4.643	2.305	1.586
	Mn^{2+}	6.48	0.625	1.953	0.755	0.781	2.356	2.5	1.473	2.589	3.518	2.862
Other ion (mg·L^{-1})	K^+	3.14	3.35	2.49	1.58	1.91	2.2	1.72	1.43	2.17	2.57	1.01
	Na^+	179.3	135.9	149.4	67.96	135.8	133.6	137.8	145.6	165.4	124.1	152.4
	Ca^{2+}	282.9	255.9	141.2	61.56	99.63	151.4	162.3	108.9	139.4	209.4	172.8
	Mg^{2+}	121.4	93.74	61.81	21.59	42.24	62.44	59.96	41.58	61.32	81.87	60.19
	Cl^-	583.6	214.6	134.5	61.32	162.7	198.6	195.1	141.8	191.6	182.9	193.4
	TDS	1675	1439	975.3	412.3	758.8	958.1	983.1	818.9	994.3	1177	1064
COD (mg·L^{-1})		4.37	1.65	4.54	6.19	56.23	162.4	268.7	56.23	337.4	40.63	31.24
pH		6.94	7.26	7.22	7.59	7.66	7.15	7.02	7.22	6.94	7.15	7.28

3.2. Alpha-Diversity Indexes

Alpha-diversity indexes of the bacterial communities in the 11 samples are shown in Table 2. The good coverage index (>0.995) showed that the obtained reads in the study were representative. The Sobs, Shannon, ACE, and Chao indexes were higher, while the Simpson index was lower, in the samples from the side-plume than in the other samples.

Table 2. Alpha-diversity indexes.

Location	Sample	Diversity Indexes					
		Sobs	**Shannon**	**Simpson**	**Ace**	**Chao**	**Coverage**
Side-plume	MW5	383	3.73	0.07	428.3	443.86	0.995
	MW4	324	3.59	0.08	356.17	358.89	0.996
	MW13	249	2.82	0.12	317.9	314.27	0.995
Upstream-source	PM7	211	2.89	0.12	260.23	262.21	0.996
	OTAW4	191	2.76	0.15	240.13	226.25	0.997
Source	MW7	197	2.79	0.12	246.43	238.44	0.996
	PM4	186	2.7	0.2	341.12	294.48	0.995
Downstream-source	MW3	190	2.61	0.17	287.61	263.75	0.995
	MW17	160	2.18	0.2	204.92	196.96	0.997
Downstream-plume	MW6	215	3.44	0.07	266.41	291.56	0.996
	MW10	204	3.05	0.09	256.63	255.11	0.996

3.3. Community Composition

The compositions of the bacterial communities in the 11 samples at the genus level are shown in Figure 2.

Sequences representing Dechloromonas were the most abundant bacterium and accounted for 23% of all bacterial sequences in MW7. Dechloromonas was also detected and accounted for between 3% and 9% of all the bacteria, in PM4 and in the samples collected from the downstream-source and downstream-plume areas.

Sequences representing *Acidovorax* were most the abundant, and accounted for 43% of all bacteria, in PW4. They were also present at abundances greater than 10% in MW7, and in samples from the downstream-source (MW17 and MW3) and downstream-plume (MW6) areas. These sequences were also present at lower abundances in other samples.

Sequences representing *Hydrogenophaga* were most the abundant in OTAW4, MW13, MW17, and MW3, where they accounted for 33%, 45%, 38%, and 37%, respectively. They were also present at abundances of between 3% and 20% in all other samples, except those from the contamination source area (samples MW7 and PM4).

Sequences representing *Comamonas,* present in all samples, were the most abundant in samples PM7 and MW4, where they accounted for 23% and 19% of all bacteria, respectively. They were most abundant in the upstream source area, and then decreased (in rank order) in the side-plume, downstream-source, downstream-plume, and were least abundant in the contaminant source areas.

Sequences representing *Rhodoferax* were the most abundant in MW5, MW6, and MW10, where it accounted for 23%, 18%, and 23% of total bacteria, respectively. Except for sample PM4, *Rhodoferax* accounted for more than 10% of all bacteria in the samples.

Sequences representing other bacteria were present at higher abundances (>5%) in certain samples. *Pseudomonas* was in all the samples, but only had abundances of more than 10% in samples from the source area (MW7 and PM4) and in MW10, for example. Treponema was detected in all samples and, apart from MW7 and MW10 where it had abundances of 7% and 10%, respectively, its abundances were less than 5%. Present in all samples, *Novosphingobium* was more abundant in the upstream-source area samples (OTAW4 (12%) and PM7 (8%)) than in other areas, where its abundances were less than 5%. *Pseudoxanthomonas* was present in samples PM4 and MW7 from the source area at abundances of 6% and 9%, respectively, and was either present at abundances of less than 1% in, or was absent from, the plume area. *Zavarzinia* was most abundant in MW4 (12%) and MW6 (8%) and was present in other samples at abundances of less than 1%. *Sulfuritalea* was detected in all samples. Apart from MW5, where its abundance was 13%, its abundances were less than 2%. *Sulfuricurvum* had abundances of 7% in MW5 and MW10, and of about 1% in MW3 and MW6. *Desulfosporosinus* was only present in MW6 and MW10 at abundances of 2% and 8%, respectively. *Nitrospira* was found in MW5 and MW4 only at abundances of 5% and 1%, respectively. Norank_p__Omnitrophica was only found in MW5 at an abundance of 5%.

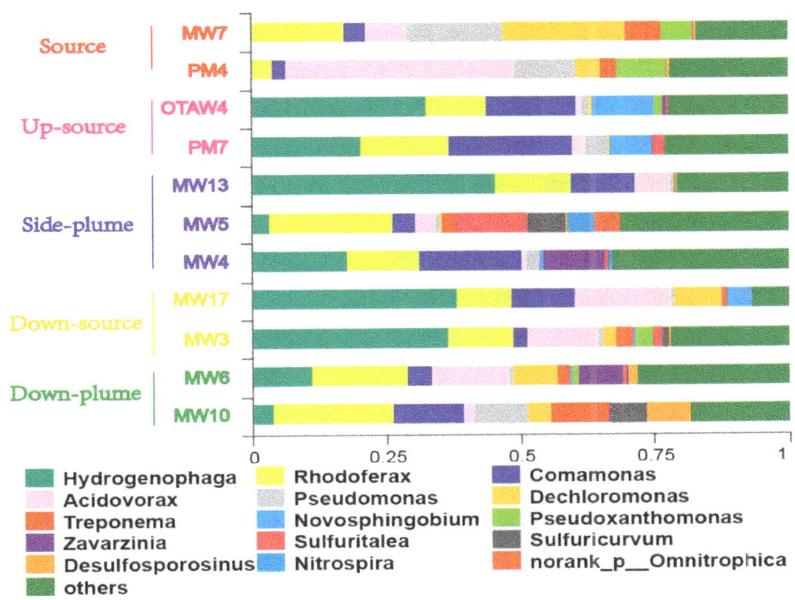

Figure 2. Bar chart of community abundances (greater than 5%) at the genus level.

3.4. Relationships between Bacterial Communities among Samples

Venn diagrams showed the number of species that were unique to, or shared between, the different groups. The species are presented according to the different degrees of petroleum contamination in Figure 3a. The 3 contamination classes shared 214 species, and the group with low contamination had 191 unique species. This figure demonstrates the reduced bacterial diversity in petroleum -contaminated wells. The species found in the different parts of the groundwater flow field are shown in Figure 3b. The different parts shared 98 species, and the groundwater from the side-plume area had the greatest number of unique species.

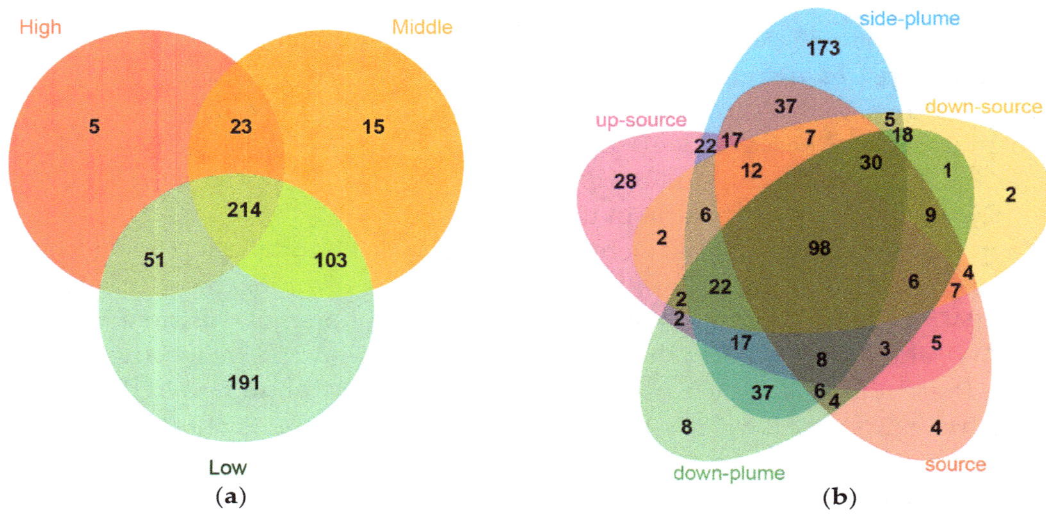

Figure 3. Venn diagrams showing the bacteria by (**a**) the degree of contamination and (**b**) locations.

The results of the PCoA are shown in Figure 4. The samples from the same groundwater flow area were grouped together. The side-plume group was grouped to the left of the graph along the first principal component (PC1), while the other samples were grouped to the right of the graph. The samples from the upstream-source, downstream-plume, downstream-source, and source areas were grouped along the second principal component (PC2).

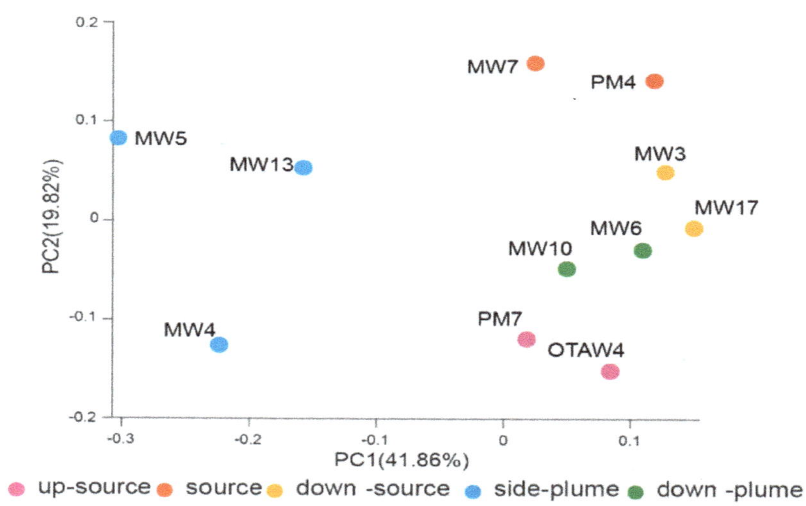

Figure 4. The PCoA plot of samples at the genus level.

The RDA plot (Figure 5) indicated that samples PM4 and MW7 from the contamination source area and sample MW6 from the downstream-plume were associated with high concentrations of COD, toluene, xylene, and ethylbenzene. Samples MW5 and MW10 were associated with high SO_4^{2-} and DO, samples MW4 and PM7 were associated with high NO_3^- and DO, sample MW13 was associated with low Mn^{2+}, and samples MW3 and MW17 were associated with high Fe^{2+} and low SO_4^{2-}. The upstream-source samples were associated with high pH values.

Most of the genera were clustering in the center of the plot, which indicates that these bacteria were present in all samples at similar abundances. The relative abundance of Acidovorax increased as the COD increased. *Pseudoxanthomonas* was positively correlated with toluene and xylene, and *Novosphingobium* and *Comamonas* were negatively correlated with toluene and xylene. *Dechloromonas* and *Pseudomonas* were positively correlated with ethylbenzene. *Hydrogenophaga* was negatively correlated with bivalent manganese (Mn^{2+}). *Rhodoferax* was negatively correlated with Fe^{2+} and positively correlated with DO.

The index of the organic content, COD, was positively correlated with various contaminants (toluene, xylene, and ethylbenzene) and HCO_3^-, Fe^{2+}, and Mn^{2+}, and was negatively correlated with SO_4^{2-}, DO, and NO_3^-. The pH was negatively correlated with TDS, Mg^{2+}, Ca^{2+}, and Cl^-.

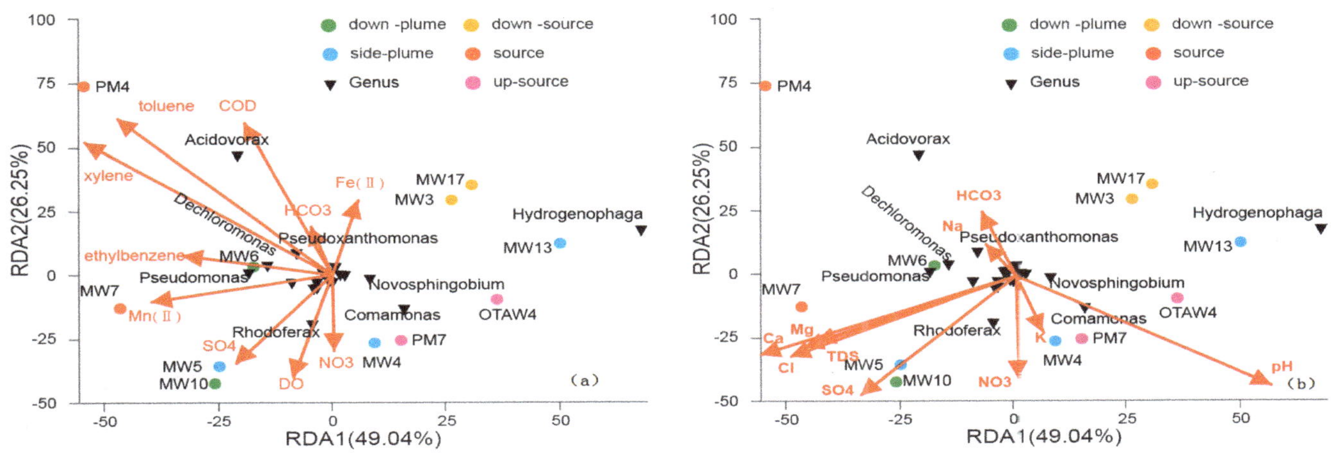

Figure 5. Redundancy analysis (RDA) of the relationship between the groundwater parameters and the relative abundance of the bacterial genus of the collected samples. The contamination indexes and the electron acceptors are mainly shown in (**a**); while the major ions, TDS, and pH are mainly shown in (**b**).

4. Discussion

4.1. Variations in Bacterial Communities with Electron Donor Concentrations

Analysis of the community diversity of all the groundwater samples showed that the bacterial communities of samples within the same groundwater flow area were similar. Most of the bacteria that were present at abundances of more than 5% in this contaminated aquifer were related to the degradation of hydrocarbons, especially aromatic hydrocarbons. The bacterial communities varied in different areas.

Hydrogenophaga was not present in the source zone (MW7 and PM4). Hydrogenophaga can metabolize various organic compounds, such as polycyclic aromatics and toluic acid, but not toluene and xylene [30]. Researches showed that high concentrations of toluene and xylene may have negatively impact it [31], *Hydrogenophaga* tends be more abundant where toluene and xylene concentrations are low and COD concentrations are high. *Comamonas* can also degrade polycyclic aromatic hydrocarbons [32], but its abundances were lowest in the source area (samples MW7 and PM4). This bacterium might be harmed by high concentrations of hydrocarbons. While *Rhodoferax* can degrade benzene [33], it was least abundant in PM4 where the toluene contamination was greatest, which might indicate that this bacterium is sensitive to toluene. *Rhodoferax* was correlated with increases in DO in the RDA plot, which suggests that its growth may also be limited by oxygen. *Acidovorax*, *Pseudomonas*, *Dechloromonas*, and *Pseudoxanthomonas*, also capable of degrading hydrocarbons, were abundant in the highly contaminated groundwater in the source zone. *Acidovorax* can use various PAHs [34]. The species of *Pseudomonas* in the study site was *Pseudomonas mendocina*, which can degrade toluene [35]. The *Dechloromonas* in the aquifer shared a 99% sequence similarity with *Dechloromonas hortensis* and *Dechloromonas denitrificans*, which can use ClO_4^-, ClO_3^-, NO_3^-, and O_2 as electron acceptors to oxidize organic compounds, such as aromatic hydrocarbons [36,37]. *Pseudoxanthomonas* can produce biosurfactants and can be used to degrade BTEX [38,39]. These four bacteria, *Acidovorax*, *Pseudomonas*, *Dechloromonas* and *Pseudoxanthomonas*, may be more tolerant to petroleum than the other bacteria, as shown in the RDA analysis. *Novosphingobium* can also degrade aromatic hydrocarbons [40,41] and was more abundant in upstream-source zone samples (OTAW4 and PM7) than in the other areas. *Treponema*, previously discovered in hydrocarbon-contaminated sediments, can degrade hydrocarbons [8,42] and was present at relatively high abundances in MW7 and MW10. *Zavarzinia*, which can degrade benzene in aerobic conditions [43], was mainly found in samples MW4 and MW6 from the plume area, where the contents of dissolved oxygen were relatively high. Apart from sample MW5 from the side-plume, these 10 hydrocarbon degrading bacteria mentioned above accounted for more than 65% of all bacteria in all the samples.

The high abundance of degradation bacteria in the groundwater samples, combined with electron acceptors-donors, indicates that natural attenuation was occurring at this site. Venn diagrams (Figure 3) and the α-diversity indexes (Table 2) showed that a considerable number of species were sensitive to petroleum organics. The values of the Sobs, Chao, ACE, and Shannon indexes were higher, while the Simpson index was lower, in the samples from the side-plume than in the other samples. The Sobs, Chao, and ACE indexes are used to reflect the richness of species, and the Shannon and Simpson indexes reflect both the richness and the evenness of the species in samples [44]. The Shannon index has been reported to be more sensitive to changes in richness, while the Simpson index is more sensitive to the evenness [45]. Therefore, it can be said that, consistent with other studies [11,46], contamination can result in a decline of diversity and increase of abundances of dominant microorganisms. The RDA plots indicated that, of all the environmental factors, the toluene, xylene, and COD had the most effect on the bacterial communities.

4.2. The Influence of Electron Acceptors on Bacterial Communities

The electron acceptors were being consumed as the petroleum was being degraded. The electron acceptors were consumed in a certain order, with O_2 consumed first, followed by NO_3^-, Mn^{4+}, Fe^{3+}, SO_4^{2-}, and HCO_3^-, during aerobic reaction, denitrification, Mn^{4+} reduction, Fe^{3+} reduction, SO_4^{2-} reduction, and methane production, respectively [16].

Samples MW3 and MW17 were associated with high Fe^{2+} and low SO_4^{2-}, which indicated that the electron acceptors Fe^{3+} and SO_4^{2-} had been consumed, these two samples probably were in the methanogenesis stage. These two samples had the highest abundances of *Hydrogenophaga*, which is always closely related with methanogenic archaea [47,48]. Other studies have suggested that Hydrogenophaga might have catalyzed hydrogen production or perhaps was an oxygen scavenger, and that it created the strictly anoxic conditions essential for the methanogenic archaea [49]. *Hydrogenophaga* can produce and use hydrogen, one of the substrate of methanogenesis, as its energy source, which suggests that there might be cooperative relationships between *Hydrogenophaga* and methanogenic archaea. Sample PM4 was associated with high Fe^{2+} and Mn^{2+}, so either Fe^{3+}, Mn^{4+}, and SO_4^{2-} reduction or methanogenesis might have dominated. However, to the best of the authors' knowledge, the two most abundant bacteria in the sample, *Acidovorax* and *Pseudomonas*, cannot use Fe^{3+}, Mn^{4+}, or SO_4^{2-} as their electron acceptors [50,51], which suggests that methanogenesis might have also dominated in this area. Other studies have shown that *Acidovorax*, a facultative aerobic microorganism that can use carboxylic acids as a carbon source, was closely associated with methanogenic archaea [49]. The sequences that represent syntrophic bacterial species, *Syntrophus*, *Syntrophomonas*, *Syntrophobacter* and *Pelotomaculum*, which usually provide hydrogen and carbon dioxide to methanogens, were found in these samples (See OTU table in Supplementary Materials).

Samples MW7 and MW6 were associated with high Mn^{2+} and SO_4^{2-}, which suggests that these samples were in the Fe^{3+} reduction phase and had the potential to reduce the SO_4^{2-}. *Dechloromonas*, *Rhodoferax*, and *Pseudomonas mendocina* were the most abundant bacteria in these two samples.

Dechloromonas often use NO_3^- and O_2 as electron acceptors, and are not known to grow in Fe^{3+} or SO_4^{2-} reducing conditions, or through syntrophic interactions with methanogenic bacteria [52]. *Rhodoferax* can use Fe^{3+}, NO_3^-, and O_2, but not SO_4^{2-} as electron acceptors [53]. *Pseudomonas mendocina* tested positive for oxidase and NO_3^- reduction (assimilatory), but negative for dissimilatory NO_3^-, Fe^{3+}, and SO_4^{2-} reduction [54]. While these bacteria cannot use SO_4^{2-} as an electron acceptor, *Treponema* [55] and *Desulfosporosinus* [56], found in samples in MW6 and MW7, can reduce SO_4^{2-}.

Samples MW5, MW10, and MW4 were associated with high SO_4^{2-}, DO, and NO_3^-, which indicated that these samples were in the aerobic or denitrification phase. *Rhodoferax,* as shown in the RDA plot, was related with these samples. The groundwater samples contained some special bacteria. *Sulfuritalea,* was detected in all samples at abundances less than 2%, apart from MW5, which had a *Sulfuritalea* content of 13%. This bacterium was isolated from the water of a freshwater lake. It can oxidize thiosulfate, elemental sulfur, and hydrogen as sole energy sources for autotrophic growth and can use NO_3^- as an electron acceptor [57]. *Sulfuricurvum,* which had relatively high abundances in MW5 and MW10, is a facultatively anaerobic, chemolithoautotrophic, sulfur-oxidizing bacterium [58]. *Desulfosporosinus,* was also abundant in MW10, which suggests that there was strong SO_4^{2-} reduction in MW10. *Nitrospira,* was only present in MW5 (5%) and MW4 (<1%). It is a ubiquitous bacterium that can oxidize the NO_2^- into NO_3^-, which is an aerobic process [59]. The relatively high oxygen contents in the two samples meant that there was enough oxygen for this bacterium to survive. Norank_p__Omnitrophica only appeared in MW5. More than 10 OTUs corresponded to this bacterium. The function of this bacterium is not well defined; it was previously identified in an anammox community [60] and might be related to Magnetotactic bacteria [61].

Samples PM7 and OTA4 were associated with high NO_3^- and low Mn^{2+}, which suggests that these samples were in the denitrification phases. The positions of the bacteria *Comamonas,* and *Novosphingobium* on the RDA plot, confirm this view. *Comamonas* can reduce Fe^{3+}, Mn^{4+}, and NO_3^- [62,63], but not SO_4^{2-}, and *Novosphingobium* can reduce NO_3^- [40]. The HCO_3^- contents in PM7 and OTA4 were lower and the pH values were higher than in the other samples. When the pH is higher, $CaCO_3$ and $MgCO_3$ precipitate more readily [64], resulting in lower TDS, Mg^{2+}, Ca^{2+} and HCO_3^- concentrations in the water.

Sample MW13 was associated with low Mn^{2+} and closely related with *Hydrogenophaga,* which can oxidize Mn^{2+} [65] also. During this process, O_2 and NO_3^- can be used as electron acceptors [48,66]. This suggests that the O_2 and NO_3^- were adequate and MW13 was in the aerobic reaction or denitrification stage.

5. Conclusions

The bacterial community diversity varied within different groundwater flow fields. Suspected hydrocarbon degrading bacteria accounted for a considerable percentage of these bacterial communities, which indicates that petroleum biodegradation potential was great. The different bacterial communities corresponded to different redox reaction stages in different locations. Generally, the side-plume and upstream-source samples were in the aerobic or denitrification stages, and the corresponding bacteria were *Rhodoferax, Novosphingobium, Hydrogenophaga,* and *Comamonas.* Samples from the source and the downstream-source areas were related to Fe^{3+}, Mn^{4+}, and SO_4^{2-} reduction and likely methanogenesis, and the corresponding bacteria were *Rhodoferax, Treponema, Desulfosporosinus, Hydrogenophaga* and *Acidovorax.* It was proposed that spatial patterns of bacterial communities were determined by groundwater flow for its influence on the distribution of electron donors and acceptors in this petroleum-contaminated aquifer.

Author Contributions: M.Z. and Z.H. Conceived and designed the experiments; M.Z., Z.H., Z.N., P.C., C.G. and P.W. collected the samples and performed the experiments; Z.N. and M.Z. analyzed the data and wrote the paper.

Acknowledgments: We thank Deborah Ballantine, PhD, from Liwen Bianji, Edanz Editing China (www.liwenbianji.cn/ac), for editing the English text of a draft of this manuscript.

References

1. Shahi, A.; Ince, B.; Aydin, S.; Ince, O. Assessment of the horizontal transfer of functional genes as a suitable approach for evaluation of the bioremediation potential of petroleum-contaminated sites: A mini-review. *Appl. Microbiol. Biotechnol.* **2017**, *101*, 4341–4348. [CrossRef] [PubMed]
2. López, E.; Schuhmacher, M.; Domingo, J.L. Human health risks of petroleum-contaminated groundwater. *Environ. Sci. Pollut. Res.* **2008**, *15*, 278–288. [CrossRef]
3. Zhang, S.; Su, X.; Lin, X.; Zhang, Y.; Zhang, Y. Experimental study on the multi-media prb reactor for the remediation of petroleum-contaminated groundwater. *Environ. Earth Sci.* **2015**, *73*, 5611–5618. [CrossRef]
4. Hunkeler, D.; Höhener, P.; Bernasconi, S.; Zeyer, J. Engineered in situ bioremediation of a petroleum hydrocarbon-contaminated aquifer: Assessment of mineralization based on alkalinity, inorganic carbon and stable carbon isotope balances. *J. Contam. Hydrol.* **1999**, *37*, 201–223. [CrossRef]
5. Hibbing, M.E.; Fuqua, C.; Parsek, M.R.; Peterson, S.B. Bacterial competition: Surviving and thriving in the microbial jungle. *Nat. Rev. Microbiol.* **2009**, *8*, 15–25. [CrossRef] [PubMed]
6. Faust, K.; Sathirapongsasuti, J.F.; Izard, J.; Segata, N.; Gevers, D.; Raes, J.; Huttenhower, C. Microbial co-occurrence relationships in the human microbiome. *PLoS Comput. Biol.* **2012**, *8*, e1002606. [CrossRef] [PubMed]
7. Schiegg, H.-O. Field infiltration as a method for the disposal of oil-in-water emulsions from the restoration of oil-polluted aquifers. *Water Res.* **1980**, *14*, 1011–1016. [CrossRef]
8. Tischer, K.; Kleinsteuber, S.; Schleinitz, K.M.; Fetzer, I.; Spott, O.; Stange, F.; Lohse, U.; Franz, J.; Neumann, F.; Gerling, S.; et al. Microbial communities along biogeochemical gradients in a hydrocarbon-contaminated aquifer. *Environ. Microbiol.* **2013**, *15*, 2603–2615. [CrossRef] [PubMed]
9. Saul, D.J.; Aislabie, J.M.; Brown, C.E.; Harris, L.; Foght, J.M. Hydrocarbon contamination changes the bacterial diversity of soil from around scott base, antarctica. *FEMS Microbiol. Ecol.* **2005**, *53*, 141–155. [CrossRef] [PubMed]
10. Abed, R.M.M.; Al-Kindi, S.; Al-Kharusi, S. Diversity of bacterial communities along a petroleum contamination gradient in desert soils. *Microb. Ecol.* **2015**, *69*, 95–105. [CrossRef] [PubMed]
11. Peng, M.; Zi, X.; Wang, Q. Bacterial community diversity of oil-contaminated soils assessed by high throughput sequencing of 16s rRNA genes. *Int. J. Environ. Res. Public Health* **2015**, *12*, 12002–12015. [CrossRef] [PubMed]
12. Lu, L.; Huggins, T.; Jin, S.; Zuo, Y.; Ren, Z.J. Microbial metabolism and community structure in response to bioelectrochemically enhanced remediation of petroleum hydrocarbon-contaminated soil. *Environ. Sci. Technol.* **2014**, *48*, 4021–4029. [CrossRef] [PubMed]
13. Fahy, A.; Lethbridge, G.; Earle, R.; Ball, A.S.; Timmis, K.N.; McGenity, T.J. Effects of long-term benzene pollution on bacterial diversity and community structure in groundwater. *Environ. Microbiol.* **2005**, *7*, 1192–1199. [CrossRef] [PubMed]
14. Zheng, Z.; Zhang, Y.; Su, X.; Cui, X. Responses of hydrochemical parameters, community structures, and microbial activities to the natural biodegradation of petroleum hydrocarbons in a groundwater–soil environment. *Environ. Earth Sci.* **2016**, *75*, 1400. [CrossRef]
15. Zhou, A.-X.; Zhang, Y.-L.; Dong, T.-Z.; Lin, X.-Y.; Su, X.-S. Response of the microbial community to seasonal groundwater level fluctuations in petroleum hydrocarbon-contaminated groundwater. *Environ. Sci. Pollut. Res.* **2015**, *22*, 10094–10106. [CrossRef] [PubMed]
16. Lueders, T. The ecology of anaerobic degraders of btex hydrocarbons in aquifers. *FEMS Microbiol. Ecol.* **2017**, *93*. [CrossRef] [PubMed]
17. Yergeau, E.; Sanschagrin, S.; Maynard, C.; St-Arnaud, M.; Greer, C.W. Microbial expression profiles in the rhizosphere of willows depend on soil contamination. *ISME J.* **2013**, *8*, 344. [CrossRef] [PubMed]
18. Main, C.E.; Ruhl, H.A.; Jones, D.O.B.; Yool, A.; Thornton, B.; Mayor, D.J. Hydrocarbon contamination affects deep-sea benthic oxygen uptake and microbial community composition. *Deep Sea Res. Part I Oceanogr. Res. Pap.* **2015**, *100*, 79–87. [CrossRef]

19. Dorer, C.; Vogt, C.; Neu, T.R.; Stryganyuk, G.; Richnow, H. Characterization of toluene and ethylbenzene biodegradation under nitrate-, iron(III)- and manganese(IV)-reducing conditions by compound-specific isotope analysis. *Environ. Pollut.* **2016**, *211*, 271–281. [CrossRef] [PubMed]
20. Acosta-González, A.; Marqués, S. Bacterial diversity in oil-polluted marine coastal sediments. *Curr. Opin. Biotechnol.* **2016**, *38*, 24–32. [CrossRef] [PubMed]
21. Gieg, L.M.; Fowler, S.J.; Berdugo-Clavijo, C. Syntrophic biodegradation of hydrocarbon contaminants. *Curr. Opin. Biotechnol.* **2014**, *27*, 21–29. [CrossRef] [PubMed]
22. Meckenstock, R.U.; Elsner, M.; Griebler, C.; Lueders, T.; Stumpp, C.; Aamand, J.; Agathos, S.N.; Albrechtsen, H.-J.; Bastiaens, L.; Bjerg, P.L.; et al. Biodegradation: Updating the concepts of control for microbial cleanup in contaminated aquifers. *Environ. Sci. Technol.* **2015**, *49*, 7073–7081. [CrossRef] [PubMed]
23. Röling, W.F.M.; van Breukelen, B.M.; Braster, M.; Lin, B.; van Verseveld, H.W. Relationships between microbial community structure and hydrochemistry in a landfill leachate-polluted aquifer. *Appl. Environ. Microbiol.* **2001**, *67*, 4619–4629. [CrossRef] [PubMed]
24. Wan, J.; Gu, J.; Zhao, Q.; Liu, Y. Cod capture: A feasible option towards energy self-sufficient domestic wastewater treatment. *Sci. Rep.* **2016**, *6*, 25054. [CrossRef] [PubMed]
25. United States Environmental Protection Agency (U.S.E.P). *Method 8260b Volatile Organic Compounds by Gas Chromatography/Mass Spectrometry (gc/ms)*; United States Environmental Protection Agency: Washington, DC, USA, 1996.
26. Standard, A. *Methods for the Examination of Water and Wastewate*; American Public Health Association: Washington, DC, USA, 1998.
27. Ye, J.; Song, Z.; Wang, L.; Zhu, J. Metagenomic analysis of microbiota structure evolution in phytoremediation of a swine lagoon wastewater. *Bioresour. Technol.* **2016**, *219*, 439–444. [CrossRef] [PubMed]
28. Wang, W.; Li, C.; Li, F.; Wang, X.; Zhang, X.; Liu, T.; Nian, F.; Yue, X.; Li, F.; Pan, X.; et al. Effects of early feeding on the host rumen transcriptome and bacterial diversity in lambs. *Sci. Rep.* **2016**, *6*, 32479. [CrossRef] [PubMed]
29. Xu, S.; Yao, J.; Ainiwaer, M.; Hong, Y.; Zhang, Y. Analysis of bacterial community structure of activated sludge from wastewater treatment plants in winter. *BioMed Res. Int.* **2018**, *2018*, 8. [CrossRef] [PubMed]
30. Song, H.; Qiu, S.; Zhang, J.; Xia, C. Study on hydrogenophaga palleronii LHJ38—A naphthalene-degrading strain with high activity. *Environ. Prot. Chem. Ind.* **2006**, *26*, 87–90.
31. Yang, Q.; Cai, S.; Dong, S.; Chen, L.; Chen, J.; Cai, T. Biodegradation of 3-methyldiphenylether (MDE) by hydrogenophaga atypical strain QY7-2 and cloning of the methy-oxidation gene mdeabcd. *Sci. Rep.* **2016**, *6*, 39270. [CrossRef] [PubMed]
32. Goyal, A.K.; Zylstra, G.J. Molecular cloning of novel genes for polycyclic aromatic hydrocarbon degradation from *Comamonas* testosteroni GZ39. *Appl. Environ. Microbiol.* **1996**, *62*, 230–236. [PubMed]
33. Aburto, A.; Peimbert, M. Degradation of a benzene–toluene mixture by hydrocarbon-adapted bacterial communities. *Ann. Microbiol.* **2011**, *61*, 553–562. [CrossRef] [PubMed]
34. Singleton, D.R.; Guzmán Ramirez, L.; Aitken, M.D. Characterization of a polycyclic aromatic hydrocarbon degradation gene cluster in a phenanthrene-degrading acidovorax strain. *Appl. Environ. Microbiol.* **2009**, *75*, 2613–2620. [CrossRef] [PubMed]
35. Yen, K.M.; Karl, M.R.; Blatt, L.M.; Simon, M.J.; Winter, R.B.; Fausset, P.R.; Lu, H.S.; Harcourt, A.A.; Chen, K.K. Cloning and characterization of a pseudomonas mendocina KR1 gene cluster encoding toluene-4-monooxygenase. *J. Bacteriol.* **1991**, *173*, 5315–5327. [CrossRef] [PubMed]
36. Liebensteiner, M.G.; Oosterkamp, M.J.; Stams, A.J.M. Microbial respiration with chlorine oxyanions: Diversity and physiological and biochemical properties of chlorate- and perchlorate-reducing microorganisms. *Ann. N. Y. Acad. Sci.* **2016**, *1365*, 59–72. [CrossRef] [PubMed]
37. Mehboob, F.; Weelink, S.; Saia, F.T.; Junca, H.; Stams, A.J.M.; Schraa, G. Microbial degradation of aliphatic and aromatic hydrocarbons with (per)chlorate as electron acceptor. In *Handbook of Hydrocarbon and Lipid Microbiology*; Timmis, K.N., Ed.; Springer: Berlin/Heidelberg, Germany, 2010; pp. 935–945.
38. Nayak, A.S.; Vijaykumar, M.H.; Karegoudar, T.B. Characterization of biosurfactant produced by *Pseudoxanthomonas* sp. PNK-04 and its application in bioremediation. *Int. Biodeterior. Biodegrad.* **2009**, *63*, 73–79. [CrossRef]
39. Kim, J.M.; Le, N.T.; Chung, B.S.; Park, J.H.; Bae, J.-W.; Madsen, E.L.; Jeon, C.O. Influence of soil components

on the biodegradation of benzene, toluene, ethylbenzene, and o-, m-, and p-xylenes by the newly isolated bacterium pseudoxanthomonas spadix BD-a59. *Appl. Environ. Microbiol.* **2008**, *74*, 7313–7320. [CrossRef] [PubMed]

40. Sohn, J.H.; Kwon, K.K.; Kang, J.-H.; Jung, H.-B.; Kim, S.-J. *Novosphingobium pentaromativorans* sp. Nov., a high-molecular-mass polycyclic aromatic hydrocarbon-degrading bacterium isolated from estuarine sediment. *Int. J. Syst. Evol. Microbiol.* **2004**, *54*, 1483–1487. [CrossRef] [PubMed]
41. Liu, Z.-P.; Wang, B.-J.; Liu, Y.-H.; Liu, S.-J. *Novosphingobium taihuense* sp. Nov., a novel aromatic-compound-degrading bacterium isolated from taihu lake, china. *Int. J. Syst. Evol. Microbiol.* **2005**, *55*, 1229–1232. [CrossRef] [PubMed]
42. Callaghan, A.V.; Davidova, I.A.; Savage-Ashlock, K.; Parisi, V.A.; Gieg, L.M.; Suflita, J.M.; Kukor, J.J.; Wawrik, B. Diversity of benzyl- and alkylsuccinate synthase genes in hydrocarbon-impacted environments and enrichment cultures. *Environ. Sci. Technol.* **2010**, *44*, 7287–7294. [CrossRef] [PubMed]
43. Rochman, F.F.; Sheremet, A.; Tamas, I.; Saidi-Mehrabad, A.; Kim, J.-J.; Dong, X.; Sensen, C.W.; Gieg, L.M.; Dunfield, P.F. Benzene and naphthalene degrading bacterial communities in an oil sands tailings pond. *Front. Microbiol.* **2017**, *8*. [CrossRef] [PubMed]
44. Feng, Y.; Li, X.; Song, T.; Yu, Y.; Qi, J. Stimulation effect of electric current density (ECD) on microbial community of a three dimensional particle electrode coupled with biological aerated filter reactor (TDE-BAF). *Bioresour. Technol.* **2017**, *243*, 667–675. [CrossRef] [PubMed]
45. Qing, X.U.; Zhang, F.; Zhong-Qi, X.U.; Jia, Y.L.; You, J.M. Some characteristics of simpson index and the shannon-wiener index and their dilution effect. *Pratacult. Sci.* **2011**, *28*, 527–531.
46. Sutton, N.B.; Maphosa, F.; Morillo, J.A.; Abu Al-Soud, W.; Langenhoff, A.A.M.; Grotenhuis, T.; Rijnaarts, H.H.M.; Smidt, H. Impact of long-term diesel contamination on soil microbial community structure. *Appl. Environ. Microbiol.* **2013**, *79*, 619–630. [CrossRef] [PubMed]
47. Guo, H.; Liu, R.; Yu, Z.; Zhang, H.; Yun, J.; Li, Y.; Liu, X.; Pan, J. Pyrosequencing reveals the dominance of methylotrophic methanogenesis in a coal bed methane reservoir associated with eastern ordos basin in china. *Int. J. Coal Geol.* **2012**, *93*, 56–61. [CrossRef]
48. Jones, E.J.P.; Voytek, M.A.; Corum, M.D.; Orem, W.H. Stimulation of methane generation from nonproductive coal by addition of nutrients or a microbial consortium. *Appl. Environ. Microbiol.* **2010**, *76*, 7013–7022. [CrossRef] [PubMed]
49. Van Eerten-Jansen, M.C.A.A.; Veldhoen, A.B.; Plugge, C.M.; Stams, A.J.M.; Buisman, C.J.N.; Ter Heijne, A. Microbial community analysis of a methane-producing biocathode in a bioelectrochemical system. *Archaea* **2013**, *2013*, 481784. [CrossRef] [PubMed]
50. Hersman, L.E.; Huang, A.; Maurice, P.A.; Forsythe, J.H. Siderophore production and iron reduction by pseudomonas mendocina in response to iron deprivation. *Geomicrobiol. J.* **2000**, *17*, 261–273.
51. Pantke, C.; Obst, M.; Benzerara, K.; Morin, G.; Ona-Nguema, G.; Dippon, U.; Kappler, A. Green rust formation during fe(ii) oxidation by the nitrate-reducing *Acidovorax* sp. Strain bofen1. *Environ. Sci. Technol.* **2012**, *46*, 1439–1446. [CrossRef] [PubMed]
52. Chakraborty, R.; Coates, J.D. Hydroxylation and carboxylation—Two crucial steps of anaerobic benzene degradation by dechloromonas strain rcb. *Appl. Environ. Microbiol.* **2005**, *71*, 5427–5432. [CrossRef] [PubMed]
53. Finneran, K.T.; Johnsen, C.V.; Lovley, D.R. *Rhodoferax ferrireducens* sp. Nov., a psychrotolerant, facultatively anaerobic bacterium that oxidizes acetate with the reduction of fe(iii). *Int. J. Syst. Evol. Microbiol.* **2003**, *53*, 669–673. [CrossRef] [PubMed]
54. Patricia, A.M.; Vierkorn, M.A.; Hersman, L.E.; Fulghum, J.E.; Ferryman, A. Enhancement of kaolinite dissolution by an aerobic pseudomonas mendocina bacterium. *Geomicrobiol. J.* **2001**, *18*, 21–35. [CrossRef]
55. Jovanović, T.; Ascenso, C.; Hazlett, K.R.O.; Sikkink, R.; Krebs, C.; Litwiller, R.; Benson, L.M.; Moura, I.; Moura, J.J.G.; Radolf, J.D.; et al. Neelaredoxin, an iron-binding protein from the syphilis spirochete, treponema pallidum, is a superoxide reductase. *J. Biol. Chem.* **2000**, *275*, 28439–28448. [CrossRef] [PubMed]
56. Ramamoorthy, S.; Sass, H.; Langner, H.; Schumann, P.; Kroppenstedt, R.M.; Spring, S.; Overmann, J.; Rosenzweig, R.F. *Desulfosporosinus lacus* sp. Nov., a sulfate-reducing bacterium isolated from pristine freshwater lake sediments. *Int. J. Syst. Evol. Microbiol.* **2006**, *56*, 2729–2736. [CrossRef] [PubMed]
57. Kojima, H.; Fukui, M. *Sulfuritalea hydrogenivorans* gen. Nov., sp. Nov., a facultative autotroph isolated from a freshwater lake. *Int. J. Syst. Evol. Microbiol.* **2011**, *61*, 1651–1655. [CrossRef] [PubMed]

58. Kodama, Y.; Watanabe, K. *Sulfuricurvum kujiense* gen. Nov., sp. Nov., a facultatively anaerobic, chemolithoautotrophic, sulfur-oxidizing bacterium isolated from an underground crude-oil storage cavity. *Int. J. Syst. Evol. Microbiol.* **2004**, *54*, 2297–2300. [CrossRef] [PubMed]
59. Koch, H.; Lücker, S.; Albertsen, M.; Kitzinger, K.; Herbold, C.; Spieck, E.; Nielsen, P.H.; Wagner, M.; Daims, H. Expanded metabolic versatility of ubiquitous nitrite-oxidizing bacteria from the genus nitrospira. *Proc. Natl. Acad. Sci. USA* **2015**, *112*, 11371–11376. [CrossRef] [PubMed]
60. Ludington, W.B.; Seher, T.D.; Applegate, O.; Li, X.; Kliegman, J.I.; Langelier, C.; Atwill, E.R.; Harter, T.; Derisi, J.L. Assessing biosynthetic potential of agricultural groundwater through metagenomic sequencing: A diverse anammox community dominates nitrate-rich groundwater. *PLoS ONE* **2017**, *12*, e0174930. [CrossRef] [PubMed]
61. Ji, B.; Zhang, S.D.; Zhang, W.J.; Rouy, Z.; Alberto, F.; Santini, C.L.; Mangenot, S.; Gagnot, S.; Philippe, N.; Pradel, N. The chimeric nature of the genomes of marine magnetotactic coccoid-ovoid bacteria defines a novel group of proteobacteria. *Environ. Microbiol.* **2017**, *19*, 1103–1109. [CrossRef] [PubMed]
62. Wu, C.-Y.; Zhuang, L.; Zhou, S.-G.; Li, F.-B.; Li, X.-M. Fe(III)-enhanced anaerobic transformation of 2,4-dichlorophenoxyacetic acid by an iron-reducing bacterium *Comamonas* koreensis Cy01. *FEMS Microbiol. Ecol.* **2009**, *71*, 106–113. [CrossRef] [PubMed]
63. Patureau, D.; Bernet, N.; Moletta, R. Study of the denitrifying enzymatic system of *Comamonas* sp. Strain $SGLY_2$ under various aeration conditions with a particular view on nitrate and nitrite reductases. *Curr. Microbiol.* **1996**, *32*, 25–32. [CrossRef]
64. Kai, X.X.; Fang, M.C. A experimental study of influence of ph on calcium carbonate crystallization fouling. *Petro-Chem. Equip.* **2004**, *33*, 11–14.
65. Marcus, D.N.; Pinto, A.; Anantharaman, K.; Ruberg, S.A.; Kramer, E.L.; Raskin, L.; Dick, G.J. Diverse manganese(II)-oxidizing bacteria are prevalent in drinking water systems. *Environ. Microbiol. Rep.* **2017**, *9*, 120–128. [CrossRef] [PubMed]
66. Mechichi, T.; Stackebrandt, E.; Fuchs, G. *Alicycliphilus denitrificans* gen. Nov., sp. Nov., a cyclohexanol-degrading, nitrate-reducing β-proteobacterium. *Int. J. Syst. Evol. Microbiol.* **2003**, *53*, 147–152. [CrossRef] [PubMed]

Permissions

All chapters in this book were first published in MDPI; hereby published with permission under the Creative Commons Attribution License or equivalent. Every chapter published in this book has been scrutinized by our experts. Their significance has been extensively debated. The topics covered herein carry significant findings which will fuel the growth of the discipline. They may even be implemented as practical applications or may be referred to as a beginning point for another development.

The contributors of this book come from diverse backgrounds, making this book a truly international effort. This book will bring forth new frontiers with its revolutionizing research information and detailed analysis of the nascent developments around the world.

We would like to thank all the contributing authors for lending their expertise to make the book truly unique. They have played a crucial role in the development of this book. Without their invaluable contributions this book wouldn't have been possible. They have made vital efforts to compile up to date information on the varied aspects of this subject to make this book a valuable addition to the collection of many professionals and students.

This book was conceptualized with the vision of imparting up-to-date information and advanced data in this field. To ensure the same, a matchless editorial board was set up. Every individual on the board went through rigorous rounds of assessment to prove their worth. After which they invested a large part of their time researching and compiling the most relevant data for our readers.

The editorial board has been involved in producing this book since its inception. They have spent rigorous hours researching and exploring the diverse topics which have resulted in the successful publishing of this book. They have passed on their knowledge of decades through this book. To expedite this challenging task, the publisher supported the team at every step. A small team of assistant editors was also appointed to further simplify the editing procedure and attain best results for the readers.

Apart from the editorial board, the designing team has also invested a significant amount of their time in understanding the subject and creating the most relevant covers. They scrutinized every image to scout for the most suitable representation of the subject and create an appropriate cover for the book.

The publishing team has been an ardent support to the editorial, designing and production team. Their endless efforts to recruit the best for this project, has resulted in the accomplishment of this book. They are a veteran in the field of academics and their pool of knowledge is as vast as their experience in printing. Their expertise and guidance has proved useful at every step. Their uncompromising quality standards have made this book an exceptional effort. Their encouragement from time to time has been an inspiration for everyone.

The publisher and the editorial board hope that this book will prove to be a valuable piece of knowledge for researchers, students, practitioners and scholars across the globe.

List of Contributors

James Reece and Mark Christenson
AirLift Environmental, LLC, 5900 N. 58th, Suite 5, Lincoln, NE 68507, USA

Ann Kambhu
Department of Civil Engineering, University of Nebraska, Lincoln, NE 68588-6105, USA
Department of Environmental Technology and Management, Kasetsart University, Bangkok 10900, Thailand

Yusong Li
Department of Civil Engineering, University of Nebraska, Lincoln, NE 68588-6105, USA

Clifford E. Harris
Department of Chemistry, Albion College, Albion, MI 49224, USA

Steve Comfort
AirLift Environmental, LLC, 5900 N. 58th, Suite 5, Lincoln, NE 68507, USA
School of Natural Resources, University of Nebraska, Lincoln, NE 68583-0915, USA

Qiaona Guo, Yahui Zhang, Zhifang Zhou and Zili Hu
School of Earth Sciences and Engineering, Hohai University, No.8 Focheng West Road, Nanjing 211100, China

Martina Mattia, Paola Tuccimei and Michele Soligo
Dipartimento di Scienze, Università "Roma Tre", 00154 Rome, Italy

Claudio Carusi
Protezione Ambiente, Mares S.r.l., 00144 Rome, Italy

Nigus Kebede Wegahita, Lei Ma, Tingwei Huan , Qiankun Luo and Jiazhong Qian
School of Resources and Environmental Engineering, Hefei University of Technology, Hefei 230009, China

Jiankui Liu
Anhui Geological Environment Monitoring Station, Hefei 230001, China

Ruben Vera, Enriqueta Anticó and Clàudia Fontàs
Department of Chemistry, University of Girona, C/ Maria Aurèlia Capmany 69, 17003 Girona, Spain

Ali Moradi
Department of Environmental Resources Engineering, Humboldt State University, Arcata, CA 95521, USA

Kathleen M. Smits
Department of Civil Engineering, the University of Texas Arlington, Arlington, TX 76019, USA

Jonathan O. Sharp
Department of Civil and Environmental Engineering, Colorado School of Mines, Golden, CO 80401, USA

Gabriele Beretta, Andrea Filippo Mastorgio, Lisa Pedrali, Sabrina Saponaro and Elena Sezenna
Department of Environmental and Civil Engineering, Politecnico di Milano, Piazza Leonardo da Vinci 32, 20133 Milan, Italy

Andrew Plymale, Jacqueline Wells, Shelby Brooks and Brady Lee
Energy and Environment Directorate, Pacific Northwest National Lab (PNNL), Richard, WA 99354, USA

Emily Graham
Earth and Biological Sciences Directorate, Pacific Northwest National Lab (PNNL), Richard, WA 99354, USA

Odeta Qafoku
Physical and Computational Sciences Directorate, Pacific Northwest National Lab (PNNL); Richard, WA 99354, USA

Martin J. Wells and Aaron R. Mittelstet
Department of Biological Systems Engineering, University of Nebraska-Lincoln, Lincoln, NE 68583, USA

Troy E. Gilmore
Department of Biological Systems Engineering, University of Nebraska-Lincoln, Lincoln, NE 68583, USA
Conservation and Survey Division, School of Natural Resources, University of Nebraska-Lincoln, Lincoln, NE 68583, USA

Daniel Snow
Nebraska Water Sciences Laboratory, Nebraska Water Center, Lincoln, NE 68583, USA

Steven S. Sibray
Conservation and Survey Division, School of Natural Resources, University of Nebraska-Lincoln, Lincoln, NE 68583, USA

Shuya Hu, Changlai Xiao and Xiujuan Liang
Key Laboratory of Groundwater Resources and Environment of Ministry of Education, Jilin University, Changchun 130021, China
College of Environmental and Resources, Jilin University, Changchun 130021, China
National-Local Joint Engineering Laboratory of In-situ Conversion, Drilling and Exploitation Technology for Oil Shale, Changchun 130021, China

Xue Jiang
College of Water Resources and Architectural Engineering, The Northeast Agricultural University, Harbin 150030, China

Jesús Alejandro Prieto-Amparán and Alfredo Pinedo-Alvarez
Facultad de Zootecnia y Ecología, Universidad Autónoma de Chihuahua, Periférico Francisco R. Almada, Km 1, Chihuahua, Chihuahua 31453, México

Beatriz Adriana Rocha-Gutiérrez, María de Lourdes Ballinas-Casarrubias and María del Rosario Peralta-Pérez
Facultad de Ciencias Químicas, Universidad Autónoma de Chihuahua. Circuito Universitario s/n, Campus II, Chihuahua, Chihuahua 31125, México

María Cecilia Valles-Aragón
Facultad de Ciencias Agrotecnológicas, Universidad Autónoma de Chihuahua, Av. Pascual Orozco s/n, Campus I, Chihuahua, Chihuahua 31200, México

Bingqing Lu
Department of Geological Sciences, University of Alabama, Tuscaloosa, AL 35487, USA

Yong Zhang
Department of Geological Sciences, University of Alabama, Tuscaloosa, AL 35487, USA
Department of Engineering Mechanics, Hohai University, Nanjing 210098, Jiangsu, China

Chunmiao Zheng
Guangdong Provincial Key Laboratory of Soil and Groundwater Pollution Control, School of Environmental Science & Engineering, Southern University of Science and Technology, Shenzhen 518055, Guangdong, China

Christopher T. Green
U.S. Geological Survey, Menlo Park, CA 94025, USA

Charles O'Neill
Department of Aerospace Engineering and Mechanics, University of Alabama, Tuscaloosa, AL 35487, USA

Hong-Guang Sun
Department of Engineering Mechanics, Hohai University, Nanjing 210098, Jiangsu, China

Jiazhong Qian
School of Resources and Environmental Engineering, Hefei University of Technology, Anhui 230009, Hefei, China

Alexander A. Haluska, Meghan S. Thiemann Jaehyun Cho and Michael D. Annable
Department of Environmental Engineering Sciences, University of Florida, 217 A.P. Black Hall, Gainesville, FL 32611, USA

Patrick J. Evans
CDM Smith, 14432 S.E. Eastgate Way, Suite 100, Bellevue,WA 98007, USA

Zhuo Ning and and Ping Wang
Institute of Hydrogeology and Environmental Geology, Chinese Academy of Geological Sciences, Shijiazhuang 050061, China
School of Chinese Academy of Geological Sciences, China University of Geosciences (Beijing), Beijing 100083, China
Key Laboratory of Groundwater Remediation of Hebei Province, Zhengding 050083, China

Min Zhang and Caijuan Guo
Institute of Hydrogeology and Environmental Geology, Chinese Academy of Geological Sciences, Shijiazhuang 050061, China

Ze He
Institute of Hydrogeology and Environmental Geology, Chinese Academy of Geological Sciences, Shijiazhuang 050061, China
Key Laboratory of Groundwater Remediation of Hebei Province, Zhengding 050083, China

Pingping Cai
Institute of Hydrogeology and Environmental Geology, Chinese Academy of Geological Sciences, Shijiazhuang 050061, China
School of Resources and Environmental Engineering, HeFei University of Technology, Hefei 230009, China

Index

www.ingramcontent.com/pod-product-compliance
Lightning Source LLC
LaVergne TN
LVHW081035210926
848485LV00004B/1030